STUDENT STUDY GUIDE

to accompany

CALCULUS

SINGLE VARIABLE **THIRD EDITION**

Deborah Hughes-Hallett
University of Arizona

Andrew M. Gleason
Harvard University

William G. McCallum
University of Arizona

et al.

Prepared by

Beverly K. Michael
University of Pittsburgh

Peg Kem McPartland
Golden Gate University

JOHN WILEY & SONS, INC.

COVER PHOTO © Eddie Hironaka/The Image Bank.

To order books or for customer service call 1-800-CALL-WILEY (225-5945).

ISBN 0-471-44194-5

Printed in the United States of America

10 9 8 7 6 5 4 3 2 1

Printed and bound by Bradford & Bigelow, Inc.

Dedicated to our husbands and children
for their patience while we burned the midnight oil.

...BKM
...PKM

Eat honey: it is good.
And just as honey from the comb is sweet on the tongue
you may be sure that wisdom is good for the soul.

...Proverbs 24:26

TO THE STUDENT

How to use this Study Guide

Welcome to the study of calculus. The purpose of this Study Guide is to supplement your instruction. We assume that you have read the text and taken notes in class. It will provide you with additional worked examples, and it follows the spirit of the *Calculus* text by emphasizing the meaning of the symbols you are using. We expect that you have a graphing calculator or computer algebra system available when using this Study Guide.

We have also tried to make this Study Guide interactive. We expect that you have a pencil in hand, and use it, while working through this Study Guide. Each chapter section begins with the objective and preparation necessary to begin your study of the chapter section. Each chapter also has some key features that are meant to aid your study of calculus. These features are:

- ✠ **Key Ideas**: This feature begins with "You should be able to..." and it lists all the key concepts of the chapter. After studying you should ask, "Do I know all the Key Ideas?"
- ✝ **Key Questions to Ask and Answer:** These are questions that a tutor or instructor might ask, or you might ask yourself. Space is provided for you to write in the answers to the questions. You may have to refer back to the text to write the answers.
- **Examples:** These examples answer the Key Questions and are meant to provide a deeper understanding of the concept.
- ✎ **Your Turn:** These are partially worked examples that require you to fill in the blanks or to provide answers. The correct answer appears to the right of the Your Turn problem. Please use your pencil to do the work.
- ✎ **On Your Own:** These are problems very similar to the examples, which you are to work "on your own", to see if you have mastered the concept. The complete solution will appear at the end of the section. Record your work in a notebook.
- **Key Notation:** When a new notation is introduced, this feature explains how to say and use the notation.
- **Key Algebra:** This feature reviews or points out algebra concepts that are needed for the section.
- 🗀 **Key Study Skills:** This is one of the most important features of this Study Guide.
    1) We suggest that you use the ✓ ? ★ system on homework problems. File these problems for review. The symbols represent:
       ✓ (check), I know and understand how to do this problem. It gave me no difficulty. I don't need to review this problem again.
       ? (question), I had trouble with this problem, but eventually I got it. I should rework this problem again at a later date.
       ★ (starred), I really didn't know how to do this problem. I need to ask about this problem in class or get help from someone to solve it. Before the next quiz or test I need to learn how to do this problem.
    2) We suggest that you choose at least two problems that represent the section, then put each on a 3 x 5 note card with the solution on the back. These problems provide a quick and readily available source of review. Take the cards out, shuffle them up, then read a problem and think about the solution. This will help you prepare for quizzes and exams. Students who use this note card system find it one of the most simple and effective learning techniques available. You can read the cards before class, while waiting for a bus, or any time you have a few minutes to spare.

Acknowledgments

To all those calculus students who take on the challenge to really think hard about problems and concepts and who challenge us to think about new ways of teaching calculus. Please feel free to contact us at our universities. --B.K. Michael and P.K. McPartland.

CONTENTS

CHAPTER ONE

A LIBRARY OF FUNCTIONS

The study of calculus begins with the study of various mathematical functions.

1.1 FUNCTIONS AND CHANGE

Objective: To be able to define a function numerically, graphically and algebraically.
To determine the properties of linear functions

Preparations: Read and take notes on section 1.1.
Suggested practice problems: 1, 4, 5, 9, 12, 15, 17, 27, 29, 35, 37.
Problems requiring substantial algebra: 21, 22.

✠ Key Ideas of the Section

You should be able to:

- Define a function.
- Recognize a function as a table, a graph, a formula, or in words.
- State the domain and range of a function by looking at the graph, table, or equation and by recognizing the algebraic properties.
- Identify independent and dependent variables.
- Identify curves that are increasing or decreasing in different ways.
- Write a linear function in three different ways
- Determine rate of change via table, graph and algebraic form
- Decide if lines are parallel, perpendicular or neither.
- Compare direct and inverse proportion.

Key Questions to Ask and Answer

1. ***What is a function?***

✍ ____

____ *(See text p. 2.)*

The concept of a functional relationship is the first big idea of higher mathematics. If you understand the definition of a function, you are starting with a good foundation. Perhaps you are used to representing functions as formulas or algebraic equations, but tables and graphs can also represent functions.

Practically speaking you have a functional relationship if *for each number you begin with (input) you have one and only one resulting number (output).*

Example 1.1.1

a) $y = x^2$, $A = \pi r^2$, $C = \pi d$, $F = \frac{9}{5}x + 32$ are formulas that represent functions. Generally they are solved for the output variable, which is raised to the power of one. The output (*dependent* variable) is written in terms of the input (*independent* variable).

b) $x = \pm\sqrt{y}$, $r = \pm\sqrt{\frac{A}{\pi}}$, $y^2 = 25 - x^2$ are formulas that **do not** represent functions. They have more than one value for the output often because the output value is raised to an even power.

c) The graph, plot and table below are **not** functions.

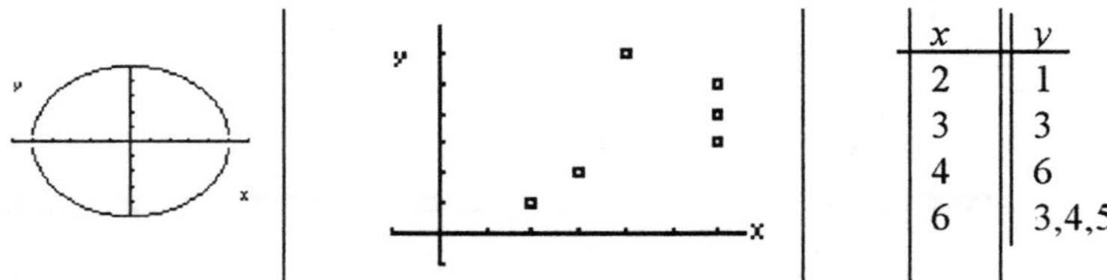

x	y
2	1
3	3
4	6
6	3,4,5

All of the above have more than one y value for an x value.

2. *What is the difference between the range and the domain of a function and when is the domain restricted?*

✍ ____

____ *(See text pp. 2-3.)*

The *domain* is always associated with the input values and the *range* with output values. The input values can be restricted by the algebraic equation, by boundaries of the graph or by the real world situation.

Example 1.1.2

a) Determine the domain of the following functions. Write the domain in words, using inequalities and interval notation[2].

Function	*Domain*
$y = x^2 - 2x - 3$	All x-values from the set of real numbers; $x = \Re$, or $x: (-\infty, \infty)$
$y = \sqrt{x}$	All x-values greater than or equal to zero; $x \geq 0$ or $x: [0, \infty)$
$y = \dfrac{1}{x}$	All x-values from the set of all real numbers, $x \neq 0$; or $x: (-\infty, 0) \cup (0, \infty)$
$y = \dfrac{1}{3}\sqrt{9 - x^2}$	All x-values between -3 and 3 inclusive, $-3 \leq x \leq 3$; or x: [-3, 3]

b) Determine the domain and range of the functions listed in *a)* above and by looking at their graphs listed below.

[2] Note: When using interval notation, brackets, [or], means include the end point, parentheses, (or), means do not include the endpoint and the union symbol, $\cup$, means "unites with" or "and".

Function	Graph	Domain and Range
$y = x^2 - 2x - 3$	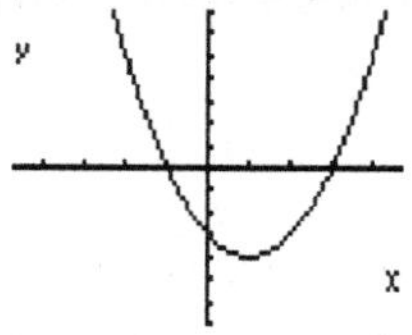	Domain: x is the set of all real numbers, or $x:(-\infty,\infty)$ Range: $y \ge -4$ or $y:[-4,\infty)$ Since the function is a quadratic, the lowest point is the vertex calculated by $x = -\frac{b}{2a}$. So x=-(-2)/(2)(1)= 1 , and $y = f(1)$ = -4
$y = \sqrt{x}$	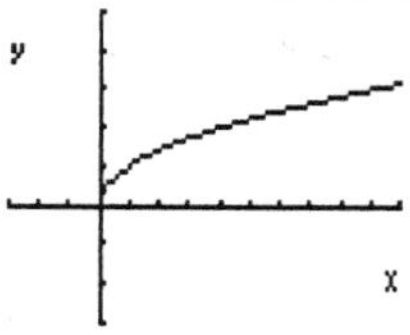	Domain: x is greater than or equal to zero, $x \ge 0$ or . $x:[0,\infty)$ Range: y is greater than or equal to zero, $y \ge 0$ or $y:[0,\infty)$.
$y = \frac{1}{x}$	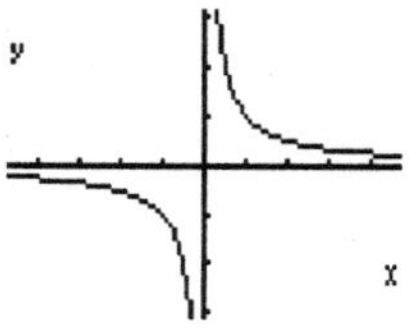	Domain: x is the set of all real numbers, $x \ne 0$, $x:(-\infty,0)\cup(0,\infty)$ Range: y is the set of all real numbers, $y \ne 0$ or $y:(-\infty,0)\cup(0,\infty)$
$y = \frac{1}{3}\sqrt{9-x^2}$	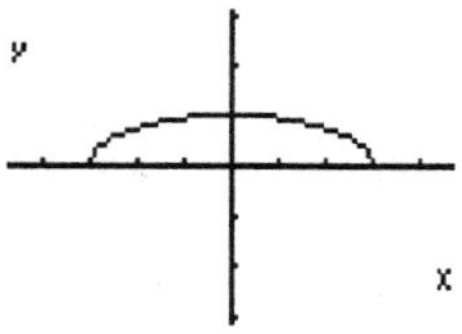	Domain: The values of x are between -3 and 3, including -3 and 3, $-3 \le x \le 3$ or $y:[-3,3]$ Range: y is greater than or equal to zero and less than or equal to one, $0 \le y \le 1$ or y: [0,1].

c) Domains are sometimes restricted by real world situations. Determine the following domains.

- The radius r of a circle. The domain: the radius is greater than zero, $r > 0$.
- The temperature, T, in degrees Fahrenheit of a town in one year. The domain of temperature is -22° ≤ T ≤ 99°.
- The number of seated students, S, in a 150 seat auditorium. The domain is $0 \le S \le 150$.

3. *What are some properties of linear functions?*

✍ List any five things you know about linear functions:

a) ______________________

b) ______________________

c) ______________________

d) ______________________

e) ______________________

Possible Answers

- $y = mx + b$
- m = slope
- b = y-intercept
- The graph is a line.
- The rate of change, slope, is constant.
- Increasing lines have positive slope, $m > 0$.
- Decreasing lines have negative slope, $m < 0$.
- Horizontal lines have slope, $m = 0$.

A linear function has the equation $y = f(x) = mx + b$, where the rate of change between any two points is constant. The graph of a linear equation is straight line. The rate of change is called the slope of the line and can be found by using any of the following instructions:

$$\text{slope} = m = \text{difference quotient} = \frac{\text{change in y}}{\text{change in x}} = \frac{\Delta y}{\Delta x} = \frac{y_2 - y_1}{x_2 - x_1}$$

$$= \frac{\text{vertical change}}{\text{horizontal change}} = \frac{f(x_2) - f(x_1)}{x_2 - x_1} = \frac{\text{rise}}{\text{run}}$$

where $(x_1, f(x_1))$ and $(x_2, f(x_2))$ are any two points on the line.

Example 1.1.3

The rate of change between the points $(1, -7)$ and $(-3, 5)$ is

$$m = \frac{5-(-7)}{-3-(1)} = \frac{12}{-4} = -3$$

4. How can rate of change be visualized?

✍ ____

____ *(See text p. 6)*

Example 1.1.4

Verify that the amount, A, one pays for a taxable item is $A = 1.05p$, if a tax rate of 5% is added to the purchase price, p. Determine if the function is a linear function.

a) Create a table of values.
b) Create a plot.
c) Show that the function goes through the graph points.

Solution:

a) The Table 1.1.1 shows that the formula takes the purchase price and adds the tax T. The rate of change is constant so the data is linear. (Note: Values have not been rounded to the nearest cent as would happen on a cash register.)

Purchase Price p	**Tax** $.05p$	**Amount** $p + .05p$	**Amount** $A = 1.05p$	**Rate of Change** $\frac{\Delta A}{\Delta P}$
\$1.00	0.050	1.050	1.050	
\$1.50	0.075	1.575	1.575	1.0500
\$2.00	0.100	2.100	2.100	1.0500
\$2.50	0.125	2.625	2.625	1.0500
\$3.00	0.150	3.150	3.150	1.0500
\$3.50	0.175	3.675	3.675	1.0500
\$4.00	0.200	4.200	4.200	1.0500
\$4.50	0.225	4.725	4.725	1.0500

Table 1.1.1

b) The plot of the data appears to be linear. The graph below shows that the function $A(p) = 1.05p$ goes through the data points

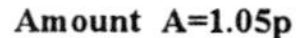

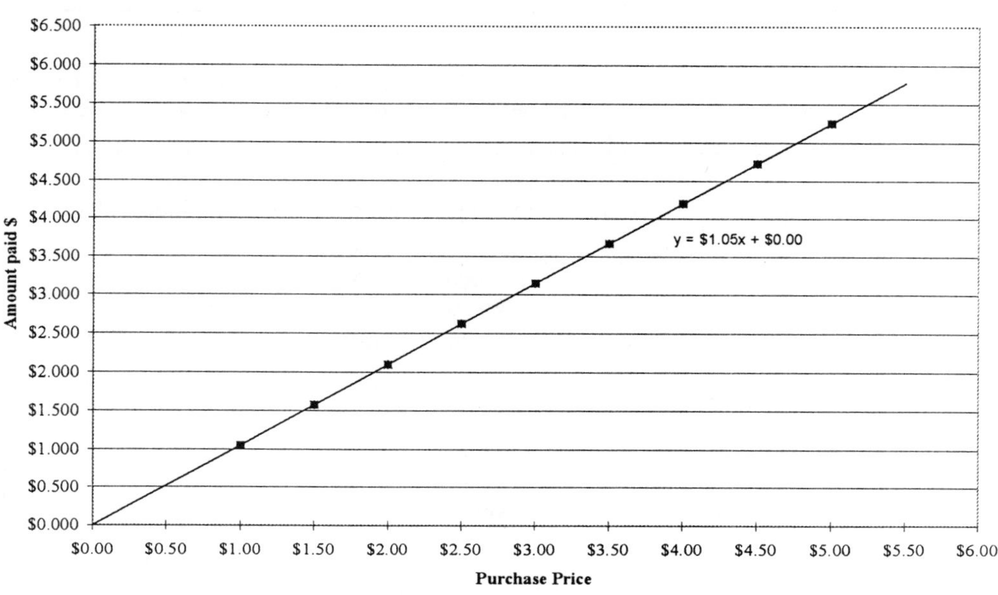

5. *What are the three forms of a linear equation?*

✍ Write the three forms of a linear equation:

	Answers
a) Slope Intercept:	$y = mx + b$
b) Point Slope:	$y - y_1 = m(x - x_1)$
c) Standard:	$Ax + By = C$

Example 1.1.5

Write the equation of a line that goes through the points (0,5) and (3,7) using three methods.

Solution:

a) Most students are familiar with the **slope intercept** form:

$$y = mx + b \quad or \quad y = b + mx$$

where m is the slope and b is the vertical or y- intercept. This form helps you to quickly draw the graph. It may also be represented in the function form $y = f(x) = b + mx$.

First find the Slope: $m = \frac{y_2 - y_1}{x_2 - x_1} = \frac{7-5}{3-0} = \frac{2}{3}$, then use the vertical intercept or any point on the line for the x and y values.

$$y = \frac{2}{3}x + b \quad \text{since x = 0 and y = 5}$$

$$5 = \frac{2}{3}(0) + b \quad \text{therefore b = 5 and the equation becomes}$$

$$y = \frac{2}{3}x + 5$$

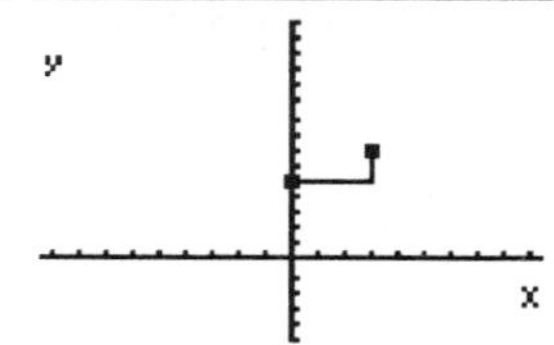	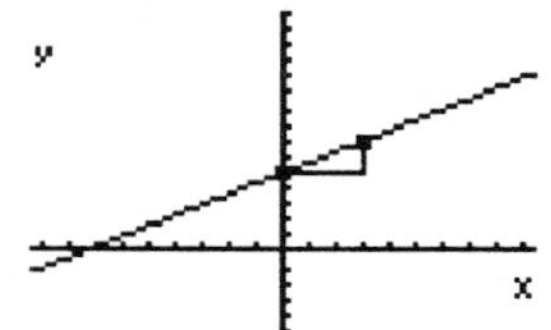
Plot the points (0,5) and (3,7) Move 3 units in a horizontal direction and 2 units in a vertical direction.	The equation is drawn through the points. $m = 2/3$ and $b = 5$ or $y = 2/3\,x + 5$

a) The **point slope** form is used more often in calculus, where m is the slope and (x_1, y_1) is a point: $y - y_1 = m(x - x_1)$, however, it is most useful in the following forms:

$$y = m(x - x_1) + y_1 \quad \text{or}$$
$$y = m(x - x_1) + f(x_1) \quad \text{where } f(x_1) = y_1.$$

In the above example the equation can be found by using either point.

$$y = \frac{2}{3}(x-0)+5 = \frac{2}{3}x+5 \quad \text{or}$$

$$y = \frac{2}{3}(x-3)+7 = \frac{2}{3}x - \frac{2}{3}(3)+7 = \frac{2}{3}x+5$$

c) The Standard form is used less often in calculus:

$$Ax + By = C$$

$$y = \frac{2}{3}x + 5 \quad \textit{becomes}$$

$$3y = 2x + 15$$

$$-2x + 3y = 15$$

6. How do parallel and perpendicular lines compare?

✍ ____

____ *(See text p. 6)*

Parallel lines have the same rate of change (slope). Parallel lines increase or decrease at the same rate, but they have different vertical intercepts.

Perpendicular lines are at a right angle to each other, so one line must increase while the other decreases or they are vertical and horizontal lines. If the lines are not horizontal or vertical, the slopes of perpendicular lines are negative reciprocals of each other.

If line L_1 is perpendicular to line L_2, where m_1 is the slope of L_1 and m_2 is the slope of L_2 then:

$$m_2 = \frac{-1}{m_1} \quad \text{or} \quad m_1 m_2 = -1$$

Example 1.1.6

Determine if the following are parallel, perpendicular or neither.

$$f(x) = \frac{2}{3}x + 5, \quad g(x) = \frac{2}{3}x - 2, \quad h(x) = \frac{-3}{2}x + 4$$

	Reason	*Graph*
f(x) is parallel to *g(x)*	The slopes are equal. $m_f = m_g = \frac{2}{3}$ They have different vertical intercepts.	
h(x) is perpendicular to both *f(x)* and *g(x)*.	The slope of *h(x)* is the negative reciprocal of the slopes of both *f(x)* and *g(x)*. $m_h = \frac{-1}{\frac{2}{3}} = -\frac{3}{2}$	

7. *What are the formulas and graphs of direct and inverse proportions?*

✍ ____

____ *(See text p. 6)*

If two variables x and y are **directly proportional** then as the magnitude of x increases, the magnitude of y also increases and as the magnitude of x decreases, the magnitude of y decreases.

a) Direct proportions are in the form

$$y = kx$$

where k is the *constant of proportionality.*

thus $k = \frac{y}{x}$

If two variables x and y are **inversely proportional** then as the magnitude of x increases, the magnitude of y decreases and as the magnitude of x decreases, the magnitude of y increases.

a) Inverse proportions are in the form

$$y = k\frac{1}{x}$$

where k is the *constant of proportionality.*

thus $k = xy$

Graph	Formula	Comments
	$C = \pi d$	• The circumference of a circle is directly proportional to the diameter, *d*. • π is the constant of proportionality.
	$V = \frac{1}{P}$	• The volume of a gas varies inversely with the pressure, *P*. • The constant of proportionality is 1.

8. *What are the formulas for distance and velocity?*

✍ ____

____ *(See text problem #39 p. 9)*

Recall that distance is the product of rate (velocity) and time:

Formula	Example
distance = (*velocity*)(*time*)	$d = 50\,\frac{mi}{hr} \times 4\text{ hr} = 200\,miles$
$Velocity = \dfrac{\text{distance}}{\text{time}}$	$v = \dfrac{200\,miles}{4\,hours} = 50\,\frac{mi}{hr}$

Key Notation

- Function notation is shorthand for many words.
- Usually the middle part of the alphabet is reserved for function symbols: *f, g, h,*. The symbol is made up of two parts, the function name and the variable. It is critical that you use and interpret function notation carefully. Be careful to include the units of the variables.

Symbol	Words	Input Variable
f(x)	*"f is a function of x" or* *"f of x"*	*x*
g(d)	*"g is a function of d"* *"g of d"*	*d*
h(t)	*"h is a function of t"* *"h of t"*	*t*

- $\dfrac{\Delta y}{\Delta x}$ means either the rate of change of y with respect to *x* or the slope of the line.
- $m = \dfrac{f(x_2) - f(x_1)}{x_2 - x_1}$ is the preferred rate of change formula in calculus. It is called the difference quotient.
- $y = m(x - x_1) + f(x_1)$ is the preferred linear equation form in calculus.
- A problem stated in words can be changed to function notation. For example if an investment of $100 at an interest rate, *r (%),* compounded annually for 10 years, yields a balance of *B* dollars , the function notation is $B = f(r)$. Here the rate *r* is the independent variable and *B* is the dependent variable. $f(8) = 215$ means that $100 invested at 8% interest for 10 years will yield a balance of $215.00.

✍ Your Turn

Write the following formulas in function form.

1. Circumference is twice the radius times π . ____________
2. The intensity of a light, *I*, is inversely proportional to the square of the distance, *d*, between the light and the observer. ____________
3. The tide of the ocean generated by the moon, *T*, varies directly with the mass of the moon, *m*, and inversely with the cube of the distance between the earth and moon, *d*. ____________

Note: the mass of the moon is constant while the distance varies.

Answers

$C(r) = 2\pi r$

1. $I(d) = \dfrac{k}{d^2}$

2. $T(d) = \dfrac{km}{d^3}$

Key Algebra

- When finding the slope, begin with the same point to find the change in y and the change in x.
- Check the graph to verify that increasing lines have positive slope and decreasing lines have negative slopes.
- Double check that you are finding the slope by dividing correctly. Slope = the vertical change divided by the horizontal change.

🗁 Study Skills

- Choose at least three problems that represent this section and put each on a 3 x 5 note card with the solution on the back.
- Use the ✔ ? ✭ system on homework problems and get answers to all ✭ problems.
- Consult an algebra text or see your instructor if you have questions.
- Learn to use the technology preferred by your instructor. The text assumes you have a graphing calculator available or a computer algebra system.

1.2 EXPONENTIAL FUNCTIONS

Objective:	To determine all of the properties of exponential functions.
Preparations:	Read and take notes on section 1.2.
	Suggested practice problems: 1, 5, 8, 10, 13,17, 19, 22, 24, 29, 36, 38.
	Problems requiring substantial algebra: 19-21, 22, 23, 31, 35, 36.

✠ Key Ideas of the Section

You should be able to:

- Draw an exponential growth and exponential decay graph.
- Identify the qualities of the family of functions, $y = ka^x$, for a > 1 and 0 < a < 1.
- Determine the common ratio, $f(x+1)/f(x)$, of an exponential function and relate it to the percent increase or percent decrease.
- Use the appropriate exponential formulas for doubling and half time.
- Identify e as an irrational number, 2.718281... , associated with continuous growth.
- Identify the continuous growth or decay functions as $Q = Q_0 e^{kt}$ or $P = P_0 e^{kt}$
- Compare and contrast linear and exponential functions.

Key Questions to Ask and Answer

1. ***What do the variables in an exponential growth equation, $P = P_0 a^t$, represent and how do they affect the graph?***

✍ ____

____ *(See text p. 10.)*

The exponential function is used often in science because it can model the growth or decay of populations. When a population increases by a constant percent per year it has a distinctive concave up shape.

All exponential functions have the algebraic form $y = ka^x$. The form that is generally used to describe a population growth or decline over time t is: $\boldsymbol{P = P_0\, a^t}$, where

P_0 = the initial value at $t = 0$ since $P = P_0\, a^0 = P_0(1) = P_0$

a = the base, it represents the factor by which you multiply. a is also know as the growth or decay factor. For each unit change in t, P will be multiplied by a units.

t = the exponent. If t is a positive integer it tells the number of factors to multiply.

Example 1.2.1

Determine the value of the exponential function for the indicated time t.

a) Let $P(t) = 100(1.5)^t$ for $t = 0, 1, 2, 3, 4$.

$P(0) = 100$

$P(1) = 100(1.5) = 150$

$P(2) = 100\ (1.5)(1.5) = 225$

$P(3) = 100(1.5)(1.5)(1.5) = 337.5$

$P(4) = 100(1.5)(1.5)(1.5)(1.5) = 506.25$

Note: Since the base is 1.5, the next value is found by multiplying by 1.5, or it is 1.5 times as large as the preceding value.

b) Let $Q = 100(.75)^t$ for $t = 3$.

$Q = 100(.75)(.75)(.75) = 42.1875$

c) Sketch the graph of the above functions on a grapher and find which values of the base a makes the exponential increase or decrease.

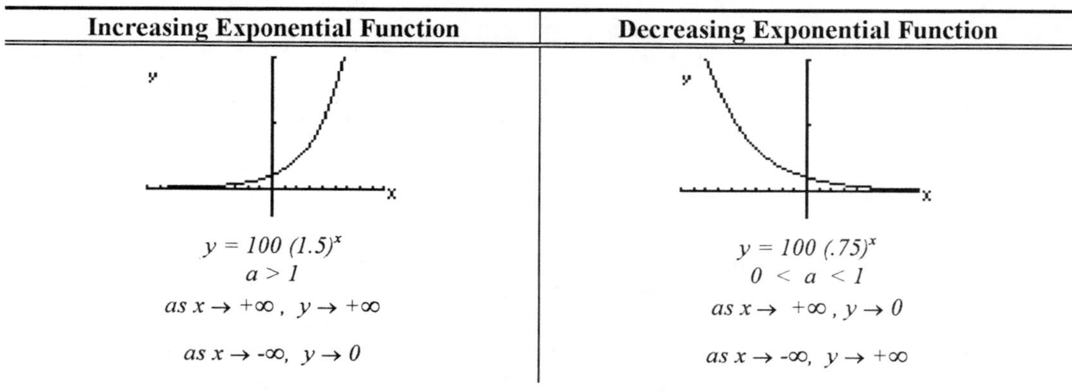

Increasing Exponential Function	Decreasing Exponential Function
$y = 100\ (1.5)^x$	$y = 100\ (.75)^x$
$a > 1$	$0 < a < 1$
as $x \to +\infty$, $y \to +\infty$	as $x \to +\infty$, $y \to 0$
as $x \to -\infty$, $y \to 0$	as $x \to -\infty$, $y \to +\infty$

2. *How is percent increase related to the base?*

✍ ____

______________________________________ *(See text p. 10.)*

Example 1.2.2

The number of reported cases of influenza increased by 50% each week over a 10 week period. At the start of the epidemic 100 people had the flu. Make a chart of the data and find the algebraic model for the number of people getting the flu. Make several observations about the data.

Week n	50% increase of flu cases	People with flu P_n (rounded)	Rate of Change $\frac{P_{n+1}-P_n}{W_{n+1}-W_n}$	Common Ratio $\frac{P_{n+1}}{P_n}$	Base of the exponential function (rounded)
0	100	100	-	-	-
1	100+.50(100)	150	50	150/100 =	1.5
2	150+.50(150)	225	75	225/150 =	1.5
3	225+.50(225)	338	113	338/225 =	1.5
4	338+.50(338)	506	169	506/338 =	1.5
5	506+.50(506)	759	253	759/506 =	1.5
6	759+.50(759)	1139	380	1139/759 =	1.5
7	1139+.50(1139)	1709	570	1709/1139 =	1.5
8	1709+.50(1709)	2563	854	2563/1763 =	1.5
9	2563+.50(2563)	3844	1281	3844/2563 =	1.5
10	3844+.50(3844)	5767	1922	5767/3844 =	1.5

- The original or initial population P_0 was 100 people.
- From the chart we see that each week the number increased by 50%, or the previous week's total was multiplied by .50 then added to the previous week amount.
- The rate of change shows that the number of flu cases is **increasing at an increasing rate.**
- The ratio of two succeeding values is 1.5. Think of this as 1 + .50, the original plus the 50% increase. This is the same as multiplying by 1.5.
- The exponential function is $\boldsymbol{P = 100\ (1.5)^{t}}$, since the initial population was 100 and the base or common ratio was 1.5.
- The function has a weekly growth factor of 1.5. The weekly rate of increase is 50%

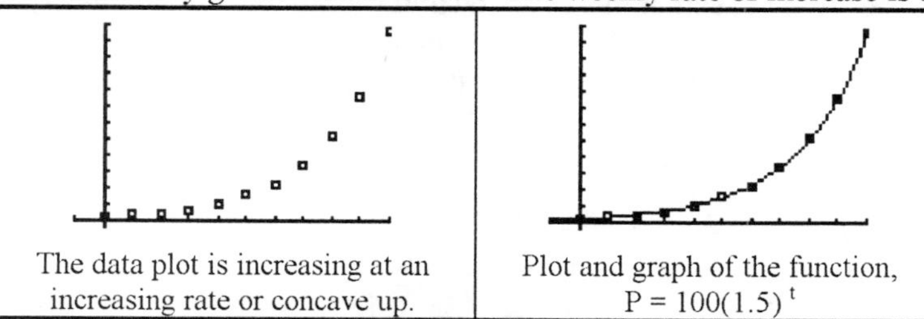

The data plot is increasing at an increasing rate or concave up. | Plot and graph of the function, $P = 100(1.5)^{t}$

3. ***Can the base of an exponential function always be thought of as a percent increase or percent decrease?***

✍ ____

____ *(See text p. 12-13)*

Recall that an *increasing exponential* function has a base that is greater than one, $a > 1$, so the base can be written as:

base = (1 + rate), where *rate* is the percent increase.

An increasing exponential is of the form: $\boldsymbol{P = P_0(1 + r)^{t}}$.

For a *decreasing exponential* function the base is less than one, $0 < a < 1$, so the base can be written as :

base = (1 - rate) where *rate* is the percent decrease.

A decreasing exponential is of the form: $\boldsymbol{P = P_0(1 - r)^{t}}$.

✎ Your Turn

Fill in the missing parts of the following table.

% Change	Base ($1 \pm r$)	Function
15% increase	(1 + .15)	$P = P_0(\ ?\ )^t$
10% decrease	?	$P = P_0(.90)^t$
?	(1 - .20)	$P = P_0(.80)^t$
?	?	$P = P_0(\ 3\ )^t$

Answers
$P = P_0(1.15)^t$
Base = (1 - .10)
20% decrease..
200% increase.
Base = (1 + 2)

4. *What are the formulas for doubling time and half life?*

✍ ____

____ *(See text p. 12)*

Doubling time of an exponentially increasing function is the time h necessary for the function to double in value:

$$P = P_0(2)^{(t/h)}$$

Half life of an exponential function is the time h necessary for the function to be reduced to half of its value.

$$Q = Q_0\left(\frac{1}{2}\right)^{(t/h)}$$

Example 1.2.3

Write an exponential equation for the following:

a) A town with population 500 initially doubles every 5 years.
b) A town of 10,000 is expected to lose half of its population every 10 years.

Solution:

a) $P = 500(2)^{\frac{t}{5}} = 500\left(2^{\frac{1}{5}}\right)^t = 500(1.18698)^t$

b) $P = 10{,}000\left(\frac{1}{2}\right)^{\frac{t}{10}} = 10{,}000\left(\left(\frac{1}{2}\right)^{\frac{1}{10}}\right)^t = 10{,}000(0.93303)^t$

5. *What is the effect of changing the value of P_0 ?*

✍ ____

____

Since P_0 represents the initial value of the population at $t = 0$. If t is represented by the horizontal axis and P is represented by the vertical axis, changing P_0 changes the *P-intercept*. Note that if $t = 0$, then $P = P_0 a^0 = P_0(1) = P_0$.

Example 1.2.4

Sketch some graphs of $P = P_0 a^t$, altering the values of P_0. Explain the effects of making $P_0 = 1,2,-2$ for both increasing and decreasing exponential functions.

Increasing exponential function $P = P_0a^t$ with $P_0 = 1,2,-2$ and $a > 1$.	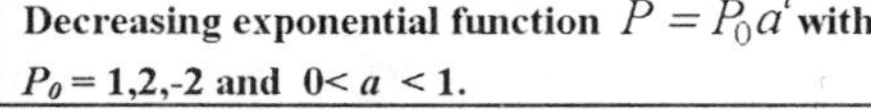 Decreasing exponential function $P = P_0a^t$ with $P_0 = 1,2,-2$ and $0 < a < 1$.
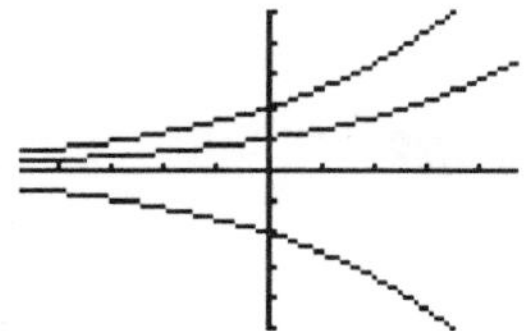	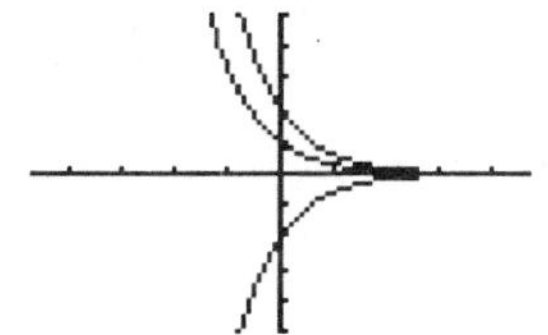
Effects $P_0 = 1$, the *y-intercept* is at 1, concave up and increasing. $P_0 = 2$, the *y-intercept* is at 2, concave up, increasing and stretched $P=a^t$ by a factor of |2|. $P_0 = -2$, the *y-intercept* is at -2, the graph has rotated $P = 2a^t$ about the *t-axis*, the graph is concave down, decreasing and stretched $P=a^t$ by a factor of |-2|.	**Effects** $P_0 = 1$, the *y-intercept* is at 1, concave up and decreasing. $P_0 = 2$, the *y-intercept* is at 2, concave up, decreasing and stretched $P=a^t$ by a factor of |2|. $P_0 = -2$, the *y-intercept* is at -2, the graph has rotated $P = 2a^t$ about the *t-axis*, the graph is concave down, increasing and stretched $P=a^t$ by a factor of |-2|.

Summary: Changing the value of P_o, changes the *P-intercept* and stretches the function by a factor of $|P_0|$. If $P_0 < 0$ it also rotates the graph about the *t-axis.*

6. *What is the number e?*

✍ ____

____ *(See text p. 13)*

The irrational number e is approximately 2.71828... ; a non-terminating, non-repeating decimal. When e is used as the base of an exponential function, $y = e^x$, we see that the graph is sandwiched between the functions 2^x and 3^x. This exponential function e^x is an important function for modeling and therefore is important in calculus.

✍ Your Turn

List some of the attributes of $y = e^x$.

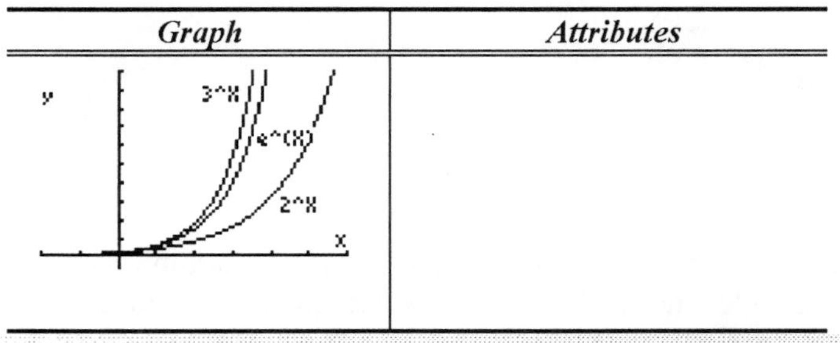

Possible Answers

- $y = e^x$ is an exponential function.
- $y = e^x$ is close to 3^x
- $y = e^x$ crosses the y-axis at (0,1).
- e^x climbs more rapidly than 2^x.

The base of an exponential function a can be written as e^k, where k is a *continuous growth rate.* So $P = P_0a^t = P_o(e^k)^t$. Use e^k when you want to express continuous growth rate.

Example 1.2.5

a) An hourly lab report shows that an initial population of 15,000 bacteria grows 5% continuously. Write the formula for the function in two forms, where t is time in hours.

$P = 15000(e^{0.05})^t = 15000(1.05127)^t$. A continuous 5% growth rate is equivalent to 5.127% hourly growth rate. $e^{.05} = 1.05127 = 1 + .05127 = 1 + 5.127\%$.

b) 50 mg of a radioactive material is expected to decay continuously at 5% over t years. Write the formula in two forms

$P = 50(e^{-.05})^t = 50(.95123)^t$. A <u>continuous</u> 5% decay rate is equivalent to an annual 4.877% decay rate. $e^{-.05} = 0.95123 = 1 - 04877 = 1 - 4.877\%$.

6. *How does exponential growth compare to linear growth?*

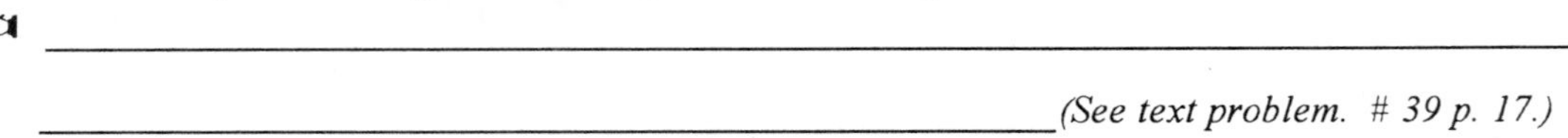

(See text problem. # 39 p. 17.)

For linear growth as x increases by one unit y is <u>added</u> by a constant. For exponential growth as x is increased by one unit, y is <u>multiplied</u> by a constant.

Example 1.2.6

Compare $y = 1.25x + 2$ to $y = 2(1.25)^x$

function	*kind of change*	*table*	*graph*
$y = 1.25x + 2$	Has an average rate of change of $m = 1.25$. 1.25 is <u>added</u> to the value of y, for each unit change in x.	X / Y1: -1 / .75; 0 / 2; 1 / 3.25; 2 / 4.5; 3 / 5.75; 4 / 7; 5 / 8.25 — Y1=1.25X+2	
$y = 2(1.25)^x$	Has a percent change of 25%. (1.25-1=.25= 25%). y is <u>multiplied</u> by 1.25 for each unit change in x,	X / Y2: -1 / 1.6; 0 / 2; 1 / 2.5; 2 / 3.125; 3 / 3.9063; 4 / 4.8828; 5 / 6.1035 — Y2=2(1.25)^X	

Both functions have the same y-intercept since when $x=0$, $y=2$, however, the graphs show that the exponential function grows at a faster rate for large x, and will eventually dominate the linear function.

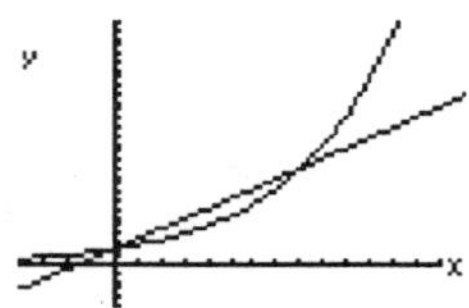

Key Notation

- $x \to \infty$, "Let the value of x assume larger and larger positive values (positive infinity)"
- $x \to -\infty$, "Let the value of x assume smaller and smaller negative values (negative infinite)"
- $f(x) \to L$, "The value of the function gets closer and closer to the value L, where L represents the horizontal line $y = L$"
- $x \to k$, "The value of x gets closer and closer to the value k, where k represents a vertical line $x = k$".

Key Algebra

- Review horizontal asymptote or saturation level (text p. 14).

Key Study Skills

- Choose at least two problems that represent this section; put each on a 3 x 5 note card with the solution on the back. Review the note cards for 1.1-1.2.
- Use the ✓ ? ★ system on homework problems and get answers to all ★ problems.

1.3 NEW FUNCTIONS FROM OLD

Objective: To transform the graphs of functions with shifts, flips and stretches and to algebraically compose new functions.
To be able to find the inverse of a function algebraically, using tables of values and using graphical methods

Preparations: Read and take notes on section 1.3 and Ready Reference in the back of the text.
Suggested practice problems: 1, 2, 4, 7, 8, 9, 15, 17, 24, 25, 28, 32, 35.
Problems requiring substantial algebra: 1-3, 5-8.

✠ Key Ideas of the Section

You should be able to:

- Recognize odd and even functions.
- Identify that multiplying a function by a non-zero constant stretches or shrinks the graph of a function.
- Identify that multiplying a function by a negative constant flips/reflects the graph about the x -axis.
- Show that a graph moves up or down when a constant is added to the function.
- Show that replacing x by $x - h$ moves the graph right or left.
- Algebraically compose a new function by recognizing the outside and the inside function.
- Recognize the notation for the inverse function, f^{-1}.
- Describe an inverse algebraically and graphically.
- Recognize when a function has an inverse that is also a function.
- Find the domains and range of f and of f^{-1} .

Key Questions to Ask and Answer.

1. *How do you stretch or shrink a function?*

✍ ____

____ *(See text p. 17.)*

In section 1.2 of this study guide, page 13, we saw that multiplying a function $f(x)$ by a non-zero constant c, $c\,f(x)$, produces three changes to the graph depending on the value of c. It is best to first think of c *without* regard to its sign, or the absolute value of c, $|c|$.

Value of c	*Change in the graph of f(x)*
$\|c\| > 1$	$f(x)$ vertically stretches
$0 < \|c\| < 1$	$f(x)$ vertically shrinks
$c < 0$	$f(x)$ reflects about x-axis

Example 1.3.1

Stretch or shrink then flip $f(x) = x^3$

Transformation	*Function*	*Graph*
This is the parent function with no stretch or shrink.	$y = x^3$	y x

Stretch vertically by a factor of 5. Note: $c = 5$ $5 > 1$	$y = 5x^3$	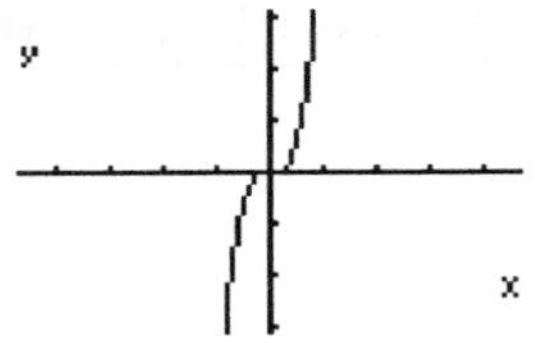
Shrink vertically by a factor of 1/10. Note: $c = 1/10$ $0 < 1/10 < 1$	$y = 0.1x^3$	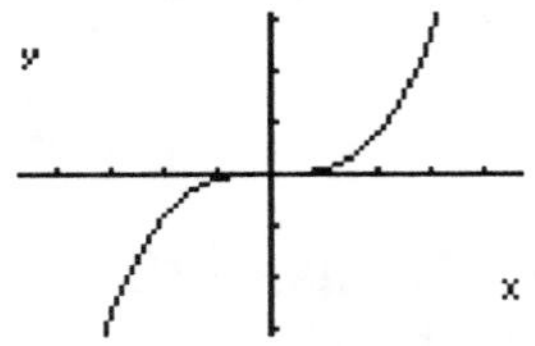
Shrink by a factor of 1/10 and reflect about the *x*-axis. Note: $c = {}^{-}1/10$ ${}^{-}1/10 < 0$	$y = -0.1x^3$	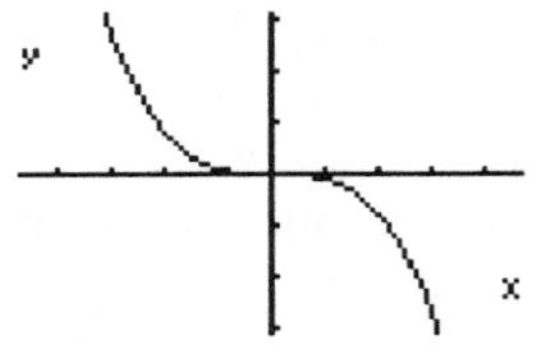

2. *How do you shift a graph up or down?*

✍ ____

____ *(See text p. 17.)*

Adding a constant k to a function moves the graph up or down $|k|$ units depending on the sign of k. This vertical translation does not change the shape of the graph. In essence you are changing the y -intercept.

Value of k	*Change in the graph of f(x)*
$k > 0$	shift k units up
$k < 0$	shift $|k|$ units down

Example 1.3.2

Shift the graph of $f(x) = x^3$ up and down.

Transformation	*Function*	*Graph*
This is the parent function with no shift.	$y = x^3$	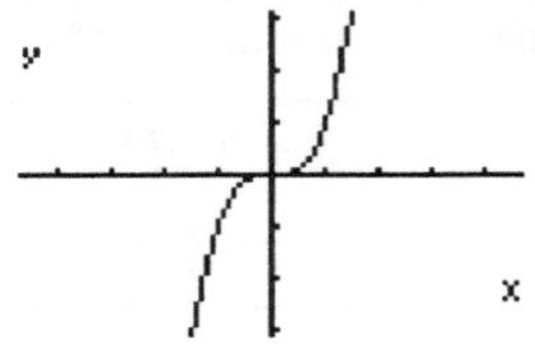
Shift up by 2 units.	$y = x^3 + 2$	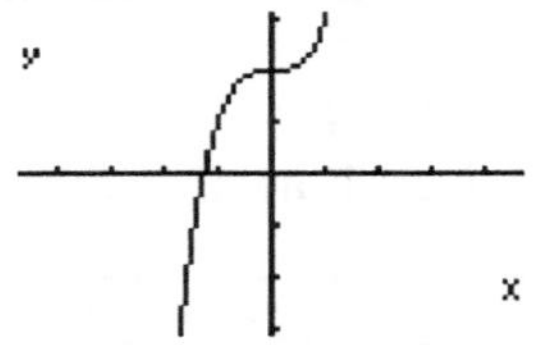
Shift down by 1.5 units.	$y = x^3 - 1.5$	

3. *How do you shift a graph left or right?*

✍ ____

____ *(See text p. 17.)*

Replacing x in $f(x)$, by $x - h$, causes a horizontal shift right or left depending on the sign of h. Think of this as $x - (h)$. You will move the graph $|h|$ units to the right for $h > 0$ or $|h|$ units to the left if $h < 0$. Instead of the graph being "centered" on the *y-axis*, the graph is "centered" on the line $x = h$. Horizontal shifts do not change the shape of the graph.

Value of h	*Change in the graph of f(x)*
$h > 0$	shift h units right
$h < 0$	shift $\|h\|$ units left

Example 1.3.3

Shift the graph of $f(x) = x^3$ horizontally.

Transformation	*Function*	*Graph*
This is the parent function with no shift.	$y = x^3$	 y x
Shift right by 3 units.	$y = (x - (3))^3 = (x - 3)^3$	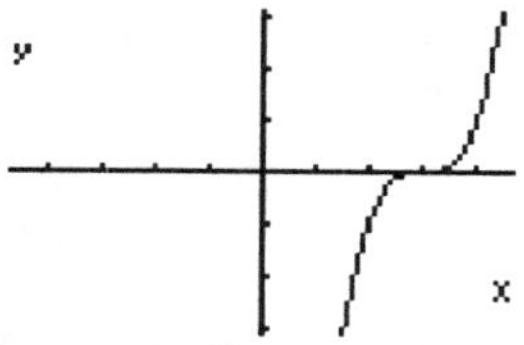 y x
Shift left by 2 units.	$y = (x - (-2))^3 = (x + 2)^3$	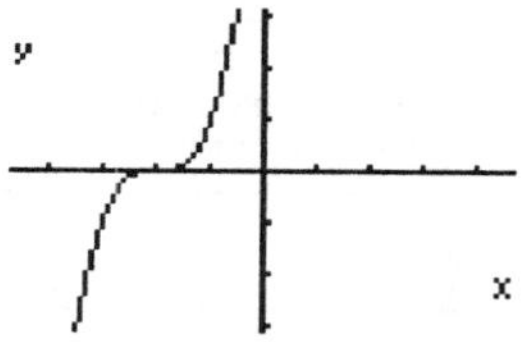 y x

4. *How do I put the transformations all together?*

✍ ____

____

Transform the function $f(x)$ into a new function $h(x)$, by using all of the above procedures:

$$h(x) = c\,f(x - h) + k$$

The order in which you perform the transformations is very important, i.e. first shift left or right, then vertically stretch or shrink, then reflect and finally shift up or down.

✎ On Your Own 1

Transform the graph of the function $g(x) = \sqrt{x}$, by using the instructions below and create a new function $h(x)$.

- Shift left by 3 units.

- Vertically stretch by 4 units.
- Reflect about the x -axis.
- Shift down by 2 units.

(The answer appears at the end of this section.)

5. *What is symmetry?*

✍ ____

______________________________________ *(See text p. 19.)*

Symmetry is present if there is a balance to the graph. If the y - axis acts as the fold line or rotation line, the right hand side will lie on top of the left side of the graph, thus an opposite value of x has the same y value.

f is an even function if it has y-axis symmetry. That is, if f(-x) = f(x).

There is symmetry about the origin if for each opposite value of x there is an opposite value of y. If two rotations occur, first about the y -axis then about the x-axis, a graph has origin symmetry.

f is an odd function if it has origin symmetry, that is, if f(-x) = -f(x).

✎ On Your Own 2

Draw the missing part of the graph for the indicated symmetry.

Symmetry	*Graph*
y-axis symmetry (even function) Hint: the points (2,0) and (3,0) reflect to (-2,0) and (-3,0). The opposite of x produces the same y.	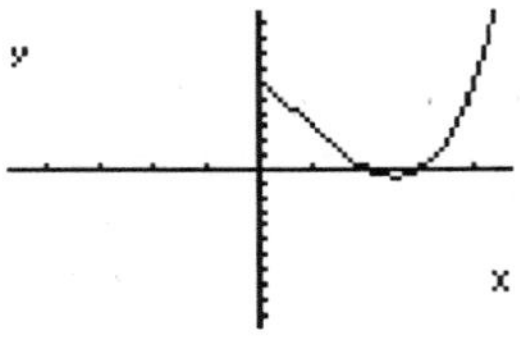 y x
Origin symmetry (odd function) Hint: the points (0,6) and (4,7) reflect to (0,-6) and (-4,-7). The opposite of x produces the opposite of y.	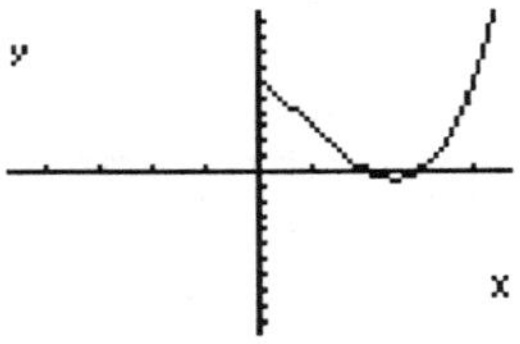 y x

(The answer appears at the end of this section.)

6. *What are composite functions?*

✍ ____

______________________________________ *(See text p. 18.)*

You can compose a new function from two others, creating a *function of a function.* $h(x) = f(g(x))$, pronounced: "*f* of *g* of *x*", means in the function $f(x)$ replace x with the value of the function $g(x)$. It is common to call g the inside function and f the outside function. The table below allows you to identify the inside function, $g(x)$ and the outside function, $f(x)$ for the composite function, $h(x)$.

h(x)=f(g(x))	*g(x), the inside function*	*f(x), the outside function*
$h(x) = (3x+1)^2$	$g(x) = 3x+1$	$f(x) = x^2$
$h(x) = 5e^{0.08x}$	$g(x) = 0.08x$	$f(x) = 5e^x$
$h(x) = \ln(5-x^3)$	$g(x) = 5-x^3$	$f(x) = \ln(x)$

On Your Own 3

If $f(t) = (t+1)^3$ and $g(t) = \frac{1}{t+1}$,

a) find *f(g(t))*,
b) find *g(f(t))*,
c) fill in the following chart.

t	f(t)	g(t)	f(g(t))	g(f(t))
-2				
0				
2				

(The answer appears at the end of this section)

7. ***How do you find the inverse of a function algebraically and graphically?***

✍ ____

____ *(See text p. 19,21.)*

A function is invertible if the independent variable for f is the dependent variable of f^{-1}, and the result is a function. Thus the domains and ranges of f and f^{-1} are interchanged.

Example 1.3.4

If the circumference of a tree, in feet, can be found by using the formula for the circumference of a circle, $C = f(r) = 2\pi r$, where r is the radius in feet, then do the following:

a) Interpret the meaning of $f^{-1}(C)$.
b) Graph $f(r)$ and $f^{-1}(C)$ on the same axes.
c) Interpret $f^{-1}(12.566)$.

Solution:

a) *f(r)* means that if you know the radius you can find the circumference. The inverse of *C*, $f^{-1}(C)$ means that if you know the circumference you can find the radius. The table below illustrates.

r in feet	$C = 2\pi r = f(r)$
0	0.000
.5	3.142
1	6.283
1.5	9.425
2	12.566
2.5	15.708
3	18.850

C in feet	$r = \frac{C}{2\pi} = f^{-1}(C)$
0.000	0
3.142	.5
6.283	1
9.425	1.5
12.566	2
15.708	2.5
18.850	3

The graph of *f(r)* shows *r* as the input or independent variable. The graph of $f^{-1}(C)$ shows *C* as the input or independent variable. The independent variable is the *x*-axis. *The domain of f^{-1} is the range of f and the range of f^{-1} is the domain of f.*

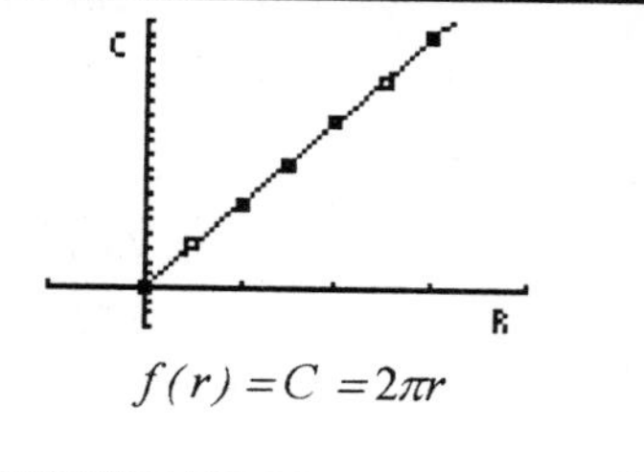

$f(r) = C = 2\pi r$

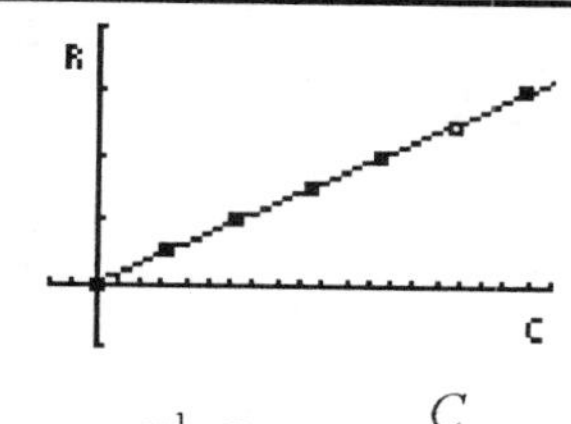

$f^{-1}(C) = r = \frac{C}{2\pi}$

b) To graph f *and* f^{-1} on the same axes, let's revert to x, y variables, with y as the dependent variable and x as the independent variable. Here y represents the circumference and x represents the radius. Find the inverse algebraically.

$Circumference = f(x)$ $y = 2\pi x$	• Rewrite using x as the radius and y as the circumference.
$x = 2\pi y$	• Switch x and y.
$radius = f^{-1}(x)$ $f^{-1}(x) = y = \dfrac{x}{2\pi}$	• Solve for y. • Now y represents the radius and x represents the circumference.

Below are the graph of f and f^{-1} on the same axis. Notice that f^{-1} is the reflection of the graph of f about the line $y = x$.

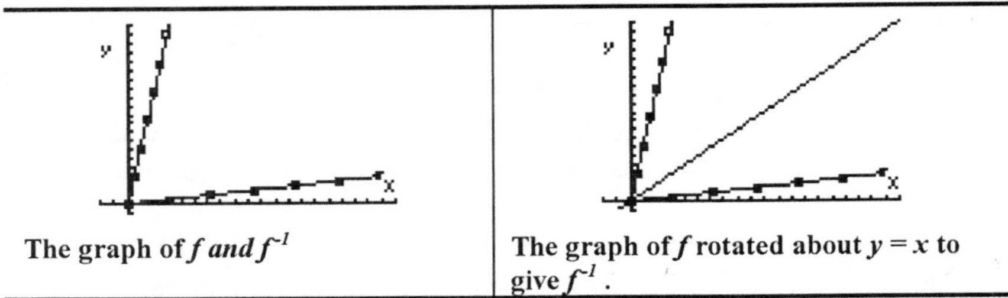

The graph of f *and* f^{-1}	The graph of f rotated about $y = x$ to give f^{-1}.

c) The interpretation of $f^{-1}(12.566)$ says that you know the circumference of a tree is 12.566 ft, find the radius. This is in contrast to $f(2)$, which says you know the radius is 2 ft.

8. *How can you sketch graph of the inverse of a function, when you don't know the formula?*

✍ ____

____

If you have a table of values you can sketch the curve. To sketch the inverse curve reverse the input and output values or rotate the graph of the original curve about the line $y = x$. The combination of the two procedures gives an accurate graph of f^{-1}.

✎ Your Turn 1

Given the table of values below for f, you fill in the table of values for f^{-1}.

x	-1	-0.5	0	0.5	1
$f(x)$	2	1.375	0	-1.375	-2

x	2	?	?	-1.375	?
$f^{-1}(x)$	-1	-.5	?	?	?

Answers: $f^{-1}(1.375) = -.5$ $\quad f^{-1}(0) = 0$ $\quad f^{-1}(-1.375) = .5$ $\quad f^{-1}(-2) = 1$

✎ Your Turn 2

Below is the graph of f, sketch the graph of f^{-1} and determine the domain and range of f^{-1}.

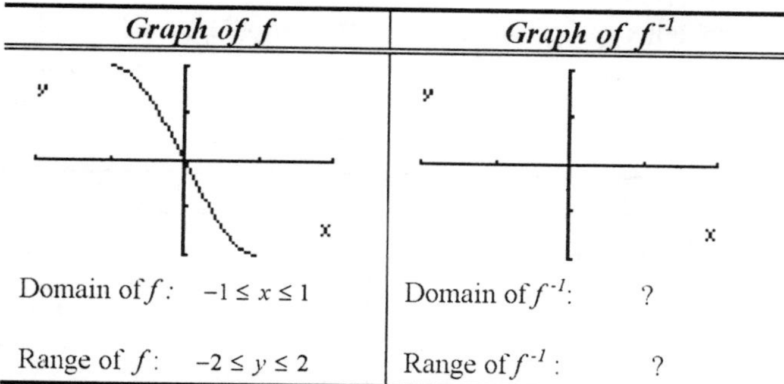

Graph of f	*Graph of f^{-1}*
Domain of f: $-1 \le x \le 1$	Domain of f^{-1}: ?
Range of f: $-2 \le y \le 2$	Range of f^{-1}: ?

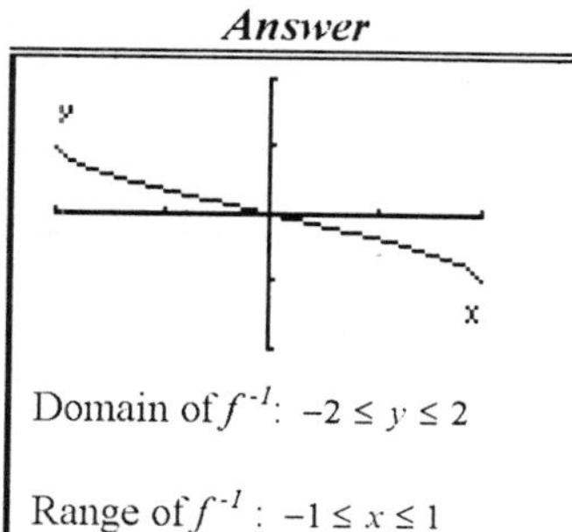

Answer

Domain of f^{-1}: $-2 \le y \le 2$

Range of f^{-1}: $-1 \le x \le 1$

9. *Which functions have inverses?*

✍ ____

____ *(See text p. 20.)*

Only functions that pass the horizontal line test have inverses. If a horizontal line can intersect the graph of a function in more than one point, the function *does not* have an inverse. Often you need to limit the domain of a function to find its inverse.

f has an inverse function, because it passes the horizontal line test	***g does not have an inverse, because it fails the horizontal line test.***
y, x	y, x
$f = x^3$ $f^{-1} = x^{1/3}$ For each x there is a unique value of y.	$g = x^3 - 3x$ ***g^{-1} can not be found*** For each x there is NOT a unique value of y.

Special Note: **Many graphers can draw the inverse of a graph. The grapher is simply reversing points and not finding the inverse algebraically.**

Key Notation

- Some instructors use the notation $f(g(x)) = (f \circ g)(x)$. Both mean replace x with $g(x)$ in the function $f(x)$.
- Order is important. In most cases $f \circ g \ne g \circ f$.

Key Algebra

- Practice the composition of functions, like the ones in On Your Own 3.

Key Study Skills

- Choose at least two problems that represent this section; put each on a 3 x 5 note card with the solution on the back. Review the cards for 1.1-1.3.
- Use the ✓ ? ☆ system on homework problems and get answers to all ☆ problems.

Answer to On Your Own 1

Transformed function	*Graph*
$h(x) = -4\sqrt{x-(-3)} - 2 = -4\sqrt{x+3} - 2$	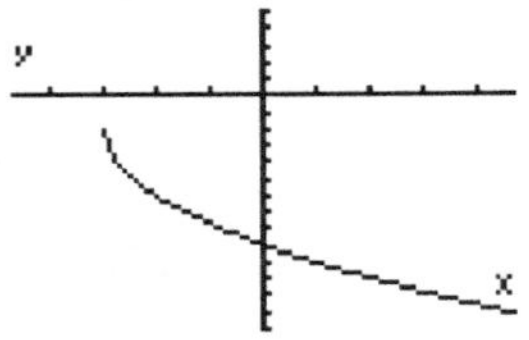

Answer to On Your Own 2

Symmetry	*Graph*
y-axis symmetry (even function)	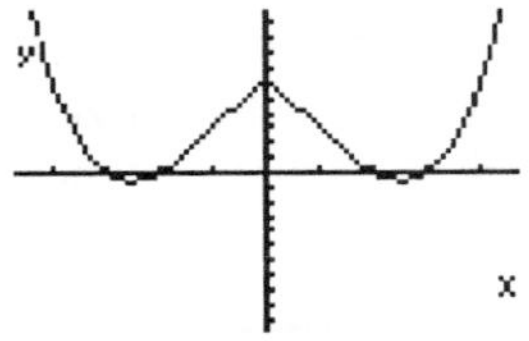
Origin symmetry (odd function)	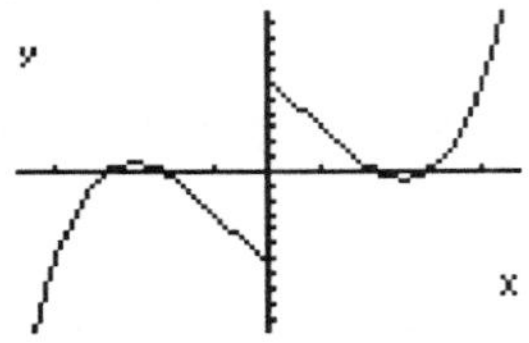

Answer to On Your Own 3

t	f(t)	g(t)	f(g(t))	g(f(t))
	$(t+1)^3$	$\frac{1}{t+1}$	$(\frac{1}{t+1}+1)^3$	$\frac{1}{(t+1)^3+1}$
-2	$(-2+1)^3 = -1$	$\frac{1}{-2+1} = \frac{1}{-1} = -1$	$(\frac{1}{-2+1}+1)^3 = 0$	$\frac{1}{(-2+1)^3+1} = \frac{1}{0} =$ undefined
0	$(0+1)^3 = 1$	$\frac{1}{0+1} = 1$	$(\frac{1}{0+1}+1)^3 = 2^3 = 8$	$\frac{1}{(0+1)^3+1} = \frac{1}{2}$
2	$(2+1)^3 = 27$	$\frac{1}{2+1} = \frac{1}{3}$	$(\frac{1}{2+1}+1)^3 = (\frac{4}{3})^3 = \frac{64}{27} = 2.370...$	$\frac{1}{(2+1)^3+1} = \frac{1}{28} =$ .03571...

1.4 LOGARITHMIC FUNCTIONS

Objective: To define a logarithm as the inverse of the exponential function.

Preparations: Read and take notes on section 1.4.
Suggested practice problems: 3, 7, 11, 17, 23, 26, 29, 33, 35, 36, 38, 42, 43, 44, 45, 47, 50.
Problems requiring substantial algebra: 2-28, 37-50.

✠ Key Ideas of the Section

You should be able to:

- Recognize that the value of a logarithm is the value of the exponent in an exponential function.
- Solve exponential equations using both properties of logarithms and a grapher.
- Notice that since a logarithm is the inverse of an exponential function; the exponential function grows very rapidly and the logarithmic function grows slowly as $x \to \infty$.
- Identify that the graph of an exponential function has a horizontal asymptote and the graph of a logarithmic function has a vertical asymptote.
- Define the natural logarithm as the logarithm to the base e :
    - $\log_e x = \ln x$.
- Identify the continuous growth or decay functions as $Q = Q_0 e^{kt}$ or $P = P_0 e^{kt}$.

Key Questions to Ask and Answer

1. *What is a logarithm?*

✍ ____

____ *(See text p. 23.)*

A logarithm is a *number*. It represents the exponent of an exponential equation. It is the inverse or "undoes" the exponential function.

Let $N = 10^c$ ***, an exponential function where*** $b = 10$ ***and*** $N>0$ ***,***
then $\log_{10} N = c$

$\log_{10} N$ means the logarithm of a number, N, to the base 10 is equal to the exponent of an exponential equation.

Example 1.4.1

$\log_{10} 1000 = 3$ because $1000 = 10^3$. (the logarithm of 1000 is 3, the exponent of 10^3).
$\log_{10} .0001 = -4$ because $.0001 = 10^{-4}$ (the logarithm of 0.0001 is -4, the exponent of 10^{-4}).

$\log_{10} N$ is called the *common logarithm* and is generally written as $\log N$. All types of scientific calculators have built in values of logarithms.

✎ Your Turn

Fill in the missing parts on the chart below.

Exponential Form	*Logarithmic Form*
$10^5 = 100{,}000$	log 100000 =?
?	log 14 = 1.46128....
$10^? = 316.227766$	log 316.227766 = ?

Answers
5
$10^{1.46128} = 14$
2.5

2. *When are logarithms used?*

✍ ____

____ (See text p. 25-27.)

Logarithms are used to solve many exponential equations. A review of the properties of logarithms appears at the end of this section. The most useful property of logarithms is: *you can take the logarithm of both sides of an equation*; just like you can add or subtract on both sides of an equation and create an equivalent equation, (as long as the domain restrictions apply).

If $a = b$ $(a, b > 0)$, then $\log a = \log b$.

Example 1.4.2

In 1990 the number of families living below the poverty line in the US was 33,585 thousand families[3]. By 1992 the number increased by 10%. If this trend continues, determine when 50,000 thousand families will be living below the poverty line.

a) Using the equation $P = P_0 a^t$ assign values to the variables, using the information from the problem.

✍

P_0 = ______________________.(the initial value)

a = ______________________. (the base or 1+ rate)

t = ______________________. (the time/interval period)

P = ______________________. (the final amount)

Answers
33,585
1.10
t/2, t=0=1990
$33585(1.1)^{t/2}$

b) Form the equation to solve.
Let P = 50,000 thousand families.

$$50,000 = 33585(1.1)^{\frac{t}{2}}$$
- Since the data is given as a two year interval, divide t by two.

$$\frac{50000}{33585} = (1.1)^{\frac{t}{2}}$$
- Divide by 33585

$$\log\left(\frac{50000}{33585}\right) = \log\left((1.1)^{\frac{t}{2}}\right)$$
- Take the log of both sides.

$$\log\left(\frac{50000}{33585}\right) = \frac{t}{2}\log(1.1)$$
- Exponent property of logs.

$$\frac{\log\left(\frac{50000}{33585}\right)}{\log(1.1)} = \frac{t}{2}$$
- Divide by $\log(1.1)$

$$t = 2(4.17525)$$
- Multiply by two, solve for t.

$$t \approx 8.4 \ \textit{years}$$

c) Solve graphically.
Graph the equations as two functions and find the point of intersection using the built in intersection finder of a grapher.

Let $f(x) = 33585(1.1)^{x/2}$ and $g(x) = 50,000$

---

[3] Bureau of the Census, US Dept. of Commerce.

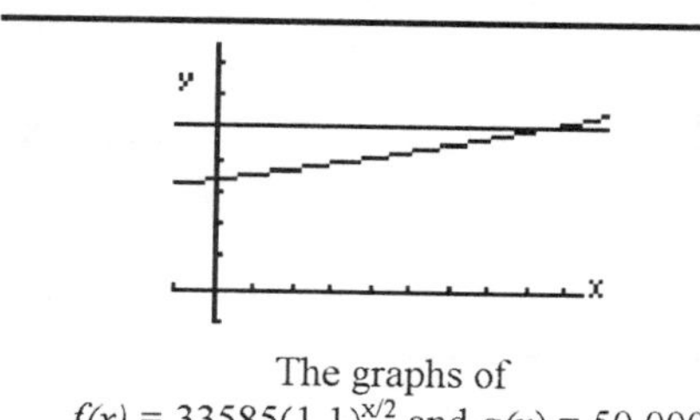 The graphs of $f(x) = 33585(1.1)^{x/2}$ and $g(x) = 50,000$	Intersection X=8.3504924 Y=50000 The point of intersection is the solution to $50000 = 33585(1.1)^{x/2}$, where $x \approx 8.4$ yr.

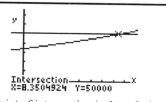

3. *How do you graph the inverse of an exponential function?*

✍ ____

____ *(See text p. 24.)*

Example 1.4.3

a) Find the inverse function of $y = 3^x$

$y = 3^x$

$x = 3^y$ • Reverse x and y.

$\log x = \log 3^y$ Take the logarithm of both sides of the equation.

$\log x = y(\log 3)$ • Exponent properties of logarithms.

$y = \dfrac{\log x}{\log 3}$ • Divide by $\log 3$, solve for y.

b) Graph the function and its inverse.

	$y = 3^x$	$y = \dfrac{\log x}{\log 3}$
Graph		
Domain	$x : (-\infty, \infty)$	$x : (0, \infty)$
Range	$y : (0, \infty)$	$x : (-\infty, \infty)$
Special features	• concave up • horizontal asymptote at $y = 0$	• concave down • vertical asymptote at $x = 0$
$x \to \infty$	y increases rapidly	y increases slowly

✎ Your Turn

Graph the function $f(x) = 2^x + 2$ and its inverse, f^{-1}, and compare their features. Fill in the appropriate blanks in the chart below.

	$f(x) = 2^x + 2$	f^{-1}	*Answers*
Graph:			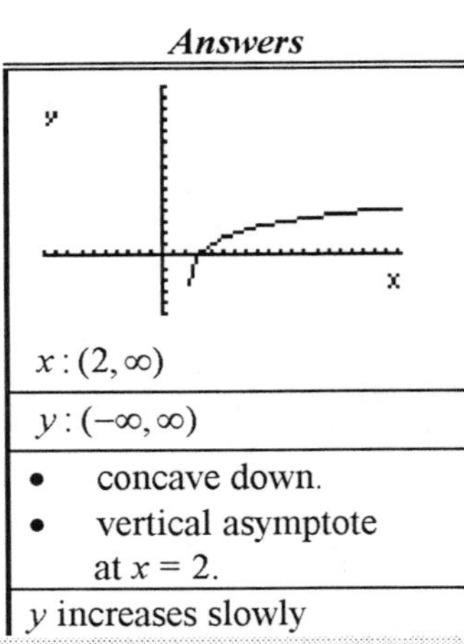
Domain:	$x:(-\infty,\infty)$	x: ?	$x:(2,\infty)$
Range:	$y:(2,\infty)$	y: ?	$y:(-\infty,\infty)$
Special features:	• concave up. • horizontal asymptote at $y = 2$.	• ? • ?	• concave down. • vertical asymptote at $x = 2$.
$x \to \infty$	y increases rapidly	y ?	y increases slowly

4. *What is the function f(x) = ln x ?*

✍ ____

____ *(See text p. 23,24.)*

The inverse function to $y = e^x$ is $y = \log_e x$, more commonly written as $y = \ln x$, "the natural logarithm". Natural logarithms are sometimes called Logarithm Napierian, thus the ln, after the Scottish mathematician John Napier who invented logarithms and was a pioneer in the use of our present decimal notation.

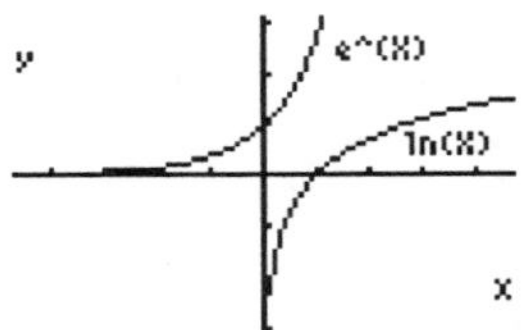

The graph of $f(x) = e^x$ and its inverse $f^{-1}(x) = \ln x$

Example 1.4.3

✍ Look at the graphs below, compare the graphs of $f(x) = \log x$ and $g(x) = \ln x$.

Graphs of $f(x) = \log x$ and $g(x) = \ln x$	**Comparison**

Possible Answers

- For $0<x<1$, $\log x > \ln x$. For $x>1$ $\ln x$, $> \log x$.
- $\ln x$ rises faster than $\log x$
- When $x=1$, $\ln x = \log x = 0$.
- as $x \to 0^+$, $\ln x \to -\infty$, $\log x \to -\infty$

3. *How are a^t and e^t related?*

✍ ____

____ *(See text p. 26.)*

In general exponential functions can be written in two ways:

$$P = P_0 a^t = P_0 e^{kt}.$$

This means that the base of the exponential function can be written as either a or e^k. Both functions have the same graph and represent the same function.

$$\text{If } a = e^k \text{, then } k = \ln a.$$

$$P = P_0 a^t = P_0 e^{(\ln a)t}$$

k can be thought of as the continuous growth rate, whereas a - 1 is the annual growth rate.

Example 1.4.4

Fill in the following chart.

$P = P_0 a^t$	$P = P_0 e^{kt}$	*Annual growth rate , (a - 1)*	*Continuous growth rate, k*		*Answers*
$P = P_0 (1.05)^t$	$P = P_0 e^{\ln 1.05\, t}$	1.05 - 1 = .05= 5%	$k = \ln 1.05 = ?$		.04879 ≈4.88% (increase)
$P = P_0 (.9)^t$	$P = P_0 e^{\ln .9\, t}$	.9 - 1 = -0.1=-10%	$k = \quad ?$		-0.10536 ≈ -10.54% (decrease)
$P = P_0 (\,?)^t$	$P = P_0 e^{.03\, t}$	?	$k = .03 = 3\,\%$		$e^{.03} \approx 1.03045$ $P_0 (1.03045\,)^t$ 1.03045-1=.03045 ≈ 3.05% (increase)

***Note:* A continuous growth rate of 4.88 % is equivalent to a 5% annual growth rate.**

On Your Own

A patient went to the hospital for a bone scan. She was told that 25 units of a radioactive isotope of molybdenum would be injected into her body so that a special machine could scan the flow of blood to the bone. She was also told that the isotope had a half life of 12 hours, and that it would be gone from her body in 24 hours. Determine if indeed the molybdenum would be out of her body after 24 hours using $Q = Q_0 e^{kt}$.

(The answer appears at the end of this section).

Key Algebra

- Remember that the value of a **logarithm is equal to the exponent** of the exponential function.
- ***If*** $10^x = 75$ ***then*** $\log_{10} 75 = x$.
- The logarithm of a number to the base ten is: $\log x$ (the base is understood to be ten).
- The logarithm of a number to the base e is: $\ln x$ (the base is understood to be e).
- Make up some examples to confirm the properties of natural logarithms and check your answers with a calculator.
- Determine why $\ln e = 1$ and $\ln 1 = 0$.
- Use the properties of logarithms to solve exponential equations. For natural logarithm just substitute *ln* into the properties.

Properties of logarithms

1. $\log (AB) = \log A + \log B$ ex: $\log 24 = \log (4 \bullet 6) = \log 4 + \log 6$
2. $\log (A/B) = \log A - \log B$ ex: $\log 24 = \log (48/2) = \log 48 - \log 2$

3. $\log (A)^p = p(\log A)$ — ex: $\log 3^2 = 2(\log 3)$
4. $\log (10) = 1$
5. $\log (10)^x = x$ — ex: $\log 10^4 = 4$
6. $10^{\log x} = x$ — ex: $10^{\log 5} = 5$
7. $\log 1 = 0$
8. $\log_b N = \log N/\log b$ — ex: $\log_7 49 = \log 49/\log 7$
9. If $y = x$ then $\log_b y = \log_b x$ — ex: $4^x = 16{,}384$

$\log 4^x = \log 16{,}384$

$x = (\log 16{,}384)/(\log 4) = 7$

Key Study Skills

- Choose at least two problems that represent this section; put each on a 3 x 5 note card with the solution on the back. Review the cards for 1.1-1.4.
- Use the ✔ ? ★ system on homework problems and get answers to all ★ problems.
- Consult an algebra text for more practice with logarithms.

Answer to On Your Own

$Q = Q_0 e^{kt}$ $\quad Q_0 = 25$ units $\quad Q = 25/2 = 12.5$ units $\quad t = 12$ hours

Find k, then determine Q for $t = 24$ hours.

$12.5 = 25e^{k(12)}$ — substitute values

$\frac{1}{2} = e^{k(12)}$ — divide by 25

$\ln\frac{1}{2} = \ln e^{k(12)}$ — take natural log of both sides

$\ln\frac{1}{2} = 12k$ — power property of logs, $\ln e = 1$

$k = \dfrac{\ln \frac{1}{2}}{12} \approx -.05776$ — divide by 12

Determine Q for t = 24

$Q = 25\,e^{-.05776\,(24)} \approx 6.25$ units of molybdenum left after 24 hours.

Note 1: There are many ways to do this problem.

Using $P = P_0 a^t$ for $a = \frac{1}{2}$, and time $= \frac{t}{12}$, since one time period is 12 hours.

$$P = 25(\tfrac{1}{2})^{\frac{t}{12}} = 25((\tfrac{1}{2})^{\frac{1}{12}})^t = 25(.94387)^t$$

for t = 24

$P = 25(.94387)^{24} \approx 6.25$ units of molybdenum after 24 hours.

Note 2: $$e^{\frac{\ln\frac{1}{2}}{12}} = \left(e^{\ln\frac{1}{2}}\right)^{\frac{1}{12}} = \left(\tfrac{1}{2}\right)^{\frac{1}{12}} \approx .94387$$

1.5 THE TRIGONOMETRIC FUNCTIONS

Objective:	To recognize the graphs of the sine, cosine and tangent functions and their inverses.
Preparations:	Read and take notes on section 1.5.
	Suggested practice problems: 5, 7, 8, 11, 13, 16, 19, 22, 24, 34, 39.

✠ Key Ideas of the Section

You should be able to:

- Determine the amplitude, period and phase shift of *sine* and *cosine* functions.
- Find the period of the tangent function.
- Define the inverse trig functions and know their restricted domains.

Key Questions to Ask and Answer

1. *Why are trig functions periodic functions?*

✍ ____

____ *(See text p. 30.)*

When trigonometric functions are defined using the unit circle, a point on the circle at t radians or $t + 2\pi$ radians or $t + 2\pi\ n$ radians (n is an integer) all share the same trigonometric values, thus the function is periodic. The graphs below show only a portion of the sine, cosine and tangent functions that continue their periodic behavior for $x \to \pm\infty$.

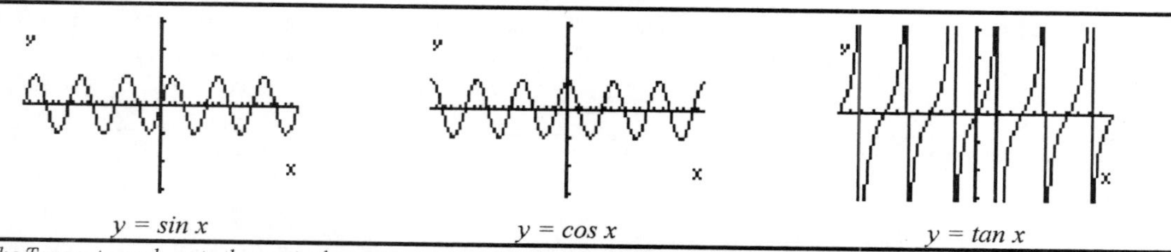

y = sin x *y = cos x* *y = tan x*

On the Tangent graph, note the vertical asymptotes every $\pi/2$ units on either side of zero. They appear as vertical lines on the grapher, but are NOT part of the graph.

Note: In calculus and most applied sciences radian measure is used instead of degree measure, because a radian is a number and calculus formulas are simpler in radians. To review radian measure, consult the Key Algebra topic at the end of this section.

Example 1.5.1

✍ Where are the maximum and minimum points for one standard period of the sine, cosine and tangent functions? Fill in the chart below.

Function	*Where Maximum Occurs*	*Where Minimum Occurs*
$y = \sin x$	1/4 of period, at $x = \pi/2$	?
$y = \cos x$	beginning and end of period, at $x=0$, and $x=2\pi$	?
$y = \tan x$	?	none

Answers
3/4 of period, at $x = \frac{3\pi}{2}$
1/2 of period, at $x = \pi$
none

2. *How do you use the rules for transformation of functions with the sine and cosine function?*

✍ ____

____ *(See text p. 31,32)*

We use the transformation rules from section 1.8:

$$h(x) = c\,f(x - h) + k\,,$$

but alter it slightly for the sine and cosine, adding a horizontal stretch or shrink factor B:

$$h(x) = A\sin(B(x - h)) + K \quad or \quad h(x) = A\cos(B(x - h)) + k$$

Variables	*Definition*	*Example*
$\lvert A\rvert$ = amplitude	• Half the distance between the maximum and minimum point. • $\lvert A\rvert$ is a stretch factor.	• For $y = \sin x$ the max. and min. values of y are +1 and -1. • For $y = 2\sin x$ the max. and min. values of y are +2 and -2.

Variables	*Definition*	*Example*
$B = 2\pi$ /period	• The value of B determines the period, which is the time of one complete cycle. • The period = $2\pi/B$ • $\lvert B\rvert$ stretches or shrinks the graph horizontally. • $B < 0$ reflects about *y-axis*.	• $y = \cos x$; $B = 1$; period = $2\pi/1 = 2\pi$. • $y = \cos 5x$; $B = 5$; period = $2\pi/5$. • $y = \sin 3\pi x$; $B = 3\pi$; period = $2\pi/3\pi = 2/3$.
h = phase shift	• A horizontal shift $\lvert h\rvert$ units left, ($h<0$) or right, ($h>0$).	• $y = \sin(x - \pi/2)$; a shift $\pi/2$ units to the right. • $y = \cos(x + 3)$; a shift of 3 units to the left.
k = vertical shift	• A vertical shift up or down $\lvert k\rvert$ units	• $y = \sin x + 10$; a shift 10 units up. • $y = \cos x - 4.5$; a shift of 4.5 units down.

Example 1.5.2

To model the cyclical nature of an animal population the following formula was used: $P(t) = 150\sin(\frac{\pi}{6}(t-4)) + 600$, where t is in months and $t = 0$ is January 1st. The graph of the function appears below.

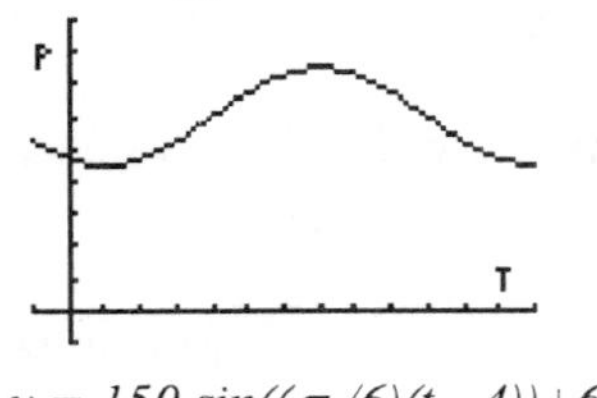

$y = 150\sin((\pi/6)(t - 4)) + 600$

- The graph shows time, t, in months with a scale of 1 month.
- The population, P, has a scale of 100 animals.
- The normal sine function has been shifted 4 units to the right, stretched vertically by 150 units, shifted up by 600 units and has a period of 12 months.

✍ Your Turn

Using the function and the graph in **Example 1.5.2**, answer the following questions.

Questions	*Your Response*
2a) What was the population on Jan. 1?	____________
2b) When did the minimum population occur?	____________
2c) When does the maximum population occur?	____________
2d) What is the period in practical terms?	____________
2e) What does the 150 in the equation represent?	____________
2f) What does the 600 in the equation represent?	____________
2g) What does the 4 in the equation represent?	____________

Answers
2a) When $t = 0$, (Jan. 1 is the beginning) $P = 150\sin((\pi/6)(0-4)) + 600 = 470.$
2b) Month 1 or February 1. $P(1) = 450.$
2c) Month 7 or August 1. $P(7) = 750.$
2d) Period = $2\pi/(\pi/6) = 12$ mo. The time from population minimum back to minimum.
2e) Half the distance from max. to min. $(750 - 450)/2 = 150$ = Amplitude stretch.
2f) The middle or median value of the function. $(450+750)/2 = 600$ units shifted up.
2g) The graph is shifted 4 units to the right. $P(4) = 600$. The middle value, 600, occurs on month 4, May 1.

3. *What are the inverse trigonometric functions for the sine, cosine and tangent and when do you use them?*

✍ ____

____ *(See text p. 33,34.)*

If you know the value of the trig function, you can use the inverse trig function to find the angle measured in radians. Recall from section 1.5 that if you know the output value of a function, the inverse function gives you the input value. For the inverse of trig functions to be defined, the domains of the trig functions themselves must be restricted.

Function	*Inverse function*	*Graph*
$y = \sin x$ for $-\pi/2 \le x \le \pi/2$	$y = \arcsin x$ for $-1 \le x \le 1$ or, $y = \sin^{-1}x$	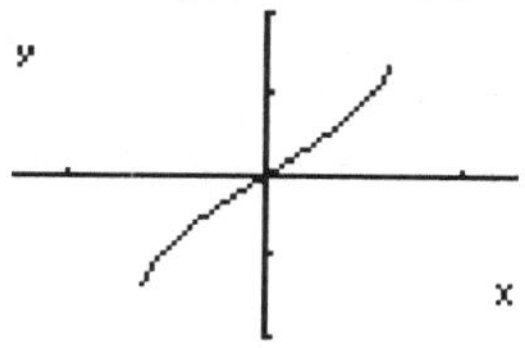 y, x
$y = \cos x$ for $0 \le x \le \pi$	$y = \arccos x$ for $-1 \le x \le 1$ or, $y = \cos^{-1}x$	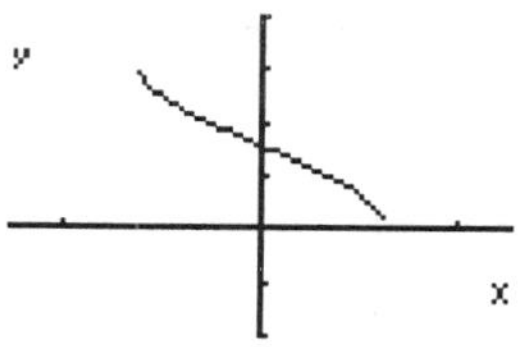 y, x
$y = \tan x$ for $-\pi/2 < x < \pi/2$	$y = \arctan x$ for $-\infty < x < \infty$ or, $y = \tan^{-1}x$	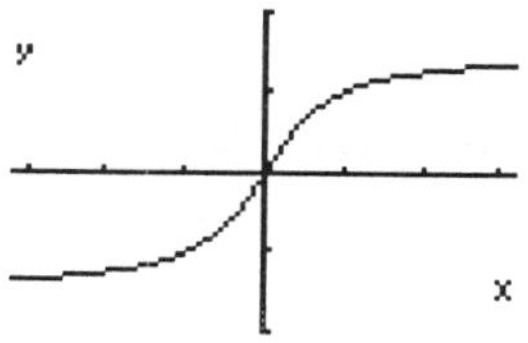 y, x

Example 1.5.3

In the **Example 1.5.2** we used the formula $P(t) = 150\sin(\frac{\pi}{6}(t-4)) + 600$ to determine the animal population. If we knew that there were 500 animals, which month would we be talking about?

Solution:

There are two methods to for solving this problem. There is the algebraic method where you will need to find the inverse of the sine function and there is the graphing method where you would use the intersection method. Ideally you will choose one method and check your answer using the second method.

Algebraic Method	*Graphical Method*
$y = 150\ sin(\frac{\pi}{6}(x-4)) + 600$ Let y = 500 $500 = 150\sin(\frac{\pi}{6}(x-4)) + 600$ $-100 = 150\ sin(\frac{\pi}{6}(x-4))$ $\frac{-100}{150} = sin(\frac{\pi}{6}(x-4))$ $\frac{-2}{3} = sin\,\theta$ let $\theta = \frac{\pi}{6}(x-4)$ $arcsin(\frac{-2}{3}) = \theta$ $\theta = -0.7297277$ $\frac{\pi}{6}(x-4) = -0.7297277$ $x \approx 2.61$ months , the second half of March.	Let $y_1 = 150\sin(\frac{\pi}{6}(x-4)) + 600$ $y_2 = 500$ 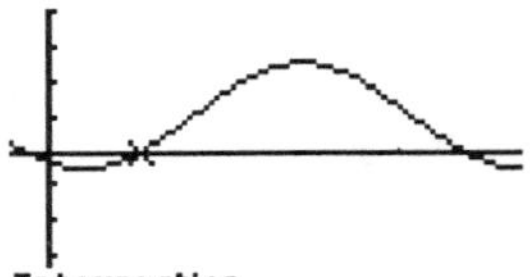 Intersection X=2.6063228 Y=500 The graph reveals a second intersection point. 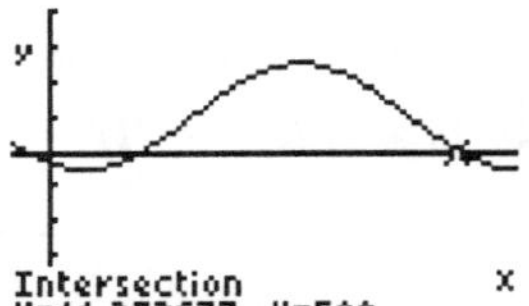 Intersection X=11.393677 Y=500

Notice that solving the problem algebraically gave only one value of θ where $-\pi/2 \le \theta \le \pi/2$.

✎ On Your Own

A population of mice in a barn over a two month period decreased from 100 to 10, thanks to Biscuit the cat. When Biscuit was removed, the mouse population was again back to 100 in another two months. Write a cosine function that would model the population over time and determine when 75 mice were present. Below is the graph of the first two month period:

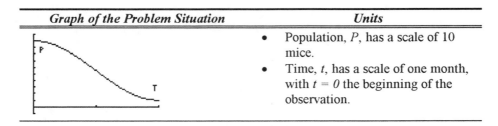

Graph of the Problem Situation	*Units*
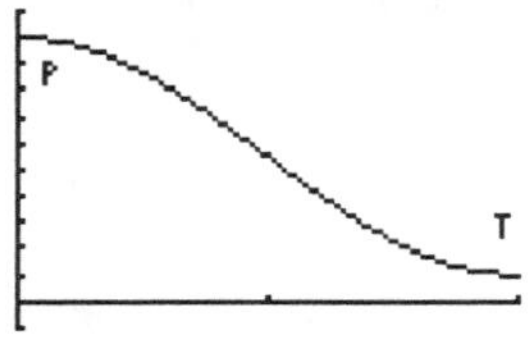 P T	• Population, *P*, has a scale of 10 mice. • Time, *t*, has a scale of one month, with *t* = *0* the beginning of the observation.

(Answer appears at the end of this section.)

Key Algebra   A review of trigonometric functions:

Right Triangle Trig		
r, y, t, x	$sin t = \frac{y}{r}$ $cos t = \frac{x}{r}$ $tan t = \frac{sin t}{cos t} = \frac{y}{x}$	If $r = 1$ $y = sin t$ $x = cos t$ $x^2 + y^2 = 1$ $sin^2 t + cos^2 t = 1$
Unit Circle Trig	***Radian /Degree***	
(cos t , sin t), s, r, t, y, (1, 0), x	1 radian = arc length of 1 unit 360° = 2 π radians 180° = π radians 1° = π /180 radians 1 radian = 180/ π degrees	arclength = s $s = rt$, where t is in radians
Trig Function	**Domain/Range**	***Graph of One Period***
$y = sin x$	$-\infty < x < \infty$ $-1 \leq sin x \leq 1$	y, x period = 2 π
$y = cos x$	$-\infty < x < \infty$ $-1 \leq cos x \leq 1$	y, x period = 2 π
$y = tan x$	x: all real numbers, except when $cos x = 0$ $x \neq \{ ...-3\pi/2, -\pi/2, \pi/2, 3\pi/2.. \}$ $-\infty < tan x < \infty$	y, x period = π
Reciprocal Functions	***Inverse Functions***	**Common Errors**
$\frac{1}{sinx} = cscx = (sin x)^{-1}$ $\frac{1}{cosx} = secx = (cos x)^{-1}$ $\frac{1}{tanx} = cot x = (tan x)^{-1}$	Find the angle if you know the value of the function. $y = arcsin x = sin^{-1} x$ for $-1 \leq x \leq 1$ $y = arccos x = cos^{-1} x$ for $-1 \leq x \leq 1$ $y = arctan x = tan^{-1} x$ for $-\infty < x < \infty$	$(\sin x)^{-1} \neq \sin^{-1} x$ $(\cos x)^{-1} \neq \cos^{-1} x$ $(\tan x)^{-1} \neq \tan^{-1} x$

Key Notation

Trig notation can be very tricky, so be careful how you use it. See the above table for the difference between reciprocal and inverse. Do not confuse the following notations:

$$(\sin x)^{-1} \neq \sin^{-1} x$$

$$(\cos x)^{-1} \neq \cos^{-1} x$$

$$(\tan x)^{-1} \neq \tan^{-1} x$$

$$\sin x^2 \neq \sin^2 x$$

$$\sin a + b \neq \sin(a+b)$$

Key Study Skills

- Choose at least two problems that represent this section; put each on a 3 x 5 note card with the solution on the back. Review the cards for 1.1-1.5.
- Use the ✓ ? ✯ system on homework problems and get answers to all ✯ problems.
- Use an algebra and trigonometry text if you are having difficulty with the trig functions.

Answer to On Your Own

A = amplitude = 45	• $\dfrac{100-10}{2} = 45$
p = period = 4 months	• $B = 2\pi/p$, so $B = 2\pi/4 = \pi/2$
h = horizontal shift = 0	• The highest point is at t=0
k = vertical shift = 55	• $\dfrac{100+10}{2} = 55$
Put it all together	•
	• let $y=75$
	• so $75 = 45\cos(\frac{\pi}{2}x) + 55$
Seventy five mice were present twice during the four month period, at .71 months and 3.3 months.	Intersection X=.70680222 Y=75 Intersection X=3.2931978 Y=75

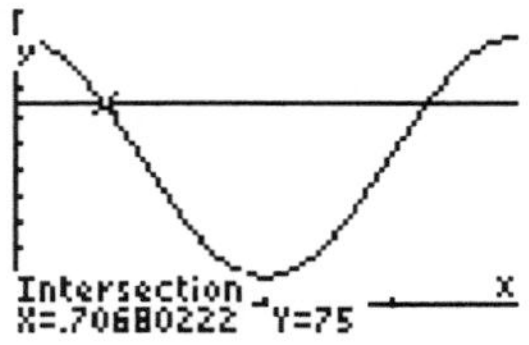

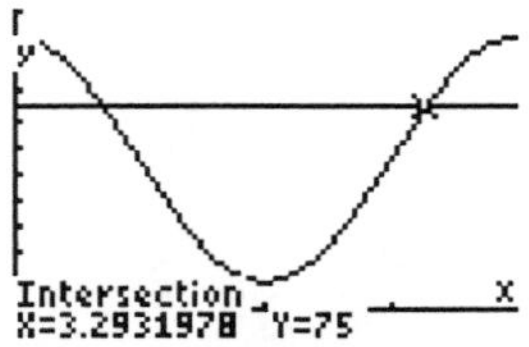

1.6 POWERS, POLYNOMIALS AND RATIONAL FUNCTIONS

Objective:	To be able to identify the graphical features of power functions on both a small and large scale. To be able to recognize the characteristics of polynomial and rational functions.
Preparations:	Read and take notes on section 1.6.

Suggested practice problems: 2, 5, 6, 8, 11, 13, 16, 17, 18, 23, 32, 34.
Problems requiring substantial algebra: 15, 24, 25, 30.

✠ Key Ideas of the Section

You should be able to:

- Identify the different graphs and global features of power functions $y = k\,x^{p}$ where k is a constant and
- p is a positive integer,
- p is a negative integer (investigate $p = 0$),
- p is a positive fraction.
- Determine the effect of the constant k on power functions $y = kx^{p}$
- Compare and contrast exponential, power and linear functions.
- Identify the degree and the maximum number of turning points for polynomials.
- Determine the end behavior of polynomials of odd and even degree.
- Determine a polynomial with given zeros and a point.
- Find vertical and horizontal asymptotes for rational functions.

Key Questions to Ask and Answer

1. What are the shapes of odd and even positive integer power functions?

✍ ____

____ *(See text p. 37)*

You will begin to notice the pattern for odd and even power functions by looking at the graphs of the functions.

Example 1.6.1

Describe the concavity of the functions by looking at the graph.

Graph	Power	Function	Shape
y, x	*$p = 2, 4, 6 \ldots.n$ where n is an even integer.*	$y = x^2$ $y = x^4$ $y = x^6$ . . .	• always concave up; $y \geq 0$. • as the power increases the graph narrows. • the graph is ∪ shaped, with ends going in the same direction.
y, x	*$p = 3,5,7 \ldots\ n$ where n is an odd integer.*	$y = x^3$ $y = x^5$ $y = x^7$ . . .	• concave down for $x < 0$. • concave up for $x > 0$. • as the power increases the graph narrows. • the graph is "seat" shaped, with ends going in opposite direction.

2. What are the shapes of odd and even negative integer power functions?

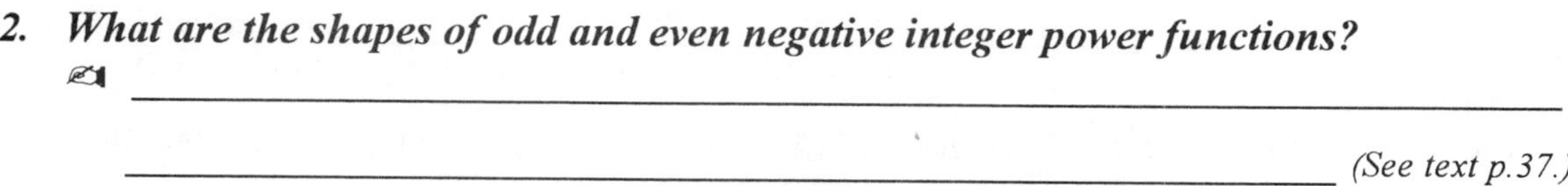

✍ ____

____ *(See text p.37.)*

Recall that a negative exponent means "take the reciprocal". The graphs of odd and even negative integer powers have similar shapes.

Example 1.6.2

Describe the concavity of the functions by looking at the graph.

Graph	Power	Function	Shape of function
	p =-1, -3, -5....*n* *where n is a negative odd integer.*	$y = x^{-1} = 1/x$ $y = x^{-3} = 1/x^3$ $y = x^{-5} = 1/x^5$	• concave down for $x < 0;\ y < 0$ • y undefined at $x = 0$ • concave up for $x > 0;\ y > 0$
	p = -2, -4, -6....*n* *where n is a negative even integer*	$y = x^{-2} = 1/x^2$ $y = x^{-4} = 1/x^4$ $y = x^{-6} = 1/x^6$	• concave up for $x < 0;\ y > 0$ • y undefined at $x = 0$. • concave up for $x > 0;\ y > 0$

3. *What are the shapes of odd and even positive fraction power functions?*

✍ ____

____

Recall that a fractional power is an alternate notation for the "root of So "*x* to the one-half power" is the same as "the square root of *x*". ". For more information see the *Ready Reference* at the back of the text.

Example 1.6.3

Describe the concavity of the functions by looking at the graph.

Graph	Power	Function	Shape of function
	p =1/2, 1/4, 1/6...1/n where n is an even integer.	$y = x^{1/2} = \sqrt{x}$ $y = x^{1/4} = \sqrt[4]{x}$ $y = x^{1/6} = \sqrt[6]{x}$	• y is undefined for $x < 0$ • concave down for $x > 0;\ y > 0$
	p = 1/3, 1/5, 1/7..1/n where n is an odd integer.	$y = x^{1/3} = \sqrt[3]{x}$ $y = x^{1/5} = \sqrt[5]{x}$ $y = x^{1/7} = \sqrt[7]{x}$	• concave up for $x < 0;\ y < 0$ • concave down for $x > 0;\ y > 0$

4. *How does the growth of exponential and power functions compare?*

✍ ____

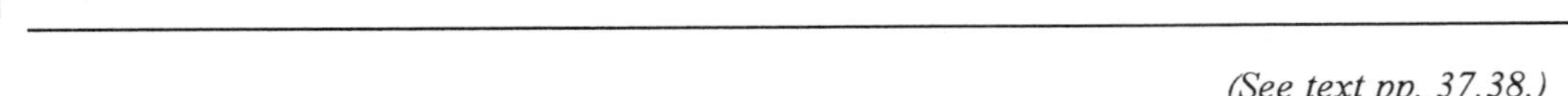

(See text pp. 37,38.)

Example 1.6.4

Compare the graphs of $f(x) = 4^x$ and $g(x) = x^4$. Which function will be on top, or above the other (or said to dominate) for different values of *x* as you move from left to right?

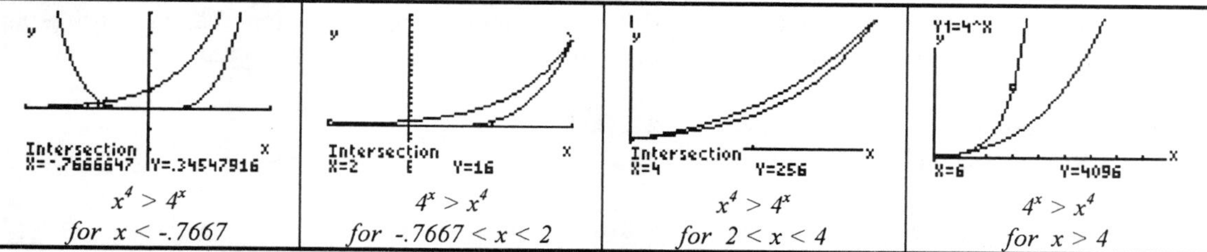

$x^4 > 4^x$ for $x < -.7667$	$4^x > x^4$ for $-.7667 < x < 2$	$x^4 > 4^x$ for $2 < x < 4$	$4^x > x^4$ for $x > 4$

Note: **Exponential growth functions will always dominate (grow faster) than power functions as $x \to \infty$.**

5. How does changing the coefficient, k, of the power function $y = kx^p$ change the shape of the graph?

✍ ____

____ *(See text p. 17)*

In general you need to consider when $k > 1$, $0 < k < 1$ and $k < 0$. Choose the simple parent graph of $y = x^2$, then observe the changes as you change the values of k.

Graph	k	Function	Shape of function
	$k = 1, 3, 7$	$y = x^2$ $y = 3x^2$ $y = 7x^2$	• The parent function $y = x^p$ gets "stretched" in a vertical direction for $k > 1$ (pulled away from x-axis).
	$k = 1, 1/2, 1/4$	$y = x^2$ $y = \frac{1}{2}x^2$ $y = \frac{1}{4}x^2$	• The parent function $y = x^p$ "shrinks" in a vertical direction for $0 < k < 1$, (pulled towards the x-axis).
	$k = -1/2, -1, -2$	$y = -\frac{1}{2}x^2$ $y = -1x^2$ $y = -2x^2$	• The parent function $y = x^p$ gets "reflected" or flipped about the x-axis.

✎ Your Turn

Use a grapher to try various power functions in the form $y = kx^p$. Try various values of k as you hold p constant. Observe the change in the graph and verify the shape of the graph according to the examples above.

6. How do you determine the degree of a polynomial?

✍ ____

____ *(See text p. 38)*

At the beginning of this section we examined the power functions in the form $f(x) = ax^n$, where the power n was a positive integer. When these particular types of power functions are added together they form a *polynomial* (many terms). The degree of the polynomial is determined by the term of the highest power.

The result of adding many power functions together can produce a curve that has turning points. The turning points represent where the graph changes from increasing to decreasing or from decreasing to increasing.

Polynomials are "well behaved" curves. They are

- continuous and
- turn around smoothly.

The degree of the polynomial can tell us something about

- the general shape of the graph of the function,
- the maximum number of turning points and
- minimum number of concavity changes.

Example 1.6.5

Determine the degree and number of turning points of the following polynomials.

Polynomial	*Degree*	*Graph*	*Turning points*
$f(x) = x^2 + 3x - 4$	second degree $n = 2$	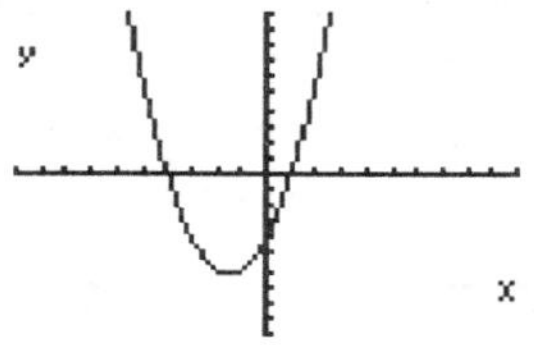	• One turning point. • Always concave up.
$g(x) = x^3 - 13x + 12$	third degree $n = 3$	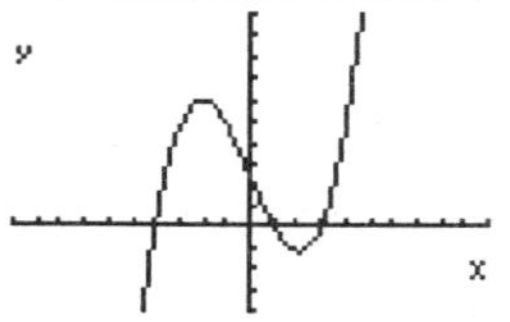	• Two tuning points. Changes from concave down to concave up, or one concavity change.
$h(x) = x^4 + 3x^3 - 13x^2 - 27x + 36$	fourth degree $n = 4$	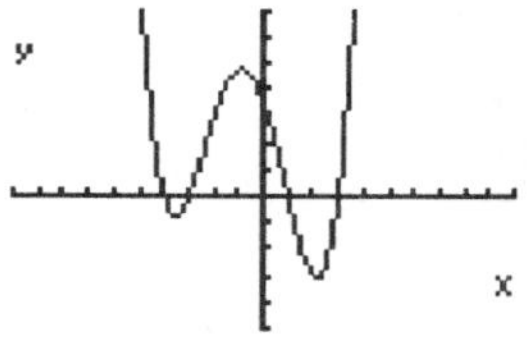	• Three turning points. • Two concavity changes.
$k(x) = x^5 - 22x^3 + 12x^2 + 117x - 108$	fifth degree $n = 5$	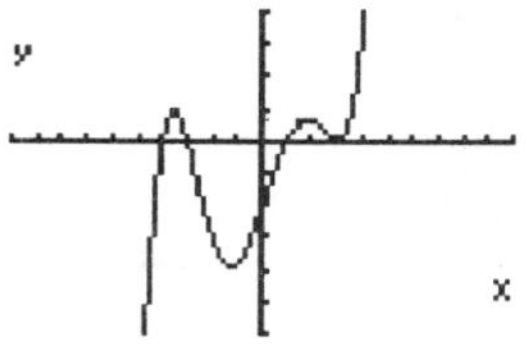	• Four turning points. • Three concavity changes.
$l(x) = a_n x^n + a_{n-1} x^{n-1} + .. a_1 x + a_0$	*nth* degree	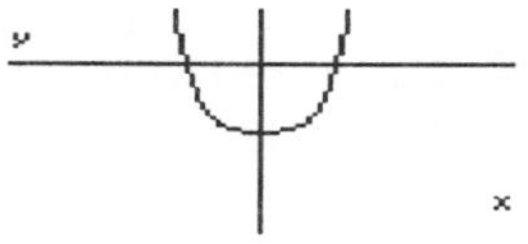 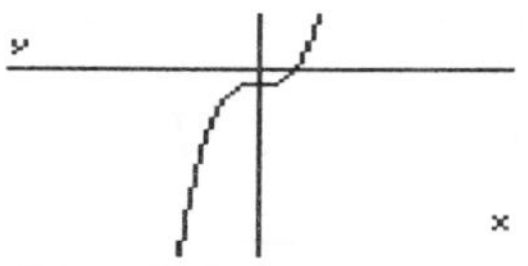	In general an *nth* degree polynomial has: • At most $n - 1$ turning points. • At least one turning point if n is even, as few as no turning points if n is odd. • At most $n - 2$ concavity changes.

Note: In the above graphs where n= 2,3,4,5, the maximum number of turning points and concavity changes are shown. Any nth degree polynomial can have fewer than n - 1 turning points.

7. *How can you find a possible formula for a function knowing the zeros and one other point?*

✍ ____

____ *(See text p. 39.)*

The factor theorem states that:

If f(x) is a polynomial, then x -c is a factor of f(x)
if and only if c is a zero of f(x).

This implies that if you know the value of a zero, c, of polynomial $f(x)$ and c is a real number then:

- c is a solution or root of $f(x) = 0$.
- c is a zero of $f(x)$.
- c is an x - intercept of the graph of $f(x)$.
- If $f(x)$ is divided by $x - c$, the remainder is zero.
- If c is a zero of $f(x)$, and the graph of $f(x)$ touches the x -axis at c, but does not cross it, then c is a multiple zero.

A polynomial with n real zeros will at *least* be an *nth* degree polynomial. The factored form of the *nth* degree polynomial would be:

$$f(x) = k(x - c_1)(x - c_2) \ldots (x - c_n)$$

where k is the real coefficient of the leading term of $f(x)$.

Example 1.6.6

Below is the graph of a polynomial with zeros at -4, -3, 1 and 3 and a y -intercept at (0, 36). Find a possible formula for the polynomial.

Graph	*Analysis*
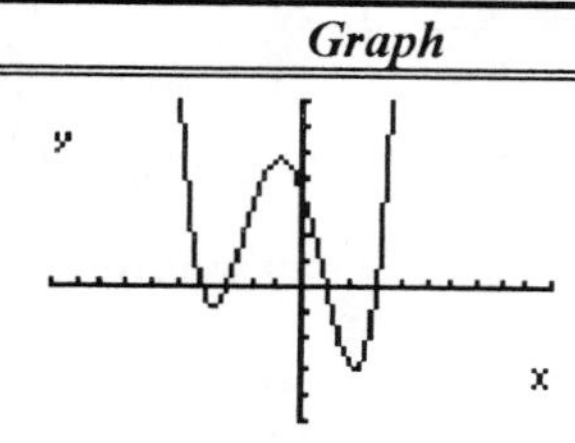	• Three turning points implies a minimum 4th degree. • Four zeros implies a minimum 4th degree with four factors in the form $(x - c)$: $f(x) = k(x-(-4))(x-(-3))(x-1)(x-3)$ • The graph opens up, so the leading coefficient should be positive.

Putting the analysis altogether a possible formula would be:

$$f(x) = k(x+4)(x+3)(x-1)(x-3)$$

Since the y -intercept is (0,36) let $x = 0$ and $y = 36$ to solve for k.

$$36 = k(0+4)(0+3)(0-1)(0-3)$$

$$36 = k(36)$$

$$k = 1$$

Expanding the formula would give:

$$h(x) = x^4 + 3x^3 - 13x^2 - 27x + 36$$

✎ On Your Own 1

a) Find a formula for a polynomial with zeros at -4, -3, 1, and 3 and a y- intercept at (0, 216) and a graph as in **Figure 1.6.1**.

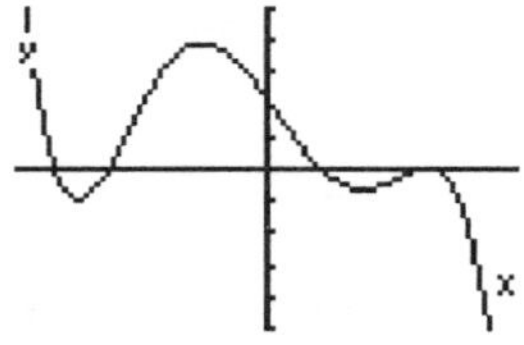

Figure 1.6.1

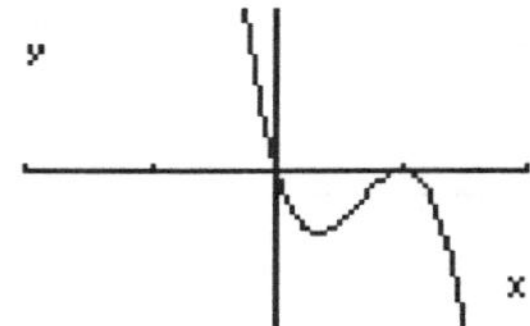

Figure 1.6.2

b) Find a formula for a polynomial with a zero at 0, a double zero at $x = 1$, and whose graph passes through the point (-1,8) and has the graph as in **Figure 1.6.2**.

(The answer appears at the end of this section)

---

8. On the very large scale, why do all polynomials have graphs that look like the graphs of their leading term, i.e. a positive integer power functions?

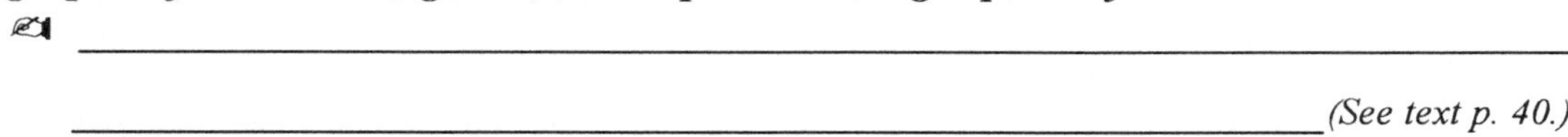

(See text p. 40.)

Earlier we learned that all positive integer functions have either a U shape if the power is even and a "seat" or lazy *S* shape if the power is odd. The graph of *f(x)* on a very large scale behaves like power functions, because the leading term of a polynomial will dominate the other terms. On a large scale the local turning points, while still present, are not seen on the graph and the end behavior of *f(x)* as $x \to \pm\infty$ is the same as the end behavior of the leading term:

- a U shape end behavior for the graphs of even degree polynomials, with both arms of the graph going in the same direction.
- a "seat" or lazy *S* shape end behavior for the graph of odd degree polynomials, with the arms of the graph going in opposite direction.

On Your Own 2

Compare the graphs of the functions $f(x) = x^3 - x + 83$ and $g(x) = x^3$ on a small scale of *x*: [-10, 10] and *y*: [-100,100], then on a large scale of *x*: [-50,50] and *y*: [-10,000, 10,000].

(The answer appears at the end of this section.)

---

9. How do you determine vertical and horizontal asymptotes for rational functions?

(See text p. 41.)

A rational function can be formed by dividing one polynomial *p(x)* by another polynomial *q(x)*, thus:

$$f(x) = \frac{p(x)}{q(x)}$$

The domain of rational functions is the set of all *x* such that $q(x) \neq 0$. Either vertical asymptotes or "missing points" occur at the values not in the domain of *f*.

To graph a rational function do the following:

- Find any vertical asymptote, let $q(x) = 0$, then solve the equation to find $x = c$, the value that makes the denominator zero. $x = c$ is either a *vertical asymptote or a "missing point". Vertical asymptotes are not part of the graph and can be drawn as dotted vertical lines.*
- Find the *y-intercept, that is, find f(0)*.

- Find *x-intercepts* , that is, the x value for $f(x) = 0$, if possible.
- Find the end behavior or any horizontal asymptotes. Let $x \to \pm\infty$, only the leading terms of $p(x)$ and $q(x)$ will matter. If the degree of $q(x)$ is greater than or equal to the degree of $p(x)$, a *horizontal asymptote* will exist. If the degree of $q(x)$ is less than the degree of $p(x)$ an *oblique asymptote* will exist
- *Horizontal and oblique asymptotes are not part of the graph and can be drawn as dashed lines.*

Example 1.6.7

Draw the graph of $f(x) = \dfrac{3x - 4}{x^2 - 8}$ indicating vertical and horizontal asymptotes, y-intercepts and x - intercepts if possible.

Solution:

The analysis of $f(x) = \dfrac{3x-4}{x^2 - 8}$

Vertical asymptote:	Let $x^2 - 8 = 0$	$x = \pm\sqrt{8} \approx \pm 2.82$
y-intercept:	Let $f(0) = \dfrac{3(0)-4}{(0)^2 - 8} = \dfrac{1}{2}$	(0, .5)
x-intercept:	Let $3x - 4=0$	$x = 4/3=1.333...$
Horizontal asymptote:	Let $x \to \pm\infty$	$f(x) = \dfrac{3x-4}{x^2-8} \approx \dfrac{3x}{x^2} = \dfrac{3}{x} \approx 0$

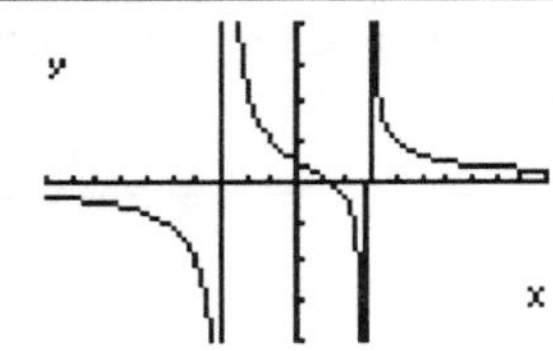

The graph of f(x).

- Vertical asymptotes at $x = \sqrt{8}$ and $-\sqrt{8}$ are not part of the graph.
- $y = 0$ is a horizontal asymptote since
  as $x \to +\infty$, $f(x) \to 0$ and
  as $x \to -\infty$, $f(x) \to 0$
- y-intercept (0, 1/2)
- x-intercept (4/3, 0)

Key Algebra

- $x^{-p} = \dfrac{1}{x^p}$ A negative power means, "take the reciprocal of the function".
- $x^{m/n} = (x^m)^{1/n} = \sqrt[n]{x^m}$ A fractional exponent means "find the nth root of the function".

Key Study Skills

- Choose at least two problems that represent this section; put each on a 3 x 5 note card with the solution on the back. Review the cards for 1.1-1.6.
- Identify and write down any key words from 1.1-1.6.
- Review the **Key Ideas** from 1.1-1.6.
- Use the ✓ ? ★ system on homework problems and get answers to all ★ problems.
- Use a grapher to help you discover the end behavior of polynomials. A grapher can also verify zeros found algebraically.

Answer to On Your Own 1a

The graph has four turning points and five roots because x=3 is a double root. Therefore a possible formula is a 5th degree polynomial, with y-intercept at (0,216). The equation is found by the following method:

$f(x) = k(x+4)(x+3)(x-1)(x-3)(x-3)$

$x = 0$ and $y = 216$

$216 = k(4)(3)(-1)(-3)(-3)$

$k = \dfrac{216}{-108} = -2$

$f(x) = -2(x+4)(x+3)(x-1)(x-3)(x-3)$

$f(x) = -2x^5 + 44x^3 - 24x^2 - 234x + 216$

Answer to On Your Own 1b

$f(x) = -2(x-0)(x-1)^2$ or

$f(x) = -2x^3 + 4x^2 - 2x$

Answer to On Your Own 2

The graph of $f(x) = x^3 - x + 83$ and $g(x) = x^3$ on a small and large scale:

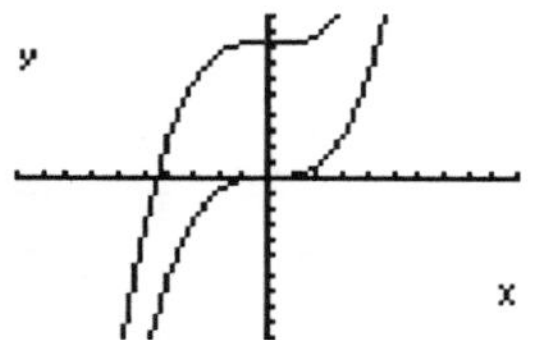

x: [-10,10] and y: [-100, 100]
On a small scale local behavior of $f(x)$ is seen.

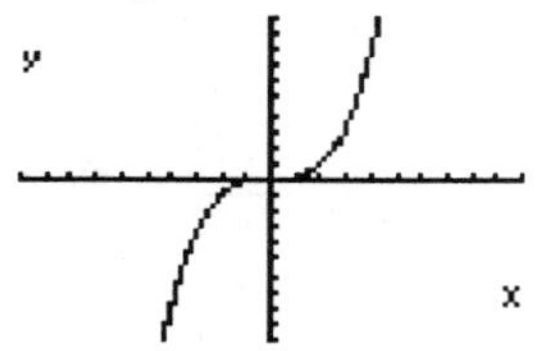

x: [-50, 50] and y: [-10000,10000]
On a large scale the graphs appear to be the same, $f(x)$ takes the shape of x^3.

1.7 INTRODUCTION TO CONTINUITY

Objective: To identify the interval over which a function is continuous.

Preparations: Read and take notes on section 1.7.
Suggested practice problems: 1, 2,6, 7, 8, 13, 16, 17, 20.

✠ Key Ideas of the Section

You should be able to:

- Identify well known functions that are continuous or not continuous.
- Recognize that continuity on an interval can be represented by a graph with no breaks, no holes and no jumps in it.
- Demonstrate numerically that for continuous functions, a small change in the independent variable produces a small change in the dependent variable.

Key Questions to Ask and Answer

1. *How do you determine if a function is continuous over an interval?*

✍ ____

____ *(See text p. 45.)*

Graphically you can tell if a function is *continuous* on an interval if its graph has no holes, no breaks and no jumps in it on that interval. *Discontinuity* can arise when the function is undefined at an x value. In the last section we saw that some rational functions had vertical asymptotes that cause a break in the graph because the function was undefined for a value of x.

- A function that contains a vertical asymptote, in an interval, is not continuous on that interval.
- A function which is a piecewise defined function for all values of x, but which has breaks or holes, is not continuous on the intervals including those breaks or holes.

Numerically, continuity means that small changes in the independent variable produce small changes in the dependent variable. See figures 1.67-1.69 in the text on page 45.

2. *Of the functions we have studied, which are continuous over the set of real numbers?*

✍ ____

____ *(See text p. 46.)*

The following are functions that are continuous on the interval $(-\infty, \infty)$:

- Polynomial functions are continuous.
- Exponential functions are continuous.
- Positive integer power functions are continuous.
- Sine and cosine functions are continuous.
- Functions created by adding , multiplying or composing any of the above are continuous.

3. *Which functions are not continuous over a set interval?*

✍ ____

____ *(See text p. 46.)*

✎ Your Turn

Determine if the functions below are continuous everywhere and supply your reason.

Function	*Graph*	*Reason*	*Answer*
$y = 5e^x$	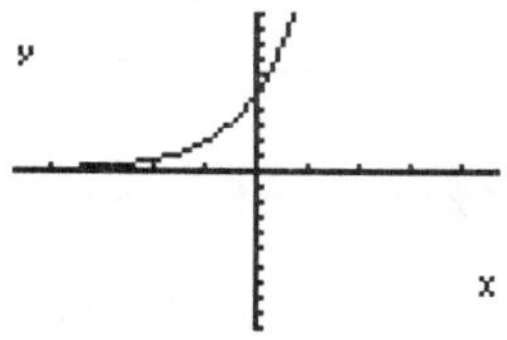 y, x		• Exponential functions are continuous for $x:(-\infty,\infty)$. • There are no breaks in the graph.
$y = \dfrac{(x-1)^2}{(x-1)} + 5$	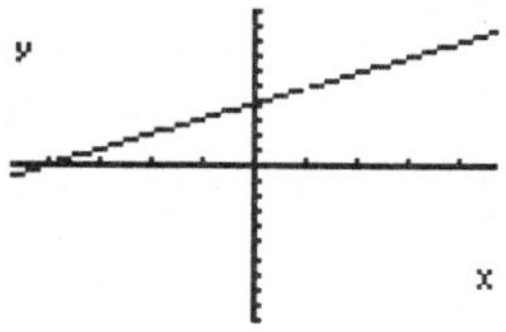 y, x		• $x \neq 1$. • Not continuous on an interval that contains $x=1$. • There is a hole in the graph at $x=1$.

Function	*Graph*	*Reason*	*Answers*
$y = \dfrac{1}{(x+2)^3} - 1$	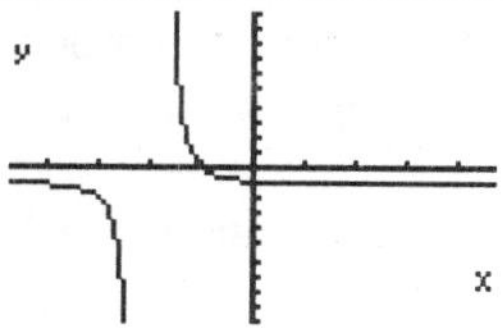 y, x		• $x \neq -2$. • Not continuous on an interval that contains $x=-2$. • Break in the graph at $x=-2$.

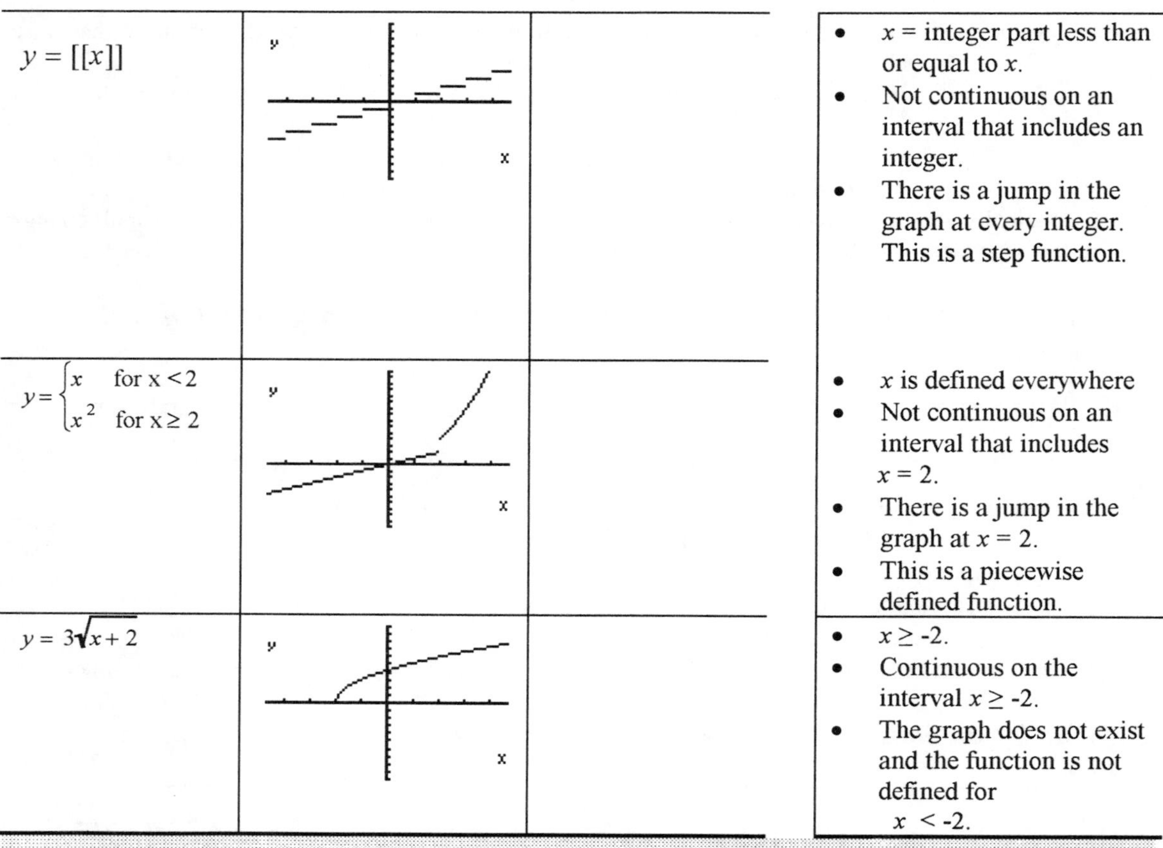

$y = [[x]]$			• x = integer part less than or equal to x. • Not continuous on an interval that includes an integer. • There is a jump in the graph at every integer. This is a step function.
$y=\begin{cases} x & \text{for } x<2 \\ x^2 & \text{for } x \geq 2 \end{cases}$			• x is defined everywhere • Not continuous on an interval that includes $x = 2$. • There is a jump in the graph at $x = 2$. • This is a piecewise defined function.
$y = 3\sqrt{x+2}$			• $x \geq -2$. • Continuous on the interval $x \geq -2$. • The graph does not exist and the function is not defined for $x < -2$.

4. *Why is the intermediate value theorem important?*

✍ ____

____ *(See text pp. 46,47.)*

We use the intermediate value theorem to help find zeros of a function (or any value between two points). For example if a continuous function has a negative value, then a positive value, it is reasonable to assume it had a zero value some where in-between.

The intermediate value theorem tells us that a continuous function over $[a,b]$, from point *(a,f(a))* to point *(b,f(b))* , then the function f takes on an intermediate value k, which is between $f(a)$ and $f(b)$ and $f(c)=k$, with c in the interval $[a,b]$. See figure 1.71 in the text page 47.

Key Algebra

- A rational function will be discontinuous on any interval that includes any values making the denominator zero. Usually the value that makes the denominator zero produces a vertical asymptote, however, if the denominator has a factor common with a factor in the numerator, the discontinuity produced may just be a missing point.
- Piecewise functions can be defined everywhere, but may have a jump or break in the graph.

🗁 Key Study Skills

- Choose at least two problems that represent this section; put each on a 3 x 5 note card with the solution on the back. Review the cards for 1.1-1.7
- The intermediate value theorem tells us that if a function f is continuous from a point $(a, f(a))$ to $(b, f(b))$, then f takes on all intermediate values between $f(a)$ and $f(b)$.
- Use the ✔ ? ✯ system on homework problems and get answers to all ✯ problems.

CUMULATIVE REVIEW CHAPTER ONE

🗁 Key Study Skills

- Review all of the **Key Ideas** from 1.1 - 1.7.
- Review the key words in the text Chapter Summary on page 48 .
- Make up your own sentences from the key words in the summary.
- Work the odd review problems at the end of the chapter one pp. 48-52 and check your answers at the back of the book.
- Answer the problems in the text from Check Your Understanding, pages 52-53.
- Do the Review Exercises in the text pages 48-52.
- Use your 3 x 5 note cards with sample problems to make up your own test.
- Turn to the **Ready Reference** in the back of the text, to review all the types of family of functions.
- After you have studied, take the Chapter One Practice Test on the next page and review the problems that gave you difficulty.

Chapter One Practice Test

1. The values of four functions, rounded to two decimal places, are given below. Find the algebraic model for the functions using the forms $y = mx + b$, $y = ka^x$ or $y = kx^p$ (One form is used twice).

x	$f(x)$	$g(x)$	$h(x)$	$k(x)$
-3	0.45	17.50	27.43	-27
-2	1.12	14.00	24.69	-8
-1	2.80	10.50	22.22	-1
0	7.00	7.00	20.00	0
1	17.50	3.50	18.00	1
2	43.75	0.00	16.20	8
3	109.38	-3.50	14.58	27
4	273.44	-7.00	13.12	64

$f(x) =$
$g(x) =$
$h(x) =$
$k(x) =$

2. In 1988, the number of reported new cases of males with AIDS was 27,049. In 1989 the number of new cases reported was 29,549 males.[4]
   a) If the rate of change in number of new cases remained constant, find the linear function that would model the number of reported new cases of males with AIDS.
   b) If the number of new cases reflected a constant percent increase, find an exponential function that would model the number of reported new cases of males with AIDS.
   c) In 1992 the actual number of reported new cases of males with AIDS was 38,917. Which of the models was a better predictor and why?
   d) Use each of the models to predict the number of reported new cases of males with AIDS in 2002.

3. The volume of a sphere of radius r is $V = \frac{4}{3}\pi r^3$. Make a table of values of the volume for various values of r. If r is measured in centimeters, what are the units of V? Sketch a graph of the function. If the radius doubles, by what factor does the volume change?

4. Compare and contrast the following functions. Construct tables of values and graphs. How do they differ? What are the characteristics of each? Over what interval(s) does each increase (or decrease)? Discuss the concavity of each.
   a. $f(x) = 2^x$
   b. $g(x) = x^2$
   c. $h(x) = 2x$
   d. $k(x) = x^{\frac{1}{2}}$
   e. $l(x) = x^{-2}$

[4] *Health United States 1993,* National Center for Health Statistics, US Dept. of Health and Human Services

5. In 1990, the annual receipts for motion pictures in the US were 39,982 million dollars.[5] In 1992 the annual receipts were 45,662 million dollars.

   a. Assume the annual receipts increase at a constant rate of change. Find a linear function that would model the annual receipts for motion pictures.
   b. If the annual receipts increased at a constant percent increase, find an exponential function that would model the annual receipts for motion pictures.
   c. The actual motion picture annual receipts in 1995 were 58,113 million dollars. Which of your models was a better predictor and why?
   d. For each model, predict when the annual receipts would be 100,000 millions of dollars.

6. Scientists have discovered two new strains of amoebas, S_1 and S_2. The two following functions give the number of thousands of cells of the two strains at time t days, where t=1 is the 1st day.

   $S_1(t) = 2^t$

   $S_2(t) = 2 * t^{100}$

   where t is time in days and S_1 and S_2 are measured in numbers of cells. Which strain has the largest population:

   a. at noon on day one?
   b. from the second to the fourth day?
   c. in one million years?

   Explain your answers.

7. Normal precipitation in San Francisco is given as follows:[6]

Month	Jan.	Feb.	Mar.	Apr.	May	June	July	Aug.	Sept.	Oct.	Nov.	Dec.
Precip. in./mo.	4.35	3.17	3.06	1.37	0.19	0.11	0.03	0.05	0.20	1.22	2.86	3.09

A possible continuous model might look like the graph that follows. Your task is to fit a *sine* or *cosine* function to model the data by finding each item that follows.

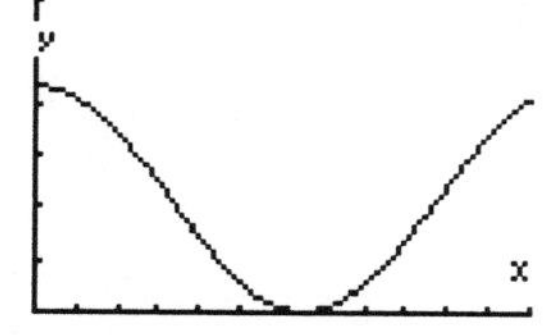

a) Find the amplitude.
b) Find the period.
c) Give a possible formula.
Use your model to predict the rainfall for July.
d) Give an estimate for the three months that the annual precipitation in San Francisco is 2 inches.

---

[5] Source: US Bureau of the Census, *Current Business Reports, Service Annual Survey*: 1995, BS/95.
[6] Source: US National Oceanic and Atmospheric Administration, *Climatography of the United States*, No. 81.

8. Using the graph of $g(x) = x^3 + 2$ below find both graphically and algebraically:

a. $-g(x + 3)$ b. $\dfrac{1}{g(x)}$ c. $g^{-1}(x)$

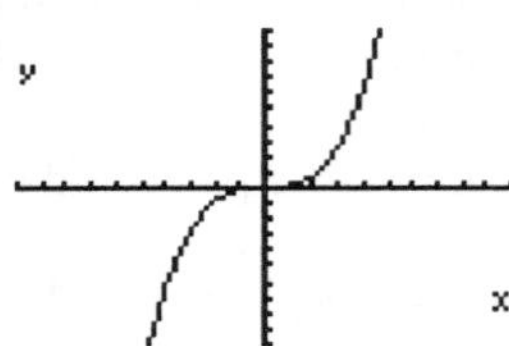

9. Given the function $f(x) = (x^3 - 6x^2 + 8x)(2 - 3^x)$ on the interval $-3 \le x \le 7$.

a. Sketch a complete graph on this interval.
b. Estimate intervals over which the function is increasing; decreasing.
c. Estimate intervals over which the function is concave down; concave up.

10. For each of the following functions:

$y_1 = 3\sin(4x - 2)$

$y_2 = \ln(x + 1)$

$y_3 = \dfrac{2x - 3}{x^2 - 4}$

a. Give rough sketches of the graphs.
b. Give any horizontal or vertical asymptotes.
c. Give domains.
d. Give any zeros.

11. Give a possible formula for a polynomial function with zeros $x =$ 2, –1, 1 and a double zero $x = 0$ passing through the point (3,–216) with the following graph.

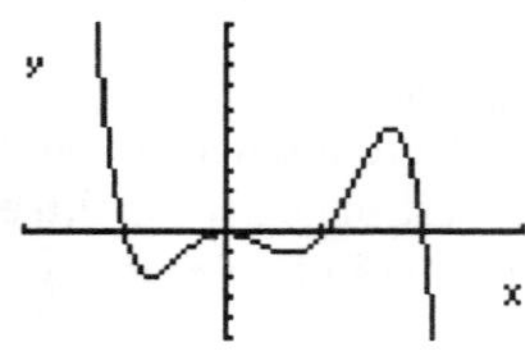

(The answers appear at the end of this study guide.)

CHAPTER TWO

KEY CONCEPT: THE DERIVATIVE

The derivative can be interpreted geometrically as the slope of the curve, and physically as a rate of change.

2.1 HOW DO WE MEASURE SPEED?

Objective: To find the average velocity between two points that represent distance over time.

Preparations: Read and take notes on section 2.1.
Suggested practice problems: 1, 3, 4, 6, 7, 8, 11, 15, 16, 18.

✠ Key Ideas of the Section

You should be able to:

- State the difference between speed and velocity.
- Compare and contrast average velocity and instantaneous velocity.
- Use the idea of a limit to find the instantaneous velocity.

Key Questions to Ask and Answer

1. How do speed and velocity differ?

✍ ____

____ *(See text p. 56,57.)*

- *Average Velocity* is the ratio of a change in position (distance) to a change in time. Velocity can be positive, zero or negative, depending on the direction traveled. If two points on the position function $s(t)$ are $(a, s(a))$ and $(b, s(b))$ then

$$\text{Average velocity} = v(t) = \frac{\text{change in position}}{\text{change in time}} = \frac{\Delta s}{\Delta t} = \frac{s(b) - s(a)}{b - a}$$

- *Speed* is the magnitude of velocity and is always positive or zero. When you drive around town you calculate your average speed because you are not concerned about the direction you are traveling, only the distance you are traveling.

Example 2.1.1

The text refers to a grapefruit being thrown straight up into the air with the following positions above the ground being recorded. Look at the average velocities, over interval t to $t + 1$. Why are some velocities positive and some negative?

time t (sec)	position s (feet)	average velocity $\frac{\Delta s}{\Delta t}\left(\frac{\text{ft}}{\text{sec}}\right)$
0	6	-
1	90	84
2	142	52
3	162	20
4	150	-12
5	106	-44
6	30	-76

Positive velocity signifies an upward direction.

Negative velocity signifies a downward direction.

2. *Compare and contrast average velocity and instantaneous velocity.*

✍ ____

____ *(See text p. 58,59.)*

- Average velocity measures the change in position over a specific time *interval* of length h units, or from $t_1 = a$ to $t_2 = a + h$.
- Instantaneous velocity refers to the velocity at a specific point in time, $t = a$ or at an *instant* in time.

Inherent in the idea of instantaneous velocity is a contradiction, because if you look at an object at an instant in time you effectively stop the motion; yet we want the answer to the question: How fast was I going at exactly 2:05 AM when I went past the state trooper with the radar?

To avoid the contradiction we use the idea of a *limit.*

- Use the average velocity to find the instantaneous velocity at $t = a$ by letting the time interval shrink in size or let h approach zero, $(h \to 0)$.

The instantaneous velocity of a function $s(t)$ at $t = a$ is:

$$\textit{instantaneous velocity} = \lim_{h \to 0} \frac{s(t_2) - s(t_1)}{(a+h) - a} = \lim_{h \to 0} \frac{s(a+h) - s(a)}{h}$$

3. *How can you visualize the instantaneous rate of change?*

✍ ____

____ *(See text p. 59)*

- The *average velocity* over a time interval $a \le t \le b$, is the same as the *slope of a line* joining the points on the graph of the function $s(t)$ at $t = a$ and $t = b$.

$$slope = \frac{\Delta s}{\Delta t} = \frac{s(b) - s(a)}{b - a}$$

(A,S(A))

(B,S(B))

- *Instantaneous velocity* is the slope of the curve at a point. A curve will look like a straight line if you "zoom in" on a section of the curve centered at a point. From this nearly linear curve, you can get better and better estimates of the instantaneous velocity by getting closer and closer to the point A.

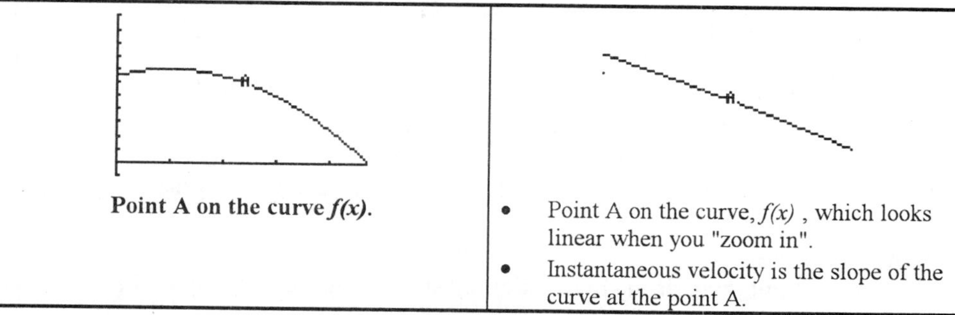

Point A on the curve *f(x)*.

- Point A on the curve, *f(x)* , which looks linear when you "zoom in".
- Instantaneous velocity is the slope of the curve at the point A.

4. *Can I look at a graph and determine if the slope of the curve at a point is positive or negative?*

✍ ____

____ *(See text p.59.)*

Recall from algebra the sign of the slope of a line:

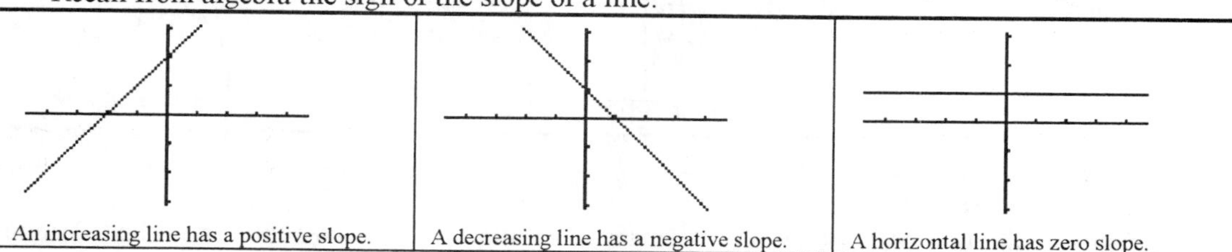

An increasing line has a positive slope. | A decreasing line has a negative slope. | A horizontal line has zero slope.

Because a curve looks linear at point, the slope of the curve at a point is:

- positive where the curve increases,
- negative where the curve decreases,
- zero where the curve is neither increasing or decreasing.

✎ Your Turn

Determine if the slope (instantaneous velocity) of the curve at the points A, B, C, D, E is positive, negative or zero.

Graph of a function f(x)	*Instantaneous velocity*
	• Slope of *f(x)* at A? ____ • Slope of *f(x)* at B? ____ • Slope of *f(x)* at C? ____ • Slope of *f(x)* at D? ____ • Slope of *f(x)* at E? ____

Answers
• positive slope • zero slope • negative slope • zero slope • positive slope

5. *What is the limit of a function at a point?*

✍ ____

____ *(See text p.60,61.)*

You can think of a limit as getting close to a point without actually being there.

- Approach the point from the left,
- Approach the point from the right

If the value of the function approaches a number L, then the limit exists.

Example 2.1.2

Estimate the limit numerically by substituting smaller and smaller values for x from the left and from the right.

Function	*Table of values*	*Graph*	*Value*
$\lim_{x \to 0} \sec x$	X | Y1 -.03 | 1.0005 -.02 | 1.0002 -.01 | 1.0001 0 | [illegible] .01 | 1.0001 .02 | 1.0002 .03 | 1.0005 Y1=1	y x	As $x \to 0$ $\sec x \to 1$
$\lim_{x \to 0} \frac{e^x - 1}{x}$	X | Y1 -.1 | .95163 -.01 | .99502 -.001 | .9995 -1E-4 | .99995 1E-4 | 1.0001 .001 | 1.0005 .01 | 1.005 Y1=1.000050002	y x	*As* $x \to 0$ $\frac{e^x - 1}{x} \to 1$ *note* : $x \neq 0$

✎ **On Your Own**

Estimate the limit numerically by substituting smaller and smaller values for x from the left and from the right for $\lim_{x \to o} \frac{\sin(3x)}{x}$

(Answer appears at the end of the section)

Key Notation

- $\lim_{h \to 0}$ "The limit as h approaches zero."
- $\lim_{h \to c}$ "The limit as h approaches c ."
- $slope\ of\ a\ line = \frac{\Delta y}{\Delta x} = \frac{change\ in\ y}{change\ in\ x} = \frac{y_2 - y_1}{x_2 - x_1} = \frac{f(x_2) - f(x_1)}{x_2 - x_1}$
- $average\ velocity = \frac{\Delta s}{\Delta t} = \frac{change\ in\ s}{change\ in\ t} = \frac{s_2 - s_1}{t_2 - t_1} = \frac{s(t_2) - s(t_1)}{t_2 - t_1}$

🗁 Key Study Skills

- Choose at least two problems that represent this section; put each on a 3 x 5 note card with the solution on the back.
- Use the ✔ ? ★ system on homework problems and get answers to all ★ problems.
- The concept of *limit* is crucial to understanding calculus; make sure you understand it well. Often we think of the limit as getting as close as you can without being there. It is a mind game you must

play. For example you can approach a wall, each time decreasing the distance by 1/2. Theoretically you will never reach the wall because there is still 1/2 of the previous distance left. However, the distance to the wall will keep getting smaller and smaller! Yet in real life you would hit the wall (limit = 0) because you can't make your steps small enough to accommodate the "mind game".

Answer to Own Your Own

Function	*Table of values*	*Graph*	*Value*
$\lim_{x \to 0} \frac{\sin(3x)}{x}$	X: -.1, -.01, 0, .1, .01, .001; Y1: 2.9552, 2.9996, ERROR, 2.9552, 2.9996, 3 Y1=2.9999955 Note: at .001 the display is rounded to 3, while the actual value is 2.9999955.....	Y1=(sin(3X))/X X=0 Y= Note: the function is undefined at x =0, but y appears to be close to 3	As $x \to 0$ $\frac{\sin(3x)}{x} \to 3$

2.2 Limits

Objective: To develop a formal definition of limit and to practice finding the limit.

Preparations: Read and take notes on section 2.2.
Suggested practice problems: 1, 5, 9, 11, 17, 21, 29, 41, 43, 46.

✠ Key Ideas of the Section

You should be able to:

- Find a sufficiently small δ given a sufficiently small ε.
- Recognize and use the properties of limits.
- Determine graphically and algebraically if $\lim_{x \to c} f(x) = L$
- Identify when a limit does not exist.
- Find limits at infinity, $\lim_{x \to \pm\infty} f(x) = L$

Key Questions to Ask and Answer

1. ***How do I define a limit formally and what does sufficiently close mean?***

✍ ____

____ *(See text p. 63.)*

The notation for a limit is $\lim_{x \to c} f(x) = L$. It means that as x gets close to the value c, $f(x)$ gets close to the number L. Sufficiently close means approach c from the left and approach c from the right without letting $x = c$.

Our authors give us the definition of a limit in words first at the top of page 63.

Your Turn

Rewrite the definition of a limit in your own words so that you understand the terminology.

____

____

2. ***What is meant by epsilon (ε) and delta (δ) in the limit definition?***

____

____ *(See text p. 63,64.)*

We say that however "close" you demand the function values to be to the number L, then you can get "close" enough to $x = c$ to meet this demand.. Think of a small vertical distance between the function, $f(x)$, and the limit. We call that small distance epsilon, ε. Thus we want $|f(x) - L|$ to be smaller than that vertical distance or in symbols: $|f(x) - L| < \varepsilon$.

Think of a small horizontal distance between x and c. We call that small distance delta, δ. Thus we want $|x - c|$ to be smaller than that horizontal distance or: $|x - c| < \delta$. Observe δ depends on ε. So no matter how small of a value given to ε, there will be a value assigned to δ which depends upon the particular value given to ε. Now the epsilon delta definition of limit can be given.

Your Turn

Write the formal definition of the limit as it appears on page 64 of the text and relate it to the graph in figure 2.13.

____

____

The *epsilon-delta* definition of limits is best understood if you look at a graph. Recall we want to specify how small to make the distance $|x - c|$ in order to make the distance $|f(x) - L|$ less than some given number. We call that given number ε, epsilon.

Example 2.2.1

Use a graph to find delta (δ) given that epsilon, $\varepsilon < 0.1$, for $f(x) = x^3 - 5x + 8$ in the region near the point (1,4).

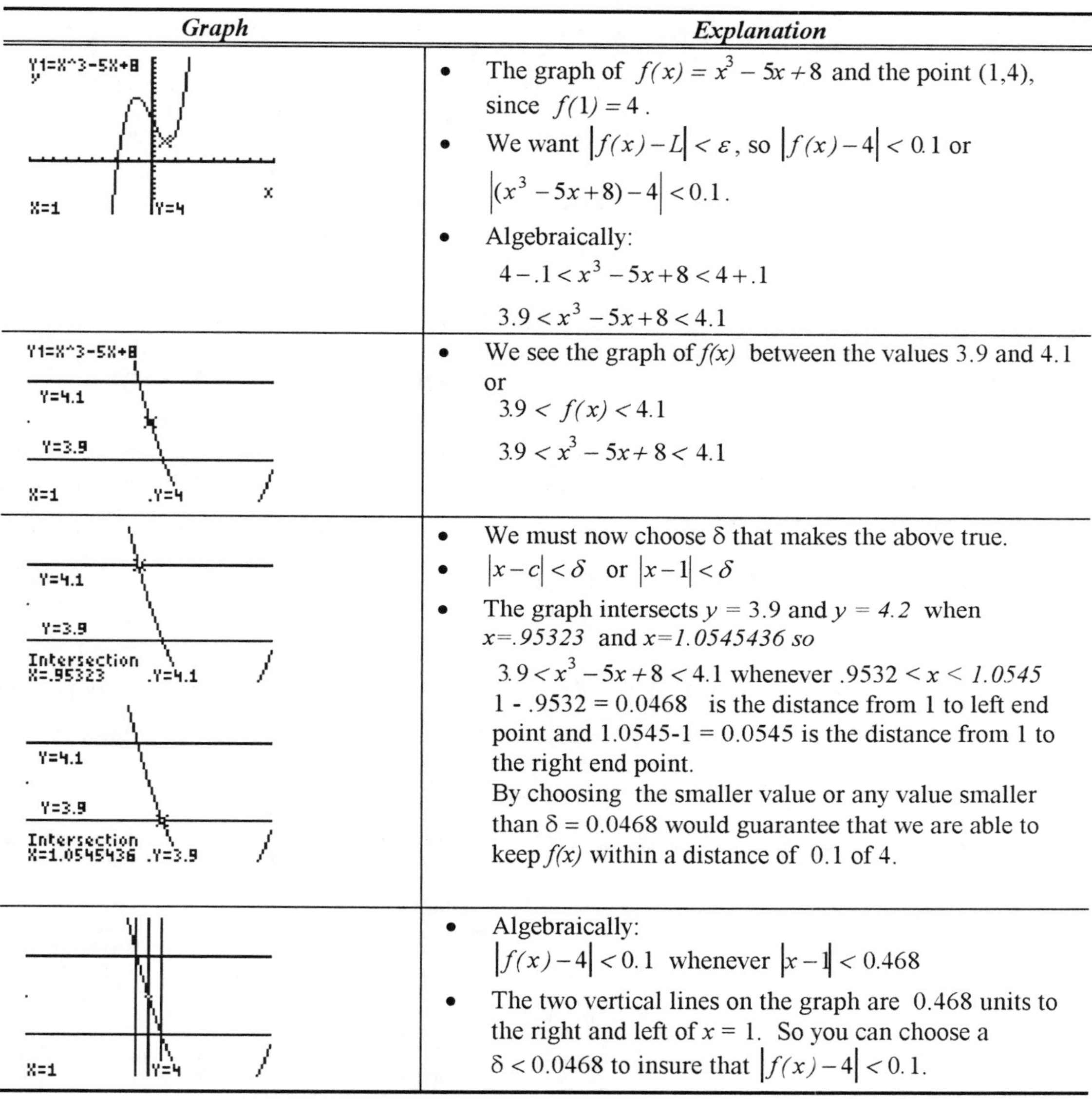

Graph	*Explanation*
Y1=X^3-5X+8 X=1 Y=4	• The graph of $f(x) = x^3 - 5x + 8$ and the point (1,4), since $f(1) = 4$. • We want $\lvert f(x) - L \rvert < \varepsilon$, so $\lvert f(x) - 4 \rvert < 0.1$ or $\lvert (x^3 - 5x + 8) - 4 \rvert < 0.1$. • Algebraically: $4 - .1 < x^3 - 5x + 8 < 4 + .1$ $3.9 < x^3 - 5x + 8 < 4.1$
Y1=X^3-5X+8 Y=4.1 Y=3.9 X=1 Y=4	• We see the graph of $f(x)$ between the values 3.9 and 4.1 or $3.9 < f(x) < 4.1$ $3.9 < x^3 - 5x + 8 < 4.1$
Y=4.1 Y=3.9 Intersection X=.95323 Y=4.1 Y=4.1 Y=3.9 Intersection X=1.0545436 Y=3.9	• We must now choose δ that makes the above true. • $\lvert x - c \rvert < \delta$ or $\lvert x - 1 \rvert < \delta$ • The graph intersects $y = 3.9$ and $y = 4.2$ when $x = .95323$ and $x = 1.0545436$ *so* $3.9 < x^3 - 5x + 8 < 4.1$ whenever $.9532 < x < 1.0545$ 1 - .9532 = 0.0468 is the distance from 1 to left end point and 1.0545-1 = 0.0545 is the distance from 1 to the right end point. By choosing the smaller value or any value smaller than δ = 0.0468 would guarantee that we are able to keep $f(x)$ within a distance of 0.1 of 4.
X=1 Y=4	• Algebraically: $\lvert f(x) - 4 \rvert < 0.1$ whenever $\lvert x - 1 \rvert < 0.468$ • The two vertical lines on the graph are 0.468 units to the right and left of $x = 1$. So you can choose a δ < 0.0468 to insure that $\lvert f(x) - 4 \rvert < 0.1$.

Our example shows that that it is possible to find a number δ no matter how small the number ε. So $\lvert (x^3 - 5x + 8) - 4 \rvert < \varepsilon$ whenever $\lvert x - 1 \rvert < \delta$. This indicates that

$$\lim_{x \to 1} (x^3 - 5x + 8) = 4.$$

Example 2.2.2

Prove that the $\lim_{x \to 1} (3x + 5) = 8$.

Let ε be a sufficiently small given positive number. We want to find δ such that:

$$\lvert f(x) - L \rvert < \varepsilon \quad \text{whenever} \quad 0 < \lvert x - c \rvert < \delta$$

$f(1) = 8$ therefore $L = 8$ and $c = 1$, we have

$$\lvert (3x + 5) - 8 \rvert < \varepsilon \quad \text{whenever} \quad 0 < \lvert x - 1 \rvert < \delta$$

but $\lvert (3x + 5) - 8 \rvert = \lvert 3x - 3 \rvert = 3\lvert x - 1 \rvert$. Therefore we want

$$3|x-1| < \varepsilon \quad \text{whenever} \quad 0 < |x-1| < \delta \text{ and}$$

$$|x-1| < \frac{\varepsilon}{3} \quad \text{whenever} \quad 0 < |x-1| < \delta .$$

This suggests we should choose $\delta \leq \frac{\varepsilon}{3}$. Therefore by the definition of a limit: $\lim_{x \to 1}(3x+5) = 8$.

3. *How can I translate the limit properties into words?*

✍ ____

____ *(See text p. 65)*

It is often easier to understand limit properties if you first translate the symbols into words. These properties can be used to prove that the limit exits in numerous examples. To help you understand them do the following.

Properties of Limits	*Word Translation*
Assuming all the limits on the right hand side exits: ✍ You write the properties that appear on p. 65 of the text. 1.	1. The limit of a constant times a function is the same as the constant times the limit of the function.
2.	2. The limit of a sum is the same as the adding the limits separately (the sum of the limits).
3.	3. The limit of a product is the same as multiplying the limits separately (the product of the limits).
4.	4. The limit of a quotient is the same as dividing the limits separately (the quotient of the limits), if the denominator $\neq 0$.
5.	5. The limit of a constant will always be that constant.
6.	6. The limit of a variable x will always be the value of the number x is approaching.

Example 2.2.3

Calculate the $\lim_{x \to 1} \frac{x-1}{x^2-1}$ in stages, using the properties of limits. Notice that the function

$f(x) = \frac{x-1}{x^2-1}$ is not defined for $x = 1$, but is defined for all values "close" to $x = 1$.

Steps	Properties
$\lim_{x \to 1} \frac{x-1}{x^2-1} = \lim_{x \to 1} \frac{x-1}{(x-1)(x+1)} = \lim_{x \to 1} \frac{x-1}{x-1} \cdot \frac{1}{x+1}$	• using algebra,
$= \lim_{x \to 1} \frac{x-1}{x-1} \cdot \lim_{x \to 1} \frac{1}{x+1}$	• property 3 , limit of a product,

$= \lim_{x\to 1} 1 \cdot \lim_{x\to 1} \frac{1}{x+1} = 1 \cdot \lim_{x\to 1} \frac{1}{x+1}$

- algebra and property 5, limit of a constant,

$= 1 \cdot \frac{\lim_{x\to 1} 1}{\lim_{x\to 1} x + 1} = 1 \cdot \frac{\lim_{x\to 1} 1}{\lim_{x\to 1} x + \lim_{x\to 1} 1}$

- property 4, limit of a quotient and
- property 2, limit of a sum,

$= 1 \cdot \frac{1}{1+1} = \frac{1}{2}$

- property 6 , limit of x and
- property 5, limit of a constant.

4. *What is the notation for left and right hand limits and when do limits NOT exist?*

✍ ____

____ *(See text p.65,66.)*

If you approach a number $x = c$ from the right we use the symbol $\lim_{x\to c+}$, or say "from the positive side of c". For example, if c =4 we allow x to get close to 4 by choosing values closer and closer to 4 from the right, such as 4.1, 4.01, 4.001, 4.0001, 4.00001 .... Thus we consider only values greater than the number.

If you approach a number $x = c$ from the left we use the symbol $\lim_{x\to c-}$, or say "from the negative side of c". For example, if $c = 4$ we allow x to get close to 4 by choosing values closer and closer to 4 from the left such as 3.9, 3.99, 3.999. 3.9999, 3.99999 .... Thus we consider only values less than the number.

The limit exists if the left and right hand limits exist and are the same.

So using our example with $c = 4$. Suppose $\lim_{x\to 4+} f(x) = 7$ and $\lim_{x\to 4-} f(x) = 7$. Since the left hand and right hand limits are the same, the limit exists and $\lim_{x\to 4} f(x) = 7$.

The limit does NOT exists if the left and right hand limits are NOT the same.

If $\lim_{x\to c+} f(x) = L_1$ and $\lim_{x\to c-} f(x) = L_2$, $L_1 \neq L_2$, then $\lim_{x\to c+} f(x) \neq \lim_{x\to c-} f(x)$, so $\lim_{x\to c} f(x)$ does not exist.

Example 2.2.4

Investigate the following limits.

a. $\lim_{x\to 0} \frac{|x|}{x}$ b. $\lim_{x\to 1} \frac{1}{x^2 - 1}$ c. $\lim_{x\to +\infty} \left(\frac{1}{x}\right) + 1$ and $\lim_{x\to -\infty} \left(\frac{1}{x}\right) + 1$

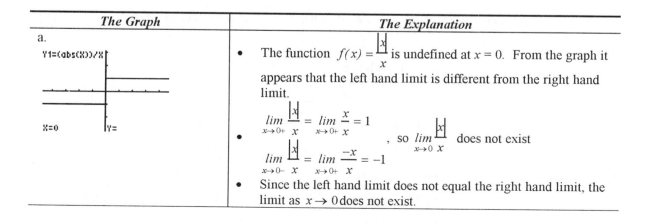

The Graph	*The Explanation*
a. Y1=(abs(X))/X X=0 Y=	• The function $f(x) = \frac{|x|}{x}$ is undefined at $x = 0$. From the graph it appears that the left hand limit is different from the right hand limit. • $\lim_{x\to 0+} \frac{|x|}{x} = \lim_{x\to 0+} \frac{x}{x} = 1$, $\lim_{x\to 0-} \frac{|x|}{x} = \lim_{x\to 0+} \frac{-x}{x} = -1$, so $\lim_{x\to 0} \frac{|x|}{x}$ does not exist • Since the left hand limit does not equal the right hand limit, the limit as $x \to 0$ does not exist.

b.

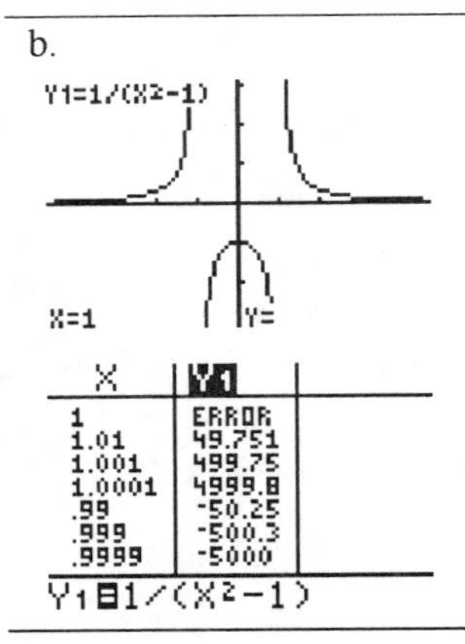

X	Y1
1	ERROR
1.01	49.751
1.001	499.75
1.0001	4999.8
.99	-50.25
.999	-500.3
.9999	-5000

Y1=1/(X²-1)

- The function $g(x) = \dfrac{1}{x^2 - 1}$ is undefined at $x = 1$. From the graph it appears that the left hand limit is different from the right hand limit.
- $$\lim_{x \to 1+} \frac{1}{x^2 - 1} = \frac{\lim\limits_{x \to 1+} 1}{\lim\limits_{x \to 1+} (x+1) \cdot \lim\limits_{x \to 1+} (x-1)} = \frac{1}{+\text{small number}} = +\infty$$
- $$\lim_{x \to 1^-} \frac{1}{x^2 - 1} = \frac{\lim\limits_{x \to 1+} 1}{\lim\limits_{x \to 1^-} (x+1) \cdot \lim\limits_{x \to 1^-} (x-1)} = \frac{1}{\text{- small number}} = -\infty$$
- The $\lim\limits_{x \to 1} \dfrac{1}{x^2 - 1}$ does not exist. The limit as $x \to 1$ does not exist because ∞ is not a number, the value of the function keeps getting larger in either a positive or negative direction. This suggests that there is a vertical asymptote at $x = 1$.

NOTE: $\lim\limits_{x \to c} f(x) = +\infty$ or $-\infty$ means the limit *does not exist*. This indicates a vertical asymptote as $x \to c$.

The Graph	*The Explanation*

c.

Y1=1/X+1

X=50 Y=1.02

- From the graph it appears that the function $h(x) = \dfrac{1}{x} + 1$. approaches a horizontal line y = 1 as $x \to +\infty$ or as $x \to -\infty$
- $$\lim_{x \to +\infty} \left(\frac{1}{x}\right) + 1 = \frac{\lim\limits_{x \to +\infty} 1}{\lim\limits_{x \to +\infty} x} + 1 = \frac{1}{+\text{very large number}} + 1 = 1$$
  $$\lim_{x \to -\infty} \left(\frac{1}{x}\right) + 1 = \frac{\lim\limits_{x \to -\infty} 1}{\lim\limits_{x \to -\infty} x} + 1 = \frac{1}{-\text{very large number}} + 1 = 1$$
- $\lim\limits_{x \to \pm\infty} \left(\dfrac{1}{x}\right) + 1 = 1$ suggesting that *h(x)* has a horizontal asymptote, $y = 1$, as $x \to \pm\infty$.

Note: If $\lim\limits_{x \to +\infty} f(x) = L$ or if $\lim\limits_{x \to -\infty} f(x) = L$, then *f(x)* has a horizontal asymptote at $y = L$.

Key Notation

- ε "Epsilon, a small vertical distance"
- δ "Delta, a small horizontal distance"
- $\lim\limits_{x \to c} f(x)$ "The limit as *x* approaches *c* of the function *f*."
- $\lim\limits_{x \to c+} f(x)$ "The limit as *x* approaches *c* from the right of the function *f*."
- $\lim\limits_{x \to c-} f(x)$ "The limit as *x* approaches *c* from the left of the function *f*."
- $\lim\limits_{x \to c} f(x) = L$ "The limit as *x* approaches *c* of the function *f* is the number *L*."
- $\lim\limits_{x \to c} f(x) = \infty$ "The limit as *x* approaches *c* of the function *f* does not exist because the value of *f(x)* keeps getting larger and larger."

Key Algebra

- If $h(x) = \dfrac{f(x)}{g(x)}$ and a common factor exists between $\dfrac{f(x)}{g(x)}$, such that $g(c) = 0$, and if $\lim\limits_{x \to c} h(x) = L$, this may suggest that $h(x)$ has a hole in the graph at $x = c$.
- For a limit to exist you must show that the left hand limit is equal to the right hand limit.
- Looking at the graph of a function and numerically substituting values into the function as $x \to c$, is helpful in determining the value of the limit. Use properties of limits however to prove that the limit exists.

🗁 Key Study Skills

- Choose at least two problems that represent this section; put each on a 3 x 5 note card with the solution on the back.
- Use the ✓ ? ✯ system on homework problems and get answers to all ✯ problems.
- Review note card problems from 2.1-2.2.
- Understanding the concept of a limit is the foundation for understanding the concept of the derivative, in the next section. Focus on understanding the limit concept. Learn the notation and symbols that go with the limit. Calculus is the mathematics of things in motion or things that are changing and the limit says, change the value of x so that you get as close to a number c, without letting $x = c$.

2.3 THE DERIVATIVE AT A POINT

Objective:	To define the derivative as the instantaneous rate of change.
Preparations:	Read and take notes on section 2.3.
	Suggested practice problems: 1, 3, 5, 8,9, 10, 13, 16, 21, 25, 30.
	Problems requiring substantial algebra: 10-19.

✠ Key Ideas of the Section

You should be able to:

- State the difference quotient and the limit definition of the derivative.
- Find the derivative of a function at a given point.
- Recognize that the derivative at a point is a number.
- Equate the derivative with the slope of the tangent line.

Key Questions to Ask and Answer

1. *How are the average rate of change and the derivative related?*

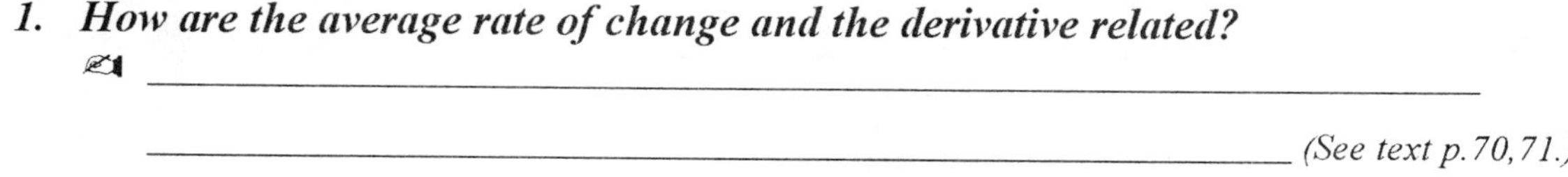

(See text p.70,71.)

The *average rate of change* of any function f from a to $a+h$ is also called the *difference quotient.* This is also the slope of the secant line between the two points, $(a, f(a))$ and $(a+h, f(a+h))$.

$$\textbf{average rate of change} = \frac{\textit{change in } f}{\textit{change in } x} = \frac{f(x_2) - f(x_1)}{x_2 - x_1} = \frac{f(a+h) - f(a)}{(a+h) - a} = \frac{f(a+h) - f(a)}{h}$$

If the interval is allowed to get very small, then you have the *instantaneous rate of change* of the function f at the point $x = a$. This is called *the derivative at point a.*

derivative of $f(a)$ = $f'(a) = \lim_{h\to 0} \frac{\text{change in } f}{\text{change in } x} = \lim_{h\to 0} \frac{f(a+h)-f(a)}{h}$

- If the limit exits, then *f* is said to be differentiable at point *a*.
- The derivative of *f* at *a* is denoted by the symbol $f'(a)$, or read "*f prime of a*".

Example 2.3.1

Find the derivative of $f(x) = \cos x$ at $x = 2$.

a) Approach 2 from the right.

$x_1 = a = 2$	*change in x* = *h*	$x_2 = a + h$	*change in y* $f(a+h) - f(a)$	$\frac{\Delta y}{\Delta x} = \frac{f(a+h)-f(a)}{h}$
2	0.1	2.1	-0.088699268	-0.886992681
2	0.01	2.01	-0.009072016	-0.907201555
2	0.001	2.001	-0.000909089	-0.909089202
2	0.0001	2.0001	-0.000091	-0.909276618
2	0.00001	2.00001	-0.000009	**-0.909295346**

The derivative *f'(2)* is close to -0.9093 when we approach 2 from the right.

b) Approach 2 from the left.

$x_1 = a = 2$	*change in x* = *h*	$x_2 = a + h$	*change in y* $f(a+h) - f(a)$	$\frac{\Delta y}{\Delta x} = \frac{f(a+h)-f(a)}{h}$
2	-0.1	1.9	0.09285727	-0.928572697
2	-0.01	1.99	0.00911363	-0.911362989
2	-0.001	1.999	0.000909505	-0.909505349
2	-0.0001	1.9999	0.0000909	-0.909318233
2	-0.00001	1.99999	0.0000091	**-0.909299508**

The derivative *f'(2)* is close to -0.9093 when we approach 2 from the left.

- The derivative of the cosine at $x=2$ is -0.9093 or *f'(2)* = -0.9093.
- Since the derivative is the slope of the function *f* and *f'(2)* is negative, the function *f* must be decreasing at $x = 2$.
- The instantaneous rate of change represents a -0.9093 vertical change for each unit horizontal change on the cosine curve at $x = 2$.

2. ***Why is the derivative of a function at point A interpreted as the slope of the tangent line at a point A?***

✍ ____

____ *(See text p. 72,73.)*

Since the graph of the function is nearly linear at point A, the tangent line is nearly the same as the function. Therefore the slope of the tangent line to point A is an approximation for the rate of change of the function at point A, and thus the derivative of the function at point A.

- Where the slope of the tangent line to the graph of the function is positive, the function is increasing.
- Where the slope of the tangent line to the graph of the function is negative, the function is decreasing.

Look at the graph of the function $f(x) = \cos x$, and the graph of the tangent line at $x = 2$ and $x=4$. Notice how the slope of the tangent line will tell you if the curve is increasing or decreasing at that point.

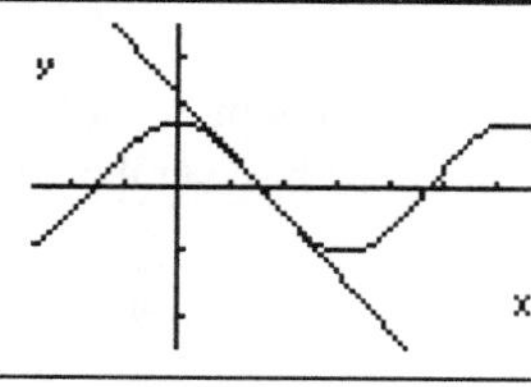	• The tangent line at $x = 2$ has a negative slope. • The function is decreasing at $x = 2$ • Near $x = 2$ the curve and the tangent line will look the same.
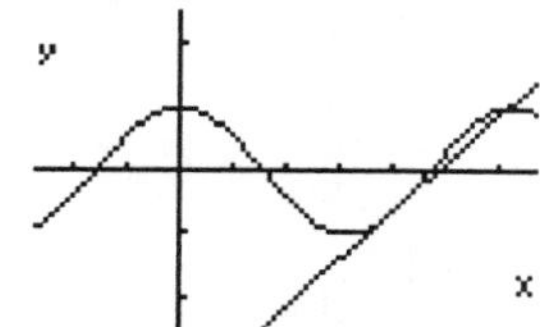	• The tangent line at $x = 4$ has a positive slope. • The function is increasing at $x = 4$ • Near $x = 4$ the curve and the tangent line will look the same.

Example 2.3.2

Find the equation of the tangent line to $f(x) = \cos x$ at $x = 2$.

a) To find the equation of a line you need the *slope* and a *point*.

- The slope of the tangent line = rate of change = derivative at $x = 2$: $m = f'(2) = -0.9093$.
- The point on the graph of the function at $x = 2$ is: $f(2) = \cos 2$, so the point is $(2, \cos 2) = (x_1, y_1)$

b) Use the point slope form for the equation of a line.

$$y = m(x - x_1) + y_1$$

$$y = f'(2)(x - 2) + \cos 2$$

$$y = -0.9093(x - 2) - 0.41615$$

On Your Own

Find the derivative of $k(t) = \frac{1}{t}$ at $t = -2$ algebraically, and then find the equation of the tangent line at $t = -2$. (Refer to the text p. 75 example 8).

(The answer appears at the end of this section.)

Key Notation

- $f'(a)$, "f prime of a" or "the derivative of f at a ".
- $f'(a) = \lim_{h \to 0} \frac{f(a+h) - f(a)}{h}$ can be interpreted as:
    - ♦ The instantaneous rate of change of f at $x = a$.
    - ♦ Slope of the tangent line of the graph of f at $x = a$.
    - ♦ The derivative of f at $x = a$.
    - ♦ The limit of the difference quotient.

Key Algebra

- $f(a + h)$ means replace x with $a + h$ in the function $f(x)$.
- Practice simplification of $f'(a) = \lim_{h \to 0} \frac{f(a+h) - f(a)}{h}$.

Key Study Skills

- Choose at least two problems that represent this section; put each on a 3 x 5 note card with the solution on the back.
- Use the ✓ ? ★ system on homework problems and get answers to all ★ problems.
- Review note card problems from 2.1-2.3.
- There are two main ideas in calculus: the derivative and the integral. You have just learned about the first main idea. Focus on understanding the concept. Learn the notation and symbols that go with the derivative.

Answer to On Your Own

Let $k(t) = \frac{1}{t}$, then

$$k'(-2) = \lim_{h\to 0}\frac{f(-2+h)-f(-2)}{h} = \lim_{h\to 0}\frac{\frac{1}{-2+h}-\frac{1}{-2}}{h} = \lim_{h\to 0}\frac{-2-(-2+h)}{-2(-2+h)}\cdot\frac{1}{h} = \lim_{h\to 0}\frac{-h}{4-2h}\cdot\frac{1}{h} = -\frac{1}{4}$$

Equation of the tangent line $T(t) = k'(-2)(t-(-2)) + k(-2) = -\frac{1}{4}(x+2) - \frac{1}{2}$

2.4 THE DERIVATIVE FUNCTION

Objective: To find the derivative function and to sketch the graph of the derivative from the graph of the function.

Preparations: Read and take notes on section 2.4.
Suggested practice problems: 5, 8, 11, 15, 22, 24, 31, 33, 38, 39.

Key Ideas of the Section

You should be able to:

- Describe the rate of increase or decrease of the graph of a function.
- Sketch the graph of the derivative given the graph of a function.
- Find the derivative of a constant, linear and power function.

Key Questions to Ask and Answer

1. ***How can you tell the rate at which a curve is changing?***

✍ ____

____ *(See text p. 78,79.)*

We use the slope of the tangent line at a point to estimate the rate of change of a function. The derivative is the slope of the tangent line to the graph at a point.

- Where the derivative is positive, the function is increasing.
- Where the derivative is negative, the function is decreasing.
- Where the derivative is zero, the function is neither increasing nor decreasing. This could represent a turning point on the function.

Example 2.4.1

Look at the graph below, determine the intervals where the derivative is positive, negative or zero.

Graph of f(x)	*Sign of the derivative, f'(x)*	*Point or Interval of x*
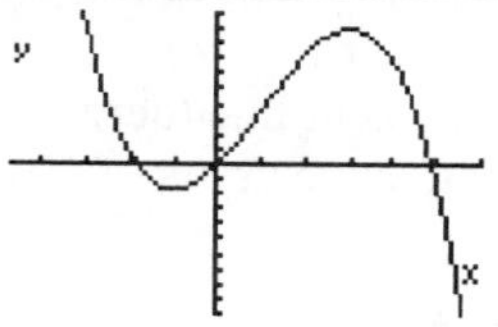	1. $f'(x) = 0$	?
	2. $f'(x) < 0$	?
	3. $f'(x) > 0$	?

Answers
1. $f'(-1) = 0$ and $f'(3) = 0$ turning points of f
2. $f'(x) < 0$ over $(-\infty, -1)$ and $(3, \infty)$ f is decreasing
3. $f'(x) > 0$ over $(-1, 3)$ f is increasing.

On Your Own 1

Look at the graph above, and arrange the following in order from least to greatest:
$f'(-2), f'(0), f'(3), f'(6)$.
(The answer appears at the end of this section.)

2. *How can I sketch the graph of the derivative, given the graph of the function?*

____

____ *(See text p.79,80)*

To sketch the derivative graph, look at the function graph, then:

- Decide what value of x makes the derivative zero.
- Determine the x intervals where the derivative is positive and negative.
- Determine where the slope of the tangent line is steepest (most negative and most positive).

Example 2.4.2

Sketch the graph of the derivative for the graph of the function given in **Example 2.4.1**.

Sketch the derivative graph	*Reason*
	• Mark on the graph where $f'(x) = 0$. • $f'(-1) = 0$ and $f'(3) = 0$.
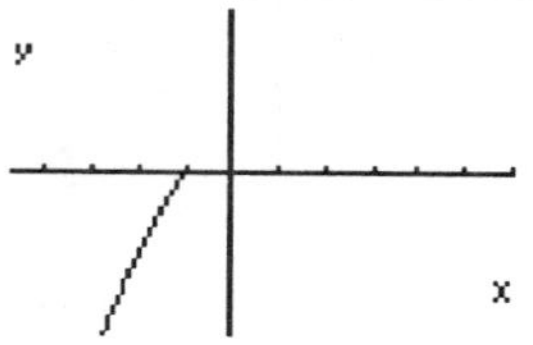	• $f'(x) < 0$ where $f(x)$ is decreasing. • $f(x)$ is decreasing over $(-\infty, -1)$. • The slope of the tangent line (derivative) is more negative where the graph of f is steepest. • The slope of the tangent line (derivative) is approaching 0 from the negative direction as x approaches -1 from the left.

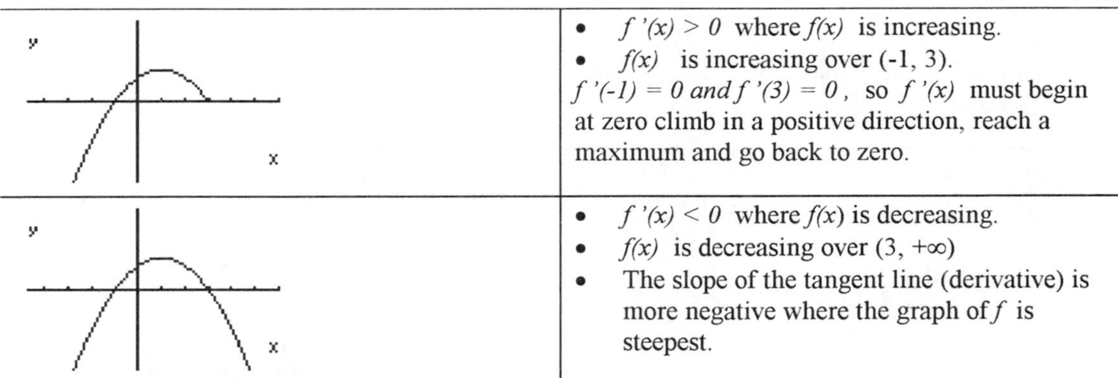

Graph	Explanation
	• $f'(x) > 0$ where $f(x)$ is increasing. • $f(x)$ is increasing over (-1, 3). $f'(-1) = 0$ and $f'(3) = 0$, so $f'(x)$ must begin at zero climb in a positive direction, reach a maximum and go back to zero.
	• $f'(x) < 0$ where $f(x)$ is decreasing. • $f(x)$ is decreasing over $(3, +\infty)$ • The slope of the tangent line (derivative) is more negative where the graph of f is steepest.

On Your Own 2

Carefully look at the graphs of the functions 1 - 6 that follows. Match the graph of the function with the graph of its derivative A-F.

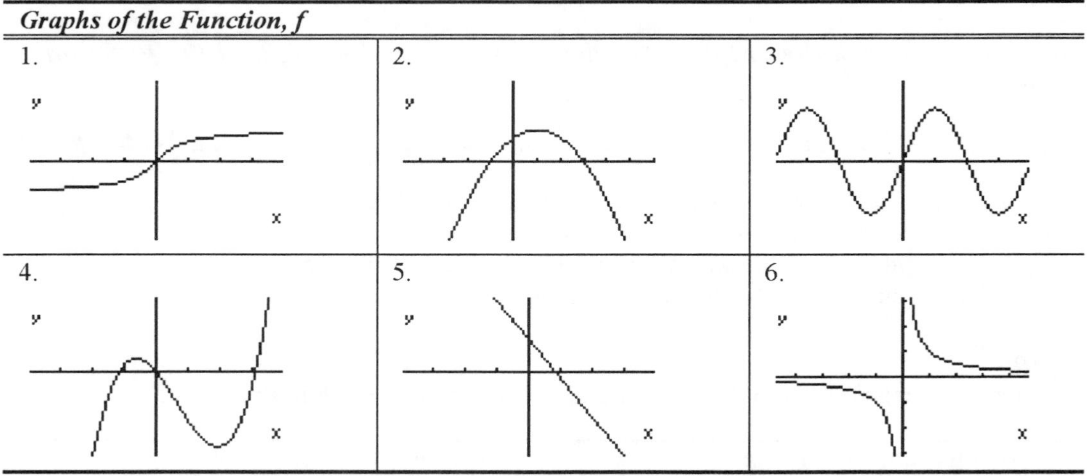

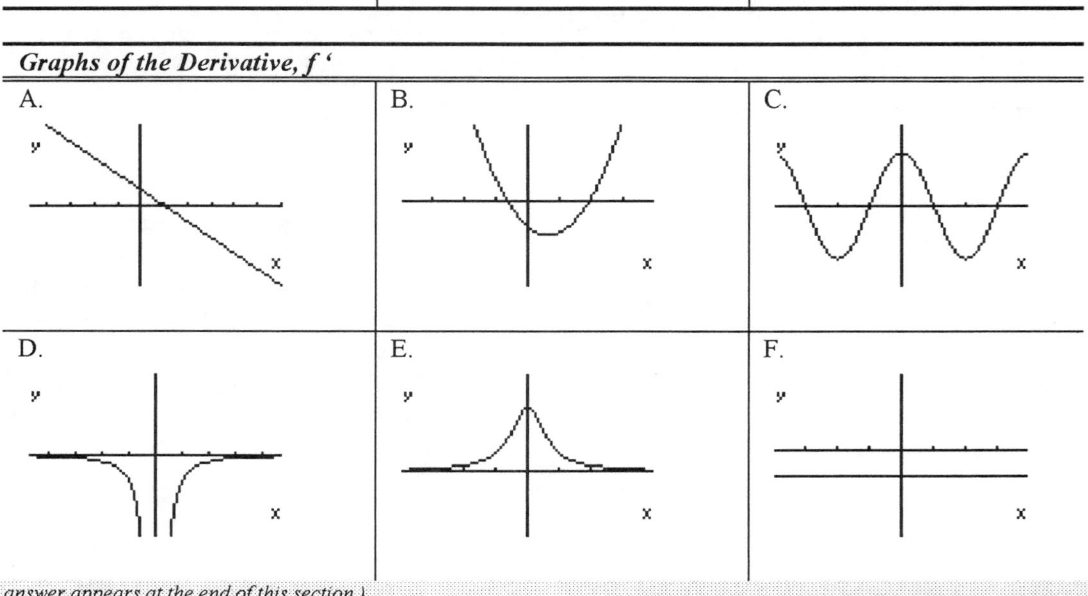

(The answer appears at the end of this section.)

2. *How can you find the derivative numerically?*

✍ ____

____ *(See text p. 80,81)*

If given a table of values, we find the derivative by finding the rate of change.

Example 2.4.3

The number of frogs (in hundreds), *N(t)* in a pond in year *t* is given below. Find the average rate of change in *frogs/year* to estimate the derivative *N'(t)*.

t *(year)*	0	0.5	1	1.5	2	2.5	3	3.5	4	4.5
N (frogs in hundreds)	23	17.5	13	9.5	7	5.5	5	5.5	7	9.5

Use the average rate of change to find the numerical derivative: $\frac{f(x_2)-f(x_1)}{x_2-x_1}$. The first value is $\frac{f(.5)-f(0)}{.5-0}=\frac{17.5-23}{.5}=\frac{-5.5}{.5}=-11$ hundred frogs/year. This means that for the first 6 months the number of frogs was *decreasing* at a rate of 1100 frogs per year. Continuing this process we would generate the following table of values:

t *years*	0	.5	1	1.5	2	2.5	3	3.5	4	4.5
N'(t) *hundred frogs/year*	-11	-9	-7	-5	-3	-1	1	3	5	---

Interpreting the meaning: the number of frogs in the pond decreases every six months until year three (*N'* is negative), when the population began to increase (*N'* is positive). *N'(t)* represents the rate of change in the number of frogs. To better estimate the change in the number of frogs in year 2 take the average of the rate of change to the left and right of year two:
$\frac{1}{2}\left(\frac{7-9.5}{2-1.5}+\frac{5.5-7}{2.5-2}\right)=\frac{-5+(-3)}{2}=-4$ hundred frogs/year. The frog population in year 2 was approximately *decreasing* at a rate of 400 frogs per year. The graph below summarizes our tables:

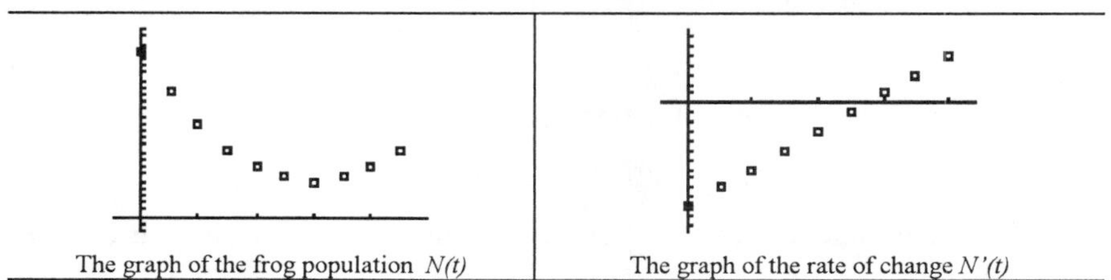

The graph of the frog population *N(t)*   The graph of the rate of change *N'(t)*

3. *How can you find the derivative function using a formula?*

✍ ____

____ *(See text p. 81-83.)*

Given a function f, its derivative function can be found using the *difference quotient*. Here we use the variable x instead of a particular point $x = a$. The result is we get the derivative function in terms of the variable x.

Rate of change of f at a *point* = the derivative of f $= f'(x) = \lim_{h\to 0} \frac{f(x+h) - f(x)}{h}$.

In addition we know the following about certain functions:

- The derivative of a horizontal line, $f(x) = c$, is zero $f'(x) = 0$.
- The derivative of a linear function, $f(x) = mx + b$, is the slope of the line, $f'(x) = m$.
- The derivative of a power function, $f(x) = x^n$, is $f'(x) = nx^{n-1}$.

Example 2.4.4

Find the derivative of the function $f(x) = \dfrac{1}{2x+3}$ using the difference quotient.

$$f'(x) = \lim_{h\to 0} \frac{\dfrac{1}{2(x+h)+3} - \dfrac{1}{2x+3}}{h} = \lim_{h\to 0} \frac{\dfrac{2x+3-(2x+2h+3)}{(2x+2h+3)(2x+3)}}{h}$$

$$= \lim_{h\to 0} \frac{-2h}{4x^2+4hx+12x+6h+9} \bullet \frac{1}{h} = \lim_{h\to 0} \frac{-2}{4x^2+4hx+12x+6h+9}$$

$$f'(x) = \frac{-2}{4x^2+12x+9} = \frac{-2}{(2x+3)^2}$$

Summary of algebraic procedure for finding the difference quotient.

- Find the value of $f(x+h)$, that is, substitute $x + h$ for x in f.
- Subtract $f(x)$, then divide by h.
- If the function is a rational function (fraction), find a common denominator.
- Simplify and factor out h, if possible.
- Let $h \to 0$, then simplify.
- If $f(x)$ is a power function, $f(x) = x^n$, check your answer by the formula $f'(x) = nx^{n-1}$.
- $f'(x)$ is a the derivative function. You can graph the function $f(x)$ and its derivative to see if the functions compare accurately.

Key Notation

- The definition of the derivative function: $f'(x) = \lim_{h\to 0} \frac{f(x+h) - f(x)}{h}$

Key Algebra

- $f(x + h)$ means replace x with $x + h$ in the function $f(x)$.
- Practice simplifying $\dfrac{f(x+h) - f(x)}{h}$ for various functions.
- $\lim_{h\to 0}$ means let h assume values closer and closer to zero until a pattern or limit is evident.

Key Study Skills

- Choose at least two problems that represent this section; put each on a 3 x 5 note card with the solution on the back.
- Use the ✓ ? ★ system on homework problems and get answers to all ★ problems.
- Review note card problems from 2.1-2.4.

- Refer to an algebra text if you are having trouble simplifying the definition of the derivative $f'(x) = \lim_{h \to 0} \frac{f(x+h) - f(x)}{h}$.

Answer to On Your Own 1

f '(-2) < 0
f '(0) > 0
f '(3) = 0
f '(6) < 0

At $x=6$ the graph of $f'(x)$ *is* negative and f *is* steeper at $x=6$ than at $x=-2$, then $f'(6) < f'(-2)$.
In order from least to greatest $f'(6) < f'(-2) < f'(3) < f'(0)$.

Answer to On Your Own 2:

1E, 2A, 3C, 4B, 5F, 6D

2.5 INTERPRETATIONS OF THE DERIVATIVE

Objective:	To use the difference quotient to interpret the meaning of the derivative.
Preparations:	Read and take notes on section 2.5.
	Suggested practice problems: 1, 2, 6, 7, 9, 12, 15, 19.

✠ Key Ideas of the Section

You should be able to:

- Find the units that describe the derivative.
- Interpret the numerical meaning of a given derivative.
- Recognize that the notation $\frac{dy}{dx}$ is equivalent to $f'(x)$.

Key Questions to Ask and Answer

1. ***Why is the notation for the derivative, $\frac{dy}{dx}$, useful in determining the units of the derivative and in interpreting the meaning of the derivative?***

✍ ____

____ (See text p. 85-87.)

The $\frac{dy}{dx}$ notation can be interpreted as :

- $\frac{dy}{dx} = f'(x)$
- the limit of the ratio: $\frac{\text{a small change in y}}{\text{a small change in x}}$ or $\frac{\text{difference in y - values}}{\text{difference in x - values}}$
- $\frac{d}{dx}(y)$, meaning "The derivative of the function y with respect to x".

Look at the units of y divided by the units of x. That ratio is the unit of the derivative, such as *miles/hour*, $/*year*, *ft/sec.*, etc..

Example 2.5.1

The area, A, effected by an oil spill in square feet over time, t in hours, is given by the function $A(t)$. Interpret:

a) $A(5) = 31{,}415$

b) $A'(t)$

c) $A'(5) = 2000$

Solution:

a) Area A, is a function of time so when $t = 5$ hours, the area effected by the oil spill is $31,415$ ft^2 (square feet).

b) This is the derivative of area with respect to time. Using the alternative notation

$A'(t) = \frac{dA}{dt}$, or $\frac{\text{a change in Area}}{\text{a change of time}}$. The units would be $\frac{\text{ft}^2}{\text{hr}}$.

c) Using the alternative notation $A'(t) = \frac{dA}{dt}$, at $t = 5$ hr. the area is changing at a rate of 2000 ft^2/hr. In other words, at hour 5 if time increases by one hour, the area effected by the spill would increase by approximately 2000 ft^2.

Key Notation

- $\frac{dy}{dx}$ An alternate notation for the derivative, $f'(x)$. It means the limit of the ratio of "a change in y divided by a change in x".

Key Study Skills

- This section has numerous examples. Read and work each example and put them in your notebook.
- Choose at least two problems that represent this section; put each on a 3 x 5 note card with the solution on the back. Review all the note cards from chapter 2.1-2.5
- Use the ✓ ? ★ system on homework problems and get answers to all ★ problems.

2.6 THE SECOND DERIVATIVE FUNCTION

Objective: To use the second derivative to determine the concavity of a curve and acceleration.

Preparations: Read and take notes on section 2.6.
Suggested practice problems: 1-7, 8, 9, 13, 15, 18, 19, 20, 22, 23..

✠ Key Ideas of the Section

You should be able to:

- Recognize that for a function f, the derivative of its derivative is called the *second derivative*.
- Use the notation $f''(x)$ or $\frac{d^2y}{dx^2}$ for the second derivative.
- Use the second derivative to find the concavity of a function.

Key Questions to Ask and Answer

1. How do I determine the rate at which a function is increasing or decreasing?

✍ ____

____ *(See text p. 90,91.)*

If the derivative tells how a function is changing, i.e. increasing or decreasing, the derivative of the derivative tells the rate at which the function is increasing or decreasing.

- The second derivative of function f is the derivative of the derivative of f.
- The second derivative tells the rate of change of the derivative of f or the rate at which f is increasing or decreasing.
- The second derivative will tell the concavity of the graph of f.
- If the first derivative tells velocity, the second derivative tells acceleration.

Example 2.6.1

Analyze the graph of $y = x^2$, using the first and second derivative graphs.

Graph of f(x)	*Graph of f '(x)*	*The value of the slope of the tangent line*	**Graph of f '(x)**	*Observations*
$y = f(x) = x^2$ 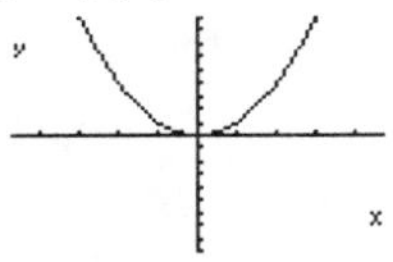 • *for x < 0 f is decreasing* • *for x > 0 f is increasing* • *f is always concave up*	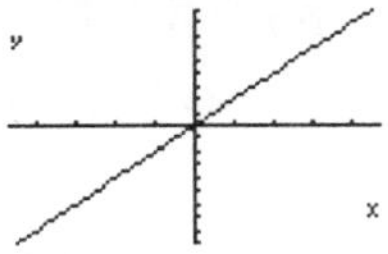 • *for x < 0 f '(x) is negative.* • *for x > 0 f '(x) is positive.* • *f '(x) is always increasing*	$f'(-2) = -4$ $f'(-1) = -2$ $f'(0) = 0$ $f'(1) = 2$ $f'(2) = 4$ *f '(x) is always increasing*	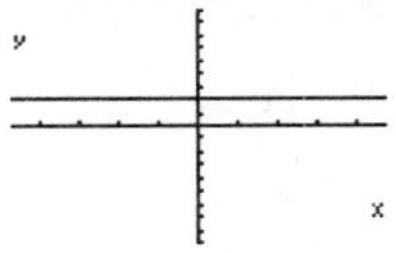 • $f''(x) > 0$	For all x: • *f(x) is concave up* • *f '(x) is increasing* • $f''(x) > 0$

In Summary:

- **If $f''(x)$ is <u>positive</u> then *f(x)* is concave up.**

Example 2.6.2

Analyze the graph of $y = -x^2$, using the first and second derivative graphs.

Graph of f(x)	*Graph of f '(x)*	*The value of the slope of tangent line*	*Graph of f ''(x)*	*Observations*
$y = f(x) = -x^2$ 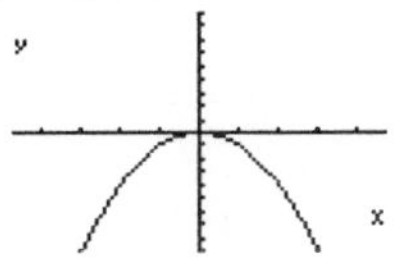 • *for x <0 f is increasing* • *for x > 0 f is decreasing* • *f is always concave down*	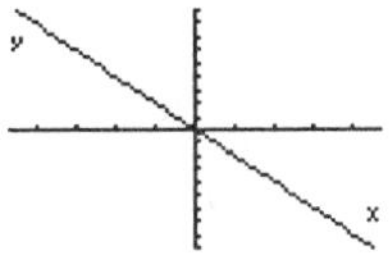 • *for x < 0 f '(x) is positive.* • *for x > 0 f '(x) is negative.* • *f '(x) is always decreasing*	$f'(-2) = 4$ $f'(-1) = 2$ $f'(0) = 0$ $f'(1) = -2$ $f'(2) = -4$ *f '(x) is always decreasing*	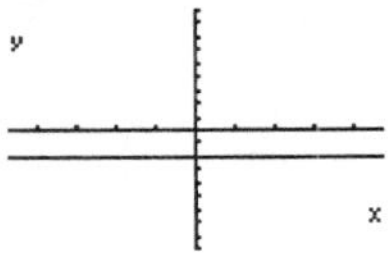 *f ''(x) < 0 for all x.*	On the same interval: • *f(x) is concave down* • *f '(x) is decreasing* • $f''(x) < 0$

In summary:

- **If $f''(x)$ is <u>negative</u>, then *f(x)* is concave down.**

Example 2.6.3

Observe the relationship between the graphs of *f(x), f '(x) and f ''(x).*

Graph of f(x)	*Graph of f'(x)*	*Graph of f''(x)*	*Observations*
f(x) changes concavity	*f '(x) decreases then increases*	*f ''(x) is negative then positive*	• *Where f ''(x) changes from negative to positive, f(x) changes concavity.* • *Where f ''(x) = 0, f(x) changes concavity.* • *Where f '(x) changes from decreasing to increasing, f ''(x) = 0.*

In summary::

- **The point at which $f''(x) = 0$, is the point where the function *may* change concavity. There are special functions where $f''(x) = 0$, but no concavity change occurs, for example $y = x^4$.**
- **The point of inflection is where a function changes concavity and where $f''(x) = 0$ or does not exist. Note that $f''(x) = 0$ does not guarantee an inflection point.**

On Your Own 1

Match the graph of each function below 1 - 6, with the graph of its second derivative A - F.

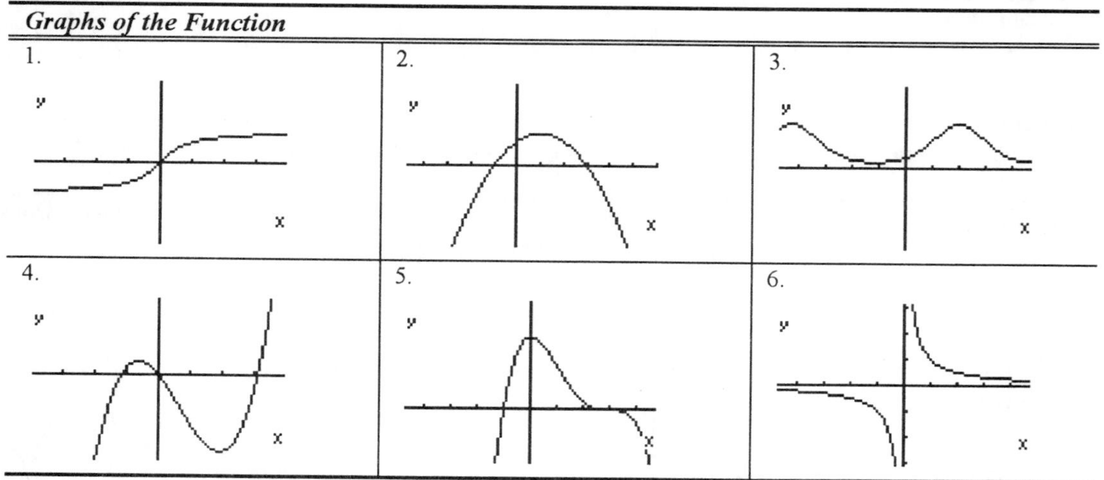

Graphs of the Second Derivative

A.	B.	C.
y x	y x	y x
D.	**E.**	**F.**
y x	y x	y x

(Answers appear at the end of the section.)

✎ On Your Own 2

Determine which of the functions graphed below is

a) increasing at an increasing rate,
b) increasing at a decreasing rate,
c) decreasing at an increasing rate or
d) decreasing at a decreasing rate.

Explain why you have chosen each.

Graphs of Increasing and Decreasing functions

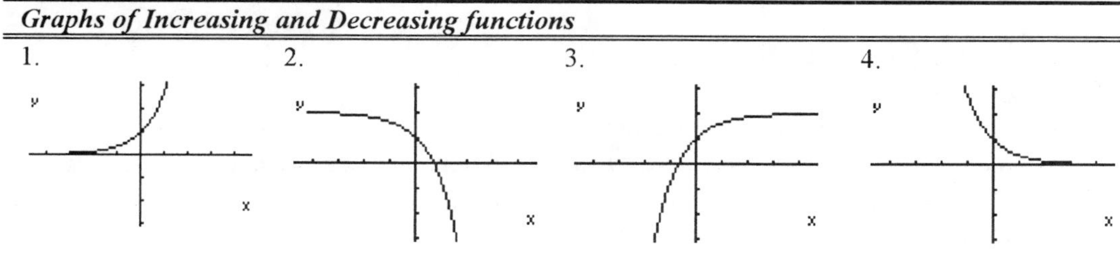

(The answer appears at the end of this section)

Key Notation

- *f ''(x), "The second derivative of f, or the derivative of f '(x)".*
- $\frac{d^2y}{dx^2} = f''(x)$, *"The second derivative of f with respect to x twice.".*
- *f ' increasing* means *f '' > 0.*
- *f 'decreasing* means *f '' < 0.*

🗁 Key Study Skills

- Choose at least two problems that represent this section; put each on a 3 x 5 note card with the solution on the back.
- Use the ✓ ? ★ system on homework problems and get answers to all ★ problems.
- Review note card problems from chapter 2.

Answers to On Your Own 1
1B, 2F, 3C, 4A, 5E, 6D

Answers to On Your Own 2

1a The graph is increasing and concave up. The slope of the tangent line is getting steeper or increasing, so the function is *increasing at an increasing rate.*

2d The graph is decreasing and concave down. The slope of the tangent line is getting more negative or decreasing, so the function is *decreasing at a decreasing rate.*

3b The graph is increasing and concave down. The slope of the tangent line is getting less steep or is decreasing, so the function is *increasing at an decreasing rate.*

4c The graph is decreasing and concave up. The slope of the tangent line is getting less steep or less negative or increasing, so the function is *decreasing at a increasing rate.*

NOTE: Be precise in your language. Avoid using the word phrases such as "it is increasing". Instead say, " The function is increasing and concave down, the derivative is positive and the second derivative is negative on the interval."

2.7 CONTINUITY AND DIFFERENTIABILITY

Objective: To determine the conditions necessary for differentiability.

Preparations: Read and take notes on section 2.7.
Suggested practice problems: 1, 2, 6, 7, 8, 9, 11, 13, 15.

✠ Key Ideas of the Section

You should be able to:

- Know the three conditions for continuity.
- State the conditions when a function is *not* differentiable at a point.

Key Questions to Ask and Answer

1. When is a function continuous?

✍ ____

____ *(See text p. 95.)*

Recall that continuous functions have no breaks, jumps or holes on an interval. A function f is continuous at a point $x = c$ if all three of the following are true:

- f is defined at $x = c$
- the limit exists, i.e. $\lim_{x \to c-} f(x) = \lim_{x \to c+} f(x) = L$
- the limit is the same as function at $x = c$, i.e. $\lim_{x \to c} f(x) = f(c)$

A function is continuous on interval $[a, b]$ if it is continuous at every point in the interval.

✎ Your Turn

What are the four continuity properties for sums, products and quotients? (See text p.95)

1.
2.
3.
4.

State the continuity conditions of composite functions. (See text p. 95)

____

____

Example 2.7.1

Let $f(x) = \begin{cases} 2x+8 & \text{for } x > -3 \\ 5x+17 & \text{for } x \leq -3 \end{cases}$ Is *f(x)* continuous at x = -3?	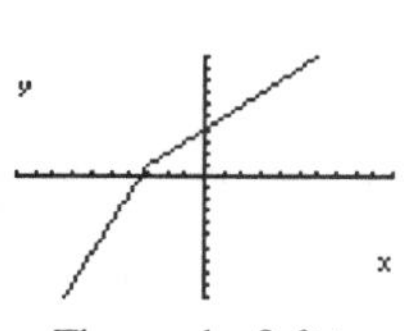 The graph of *f(x)*.

Applying the rules of continuity

1. *f(-3)* = 2 , so f is defined at $x = -3$.
2. The left and right hand limits are the same: $\lim_{x \to -3^-} 5x+17 = 2$ and $\lim_{x \to -3+} 2x+8 = 2$, so the limit exists at $x = -3$..
3. $\lim_{x \to -3} f(x) = f(-3) = 2$

f(x) is continuous at x = -3 since all three conditions are met

4. *When is a function differentiable at a point?*

✍ ____

____ *(See text p. 96)*

A function is differentiable at a point, if it has a derivative at that point. A function is NOT differentiable at a point if:

- The function is NOT continuous (there are jumps, breaks or holes).
- The graph has a sharp corner at the point and cannot be linearized.
- The graph has a vertical tangent line at the point.

Example 2.7.2

The following graphs show functions that are not differentiable at point A

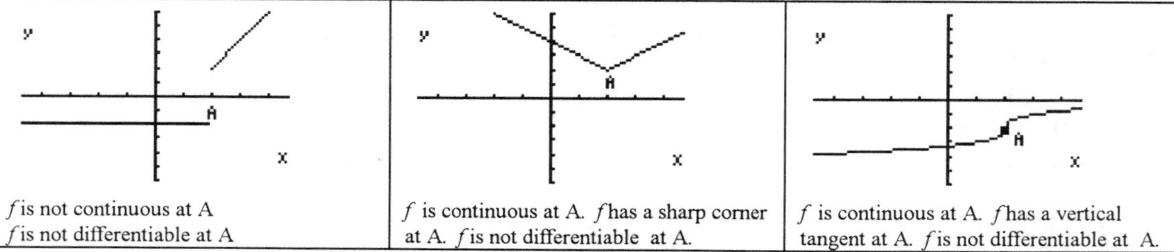

f is not continuous at A *f* is not differentiable at A	*f* is continuous at A. *f* has a sharp corner at A. *f* is not differentiable at A.	*f* is continuous at A. *f* has a vertical tangent at A. *f* is not differentiable at A.

Example 2.7.3

The graph of a function is defined by

$f(x)=\begin{cases} 2+cos(\frac{\pi x}{2}) & \text{for } -2 \le x \le 2 \\ 1 & \text{for } x<-2 \text{ or } x>2 \end{cases}$	y x The graph of *f(x)*.

a. Is *f(x)* continuous at *x* = -2?

Yes because $f(-2)=\lim\limits_{x\to -2} f(x)=1$

b. Is *f(x)* differentiable at *x* = -2?

At *x* = -2 *f(x)* is at the bottom of it's cycle and $f(-2)=2+cos\left(\frac{-2\pi}{2}\right)=2-1=1$. At *f(-2)* the tangent line would be horizontal and *f'(-2)* = 0. . From the left of *x* = -2, *f(x)* = 1 or a horizontal line, so as $x \to -2^-$, *f'(x)* = 0. Since the slopes of the tangent lines are the same from the left and the right and the function is continuous then derivative exists at *x* = -2.

🗁 Key Study Skills

- Choose at least two problems that represent this section; put each on a 3 x 5 note card with the solution on the back. Review the note card problems from sections 2.1-2.7.
- Use the ✓ ? ★ system on homework problems and get answers to all ★ problems.

CUMULATIVE REVIEW CHAPTER TWO

🗁 Key Study Skills

- Review all of the **Key Ideas** from 2.1 - 2.7.
- Review the key words in the chapter two summary of the text and make up your own sentences from the key words in the summary.
- In the text, work the odd review problems at the end of the chapter and check your answers at the back of the book.
- Use your 3 x 5 note cards with sample problems to make up your own test.
- From this Study Guide Chapter 2, review all the Key Ideas.
- From the text do the *Review Problems* on pages 99-102 and the *Check Your Understanding* page 103
- After you have studied, take the Chapter Two Practice Test, check your answers and review the problems that gave you difficulty.

Chapter Two Practice Test

1a. Below is the graph of $f(x)$. Sketch the graph of the derivative on the axis provided.

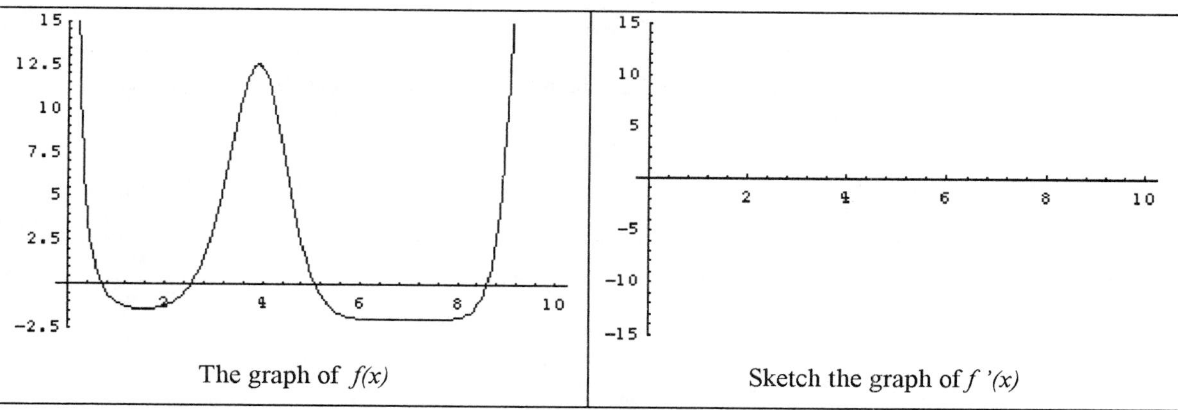

The graph of $f(x)$ | Sketch the graph of $f'(x)$

1b. Arrange the following in order from smallest to largest value (in increasing order): $f'(3), f'(5), f'(7)$. Explain your reasoning.

1c. Which is larger $f''(2)$ or $f''(4)$? Please explain.

2. Below are the graphs of three functions. Label the graphs of $f(x)$, $f'(x)$ and $f''(x)$. Explain in detail why you have chosen each using intervals to compare the three functions.

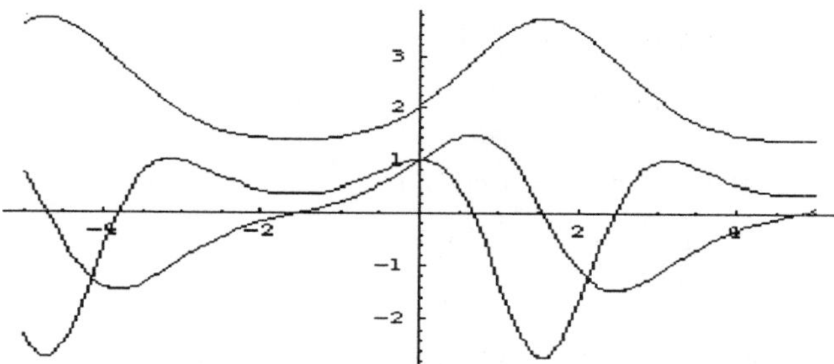

Label $f(x)$, $f'(x)$, and $f''(x)$

3 Below is the graph of $f(x) = (\log\ (x - 1))/\log 2$

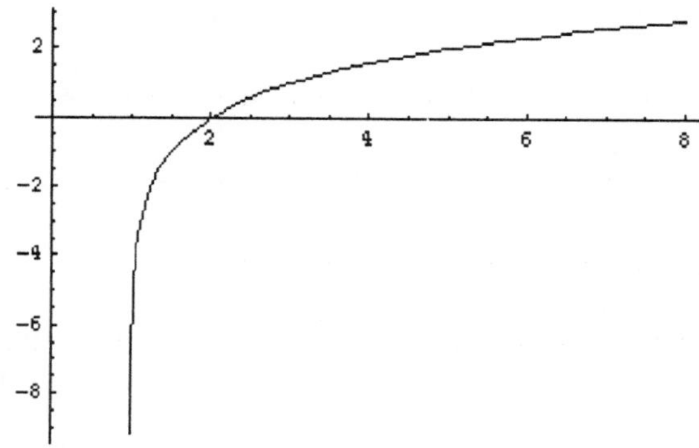

3a. For the above function $f(x)$, write the definition of the derivative, $f'(a)$. Do Not Simplify. What does $f'(a)$ represent?

3b. What is the meaning of $f'(2) = 1.4427$?

3c. Use the value of $f'(2)$ to approximate the value of $f(2.5)$. Will this be an over or under approximation? Explain.

3d. Sketch the graph of $f'(x)$.

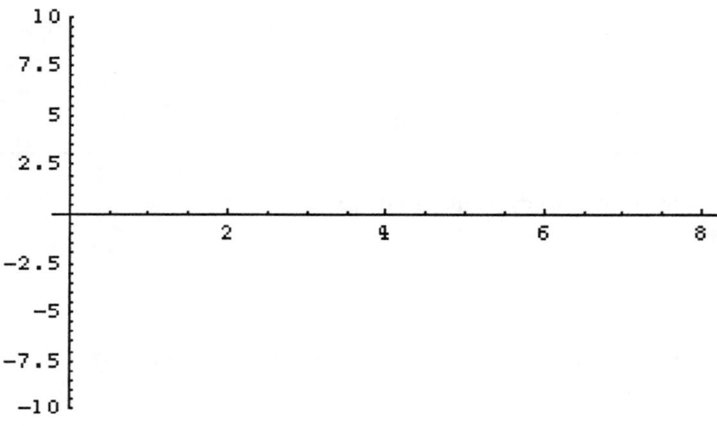

4. Let $f(x) = x^{2/3}$

4a. Use the definition of the derivative to find $f'(0)$.

4b. Is $f(x)$ continuous at $x = 0$?

4c. Is $f(x)$ differentiable at $x = 0$?

(Answers appear at the end of this Study Guide.)

CHAPTER THREE

SHORT-CUTS TO DIFFERENTIATION

In this chapter we take a systematic approach to finding the derivative of functions using given formulas.

To The Student

In this section we deviate slightly from a section by section coverage. We will cover a few sections at a time in order that you can begin to recognize when to use and when not to use the various rules and formulas for finding the derivative of a function.

Read each section in the text carefully, in order to understand the justification of each derivative rule. This Study Guide chapter will focus on helping you to use the derivative rules and to practice the techniques of differentiation.

You should work and rework as many problems as possible. Continued practice makes perfect derivatives.

3.1 Powers and Polynomials; 3.2 The Exponential Function; 3.3 The Product and Quotient Rule

Objective: To distinguish between the power and exponential rules for the derivative and to use the product and quotient rules for the derivative.

Preparations: Read and take notes on section 3.1, 3.2, and 3.3.
Suggested practice problems:
Section 3.1: 5,6,9,10,21,22,25,27,40,42,43,44,47,55,59
Section 3.2: 5, 6, 12, 16, 24, 25, 28, 35, 38, 42-44.
Section 3.3: 2, 4, 11, 20, 28, 29, 32, 41, 42, 47.

✠ Key Ideas of the Section

You should be able to:

- Recall from Chapter 2 the derivatives of a constant, linear and power functions.
- Recognize the difference between a power and an exponential function.
- Find the derivative of a power function and an exponential functions and recognize their differences..

- Use various notation for finding the derivative, i.e. $f'(x)$ or $\frac{d}{dx}(f)$.
- Find the equation of the tangent line at a given point on the graph of a function.
- Recognize that $y = e^x$ is a special case of $f(x) = a^x$.
- Memorize and use both the product and quotient rules for finding derivatives.

Key Questions to Ask and Answer

3.1 Powers and Polynomials

1. ***What are the rule for the derivative of a) a constant function b) a constant multiple of a function, c) a sum and difference of two functions and d) a power function?*** *(See text p. 106-109 .)*

Write the derivative rules in symbols an in your own words:

a) ____

b) ____

c) ____

d) ____

From Chapter Two we know that the slope (rate of change) of a horizontal line is zero. So the derivative of a constant function is zero.
Let c be any real number:

- $\frac{d}{dx}(c) = 0$

If f is a differentiable function and c is a constant then the derivative of a constant times the function is the constant times the derivative of f.

- $\frac{d}{dx}(cf(x)) = c\frac{d}{dx}f(x) = cf'(x)$

If f and g are differentiable functions then the derivative of the sum (difference) of f and g is the sum (difference) of their derivatives.

- $\frac{d}{dx}[f(x)+g(x)] = \frac{d}{dx}(f(x)) + \frac{d}{dx}(g(x)) = f'(x)+g'(x)$

and

- $\frac{d}{dx}[f(x)-g(x)] = \frac{d}{dx}(f(x)) - \frac{d}{dx}(g(x)) = f'(x)-g'(x)$

To find the derivative of any power function where n is a constant real number you multiply the function by the power and reduce the power by one:

- $\frac{d}{dx}(x^n) = nx^{n-1}$
-

Your Turn

Write the derivative of the function $h(t)$ in a step by step fashion below and supply the missing terms :

$$h(t) = 5t^4 + 12\sqrt[3]{t} - \frac{1}{7t^2} + \pi^2$$

Step1. Write using derivative notation:

$$h'(t) = \frac{d}{dt}(5t^4 + 12\sqrt[3]{t} - \frac{1}{7t^2} + \pi^2)$$

Step 2. Rewrite as a sum and difference in power form :

$$h'(t) = \frac{d}{dx}(5t^4 + 12t^{\frac{1}{3}} - \frac{1}{7}t^{-2} + \pi^2)$$

Step 3. Write as a sum using derivative notation.

$$h'(t) = 5\frac{d}{dt}(t^4) + 12\frac{d}{dt}(t^{\frac{1}{3}}) - \frac{1}{7}\frac{d}{dt}(\_\_?\_\_) + \frac{d}{dt}(\_\_?\_\_)$$

Answer

t^{-2} , π^2

Step 4. Apply the power rule:

$$h'(t) = 5(4t^{4-1}) + 12(\frac{1}{3}t^{\ -?-}) - \frac{1}{7}(-2t^{-2-1}) + (\_?\_)$$

Answer

$t^{\frac{1}{3}-1} = t^{\frac{-2}{3}}$, 0

Step 5. Simplify algebraically:

$$h'(t) = (\_?\_\_)t^3 + (\_?\_\_)t^{-\frac{2}{3}} + \frac{2}{7}(\_?\_)$$

Answer

$20t^3 + 4t^{-2/3} +$ $(2/7)(t^{-3})$

Two other forms of the correct answer are:

$$h'(t) = 20t^3 + \frac{4}{\sqrt[3]{t^2}} + \frac{2}{7t^3} = 20t^3 + \frac{4\sqrt[3]{t}}{t} + \frac{2}{7t^3}$$

A special note: In $h(t)$ above, since $t \neq 0$ the derivative does not exist at $t = 0$.

✎ On Your Own 1

Find the derivative of :

$$f(x) = \frac{6x^2 + 4x^5 - 7}{3x^4}$$

Hint: change the quotient into a sum of fractions.

The answer appears at the end of section 3.3.

3.2 The Exponential Function

2. *Why does the derivative of an exponential function look like itself?*

✍ ____

____ *(See text p. 113-114.)*

We know that an exponential function $f(x) = a^x$ is an increasing function when $a > 1$. In fact it increases at an increasing rate. So the graph of the derivative of $f(x)$ must be positive and increasing.

 The graph of $f(x) = a^x$ for x>0.	 The graph of $f'(x)$.	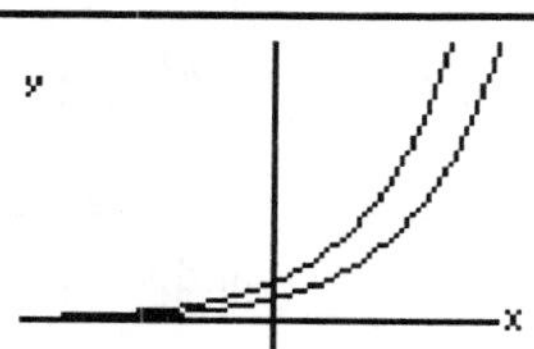 The graph of both $f(x)$ and $f'(x)$.

The derivative of $f(x) = a^x$ is itself times a constant (the natural logarithm of the base):

- $\frac{d}{dx}(a^x) = (\ln a)a^x$

The derivative of *f(x)* = e^x is exactly itself and is a special case of the exponential function derivative rule:

- $\frac{d}{dx}(e^x) = (\ln e)e^x = e^x$

Students often confuse the power function derivative rule and the exponential derivative rule, especially when constant numbers like *ln a*, π, or *e* are used.

Your Turn

Fill in the chart below. Indicate with a ✓, whether the following functions are power or exponential functions, then find their derivatives.

Function	*Power*	*Exponential*	*Derivative*	*Answer*
$y=(\ln 5)5^x$				$y' = (\ln 5)(\ln 5)5^x$ *(exp)*
$z = x^{\ln 5}$				$z' = (\ln 5)x^{\ln 5 - 1}$ *(power)*
$t = \pi^m$				$t' = (\ln \pi)(\pi^m)$ *(exp)*
$n = e^\pi$				$n' = 0$. This is a constant *(power)*
$s = t^\pi$				$s' = \pi t^{\pi-1}$ *(power)*
$f(\theta) = \pi e^\theta$				$f' = \pi e^\theta$ *(exp)*

On Your Own 2

Find the derivative of the following: $s(p) = ap^3 - (\ln b)3^p + \sqrt{3}(\ln 6)^p - ce^p + k\pi^3$

The function *s(p)* is in terms of the variable *p*, other letters are considered constants.

(The answer appear at the end of section 3.3.)

On Your Own 3

Do the odd practice problems 1 -35 in the text on p. 116,117.

(Check your answers at the back of the text)

3.3 The Product and Quotient Rule

3. *What are the two forms of the product rule for derivatives?*

If *u* = *f(x)* and *v* = *g(x)* are differentiable , then

- ____
- ____ *(See text p. 119)*

In words:

The derivative of a product is the derivative of the first function times the second function, plus the first function times the derivative of the second function: *(fg)'* = *f'g* + *fg'*.

You should be aware that some texts or instructors use the form: $(fg)' = fg' + gf'$ for the product rule, but as you can see by the commutative properties of addition and multiplication the expression is equivalent to $(fg)' = f'g + fg'$. Also be aware that many functions at first appear not to be a product but can be rewritten to use the product rule.

Example 3.3.1

Find the derivative of each function. You may have to rewrite the problem first. Identify what you are calling the 'first' and 'second' function.

a)

$$k(w) = (5w^3 - \sqrt[3]{w})(5^w) = f(w) \cdot g(w)$$

$$f(w) = 5w^3 - w^{\frac{1}{3}}, \qquad g(w) = 5^w$$

$$f'(w) = 15w^2 - \tfrac{1}{3}w^{-\frac{2}{3}}, \qquad g'(w) = (\ln 5)5^w$$

$$k'(w) = f'g + fg'$$

$$k'(x) = (15w^2 - \tfrac{1}{3}w^{-\frac{2}{3}})(5^w) + (5w^3 - w^{\frac{1}{3}})(\ln 5)(5^w)$$

The calculus is done at this point and the rest is algebraic simplification, which is a matter of style or preference. Ask your instructor if he/she expects simplification and if you are to rationalize the denominator.

$$k'(w) = 5^w\left[\left(15w^2 - \frac{1}{3\sqrt[3]{w^2}}\right) + (\ln 5)\left(5w^3 - \sqrt[3]{w}\right)\right]$$

$$k'(w) = 5^w\left(5(\ln 5)w^3 + 15w^2 - (\ln 5)\sqrt[3]{w} - \frac{\sqrt[3]{w}}{3w}\right)$$

b) Find $w'(t)$

$$w(t) = \frac{3^t}{4t^3} \qquad \text{rewrite as a produc}$$

$$w(t) = (3^t)(\frac{1}{4}t^{-3}) = f(t) \cdot g(t)$$

$$f(t) = 3^t, \qquad g(t) = \frac{1}{4}t^{-3}$$

$$f'(t) = (\ln 3)(3^t), \qquad g'(t) = \frac{-3}{4}t^{-4}$$

$$w'(t) = f'g + fg'$$

$$w'(t) = [(\ln 3)(3^t)](\frac{1}{4}t^{-3}) + (3^t)(\frac{-3}{4}t^{-4})$$

The calculus is done at this point and the rest is algebraic simplification.

$$w'(t) = \frac{(\ln 3)3^t}{4t^3} + \frac{(-3)3^t}{4t^4} = \frac{3^t t(\ln 3) + (-3)3^t}{4t^4} = \frac{3^t(t\ln 3 - 3)}{4t^4}$$

4. *What are the two forms of the quotient rule for derivatives?*

✍ If $u = f(x)$ and $v = g(x)$ are differentiable, then

- ____
- ____ *(See text p. 120.)*

In words:

The derivative of the quotient is the derivative of the numerator times the denominator minus the numerator times the derivative of the denominator, all over the denominator squared.

In symbols:

$$\left(\frac{u}{v}\right)' = \frac{u'v - uv'}{v^2} \quad \text{or} \quad \left(\frac{f(x)}{g(x)}\right)' = \frac{f'(x)g(x) - f(x)g'(x)}{(g(x))^2} = \frac{f'g - fg'}{g^2}$$

Example 3.3.2

Use the quotient rule to find the derivative of the function in **Example 3.3.1b** above

$$w(t) = \frac{3^t}{4t^3}$$

$$f(t) = 3^t, \qquad g(t) = 4t^3$$

$$f'(t) = (\ln 3)3^t, \qquad g'(t) = 12t^2$$

$$w'(t) = \left(\frac{f}{g}\right)' = \frac{f'g - fg'}{g^2}$$

$$w'(t) = \frac{(\ln 3)(3^t)(4t^3) - 3^t(12t^2)}{(4t^3)^2} = \frac{4t^2(3^t)(t\ln 3 - 3)}{16t^6} = \frac{3^t(t\ln 3 - 3)}{4t^4}$$

This answer is the same as when we simplified **Example 3.3.1b.**

On Your Own 4

Find the derivative of $y = \dfrac{5 - 2t^2}{3\sqrt{t} + 4t^4}$

(The answer appears at the end of the section.)

Key Notation

- $\dfrac{d}{dx}(f)$: "The derivative of the function f with respect to the variable x ".
- $\lim_{h\to 0}(1+h)^{\frac{1}{h}} = e$. A definition of the number e . Try putting larger and larger values for h into this expression, then compare it to the value of e in your calculator.

Key Algebra

- Review rules of exponents.

Key Study Skills

- Choose at least two problems that represent each section; put each on a 3 x 5 note card with the solution on the back. Review all the note card problems 3.1, 3.2, and 3.3.
- Use the ✓ ? ★ system on homework problems and get answers to all ★ problems.
- Use a computer algebra system to find the simplified answer to each derivative problem in sections 3.1-3.3, then practice algebraic simplification on the problems in section 3.1-3.3.
- Make up your own problems for this section and give them to friends to solve. Have each person, including yourself, make up an answer key then compare your answers.
- Practice all quotient problems first by using the quotient derivative rule, then by changing the quotient to a product and use the product derivative rule. Compare the answers.

Answer to On Your Own 1

$$f(x)=\frac{6x^2+4x^5-7}{3x^4}=\frac{6x^2}{3x^4}+\frac{4x^5}{3x^4}-\frac{7}{3x^4}=\frac{2}{x^2}+\frac{4x}{3}-\frac{7}{3x^4}=2x^{-2}+\frac{4}{3}x^1-\frac{7}{3}x^{-4}$$

$$f'(x)=2\frac{d}{dx}(x^{-2})+\frac{4}{3}\frac{d}{dx}x^1-\frac{7}{3}\frac{d}{dx}x^{-4}=2(-2x^{-3})+\frac{4}{3}(1x^0)-\frac{7}{3}(-4x^{-5})$$

$$f'(x)=-4x^{-3}+\frac{4}{3}(1)+\frac{28}{3}(x^{-5})=-\frac{4}{x^3}+\frac{28}{3x^5}+\frac{4}{3}$$

Answer to On Your Own 2

- Set up the problem using derivative notation recognizing that *a, b, c,* and *d* are constants.

$$s'(p)=a\frac{d}{dp}(p^3)-\ln b\frac{d}{dp}(3^p)+\sqrt{3}\frac{d}{dp}\left((\ln 6)^p\right)-c\frac{d}{dp}(e^p)+k\frac{d}{dp}(\pi^3)$$

- Recognize which functions are power and which are exponential and that π^3 is a constant.

$$s'(p)=3ap^{3-1}-(\ln b)(\ln 3)3^p+(\sqrt{3})(\ln(\ln 6))(\ln 6)^p-ce^p+0$$

- Simplify algebraically.

$$s'(p)=3ap^2-(\ln b)(\ln 3)3^p+(\sqrt{3})(\ln(\ln 6))(\ln 6)^p-ce^p$$

Answer to On Your Own 3:

Check the answers to odd problems for section 3.2 in the back of the text.

Answer to On Your Own 4

$$y=\frac{5-2t^2}{3\sqrt{t}+4t^4}\Rightarrow\quad y'=\frac{\left(\frac{d}{dt}(5-2t^2)\right)(3t^{\frac{1}{2}}+4t^4)-(5-2t^2)\left(\frac{d}{dt}(3\sqrt{t}+4t^4)\right)}{(3t^{\frac{1}{2}}+4t^4)^2}$$

$$y'=\frac{(-4t)(3t^{\frac{1}{2}}+4t^4)-(5-2t^2)(\frac{3}{2}t^{-\frac{1}{2}}+16t^3)}{(3t^{\frac{1}{2}}+4t^4)^2}$$

3.4 The Chain Rule; 3.5 The Trigonometric Functions; 3.6 Applications of the Chain Rule

Objective: To recognize when and how to use the chain rule with all of the functions studied thus far including trigonometric and logarithmic functions.

Preparations: Read and take notes on section 3.4, 3.5, and 3.6.
Suggested practice problems:
Section 3.4: 3, 4, 5, 11, 16, 18, 21, 20, 25, 27, 31, 37, 46, 47, 48.
Section 3.5: 3, 4, 5, 11, 18, 20, 25, 27, 37, 46, 47, 48.
Section 3.6: 3, 4, 7, 8, 16, 17, 20, 21, 22, 23, 25, 32, 41, 45, 47.
Practice as many problems as possible.

✠ Key Ideas of the Sections 3.4-3.6

You should be able to:

- Recognize when a function is a composition of two functions.
- Use the chain rule to find the derivative of compositions of functions, identifying the outside and inside functions.
- Recognize when the chain rule is to be used and practice using it.
- Give a justification for the derivative of the sine and cosine from the graph of each function.
- Derive the derivative of the tangent function using the basic definition $\tan x = \dfrac{\sin x}{\cos x}$
- Use the chain rule to find the derivative of the inverse trigonometric functions.
- Use alternative notations for the derivative: y', $f'(x)$, $\dfrac{d}{dx}(f)$, $(f(x))'$.

Key Questions to Ask and Answer

3.4 The Chain Rule

1. What is the Chain Rule for the derivative and when do you need to use the chain rule?

✍ ____

____ *(See text p. 123)*

We have discussed derivative rules for a function that depends directly on *x*, or a single variable. Now we look at the rules when the function depends upon another function. This is called composition of functions and it is important to recognize the expression for the outside function and the expression for the inside function. The inside function is generally the more complicated function.

The text uses the substitution method to identify the inside function.

If $y = (5x^3 - 3x^2 + 4x - 3)^3$, we recognize that this is a complicated function raised to a power. The inside function is the more complicated so we rewrite *y* as $y = z^3$ where $z = 5x^3 - 3x^2 + 4x - 3$.

For $y = e^{\sin x}$, we recognize this as an exponential function with a complicated exponent, the inside function is $z = \sin x$. So we can rewrite *y* as $y = e^z$ where $z = \sin x$.

The *chain rule* states that since *y* depends upon <u>both</u> *z and x*, the derivative $\dfrac{dy}{dx}$ is given by the formula:

$$\frac{dy}{dx} = \frac{dy}{dz} \cdot \frac{dz}{dx}$$

In other words <u>we chain the derivatives together</u>. This notation is useful because, although we can not literally do it, we can think of *dz* canceling, leaving only $\dfrac{dy}{dx}$.

The Chain Rule states:

"Take the derivative of the outside function without changing the inside function and multiply this result by the derivative of the inside function".

If *f* and *g* are functions then *f[g(x)]* is a function of *x*. Using an alternative notation the *chain rule* is given by the formula:

$$\frac{d}{dx} f(g(x)) = f'(g(x)) \cdot g'(x)$$

✎ Your Turn

Identify the inside function by calling it z, then set up the derivative. You supply the missing parts. Part a is completely solved for you.

Function	***Identify the Chain***	***Compute the Derivative of the chain***	***Compute the Derivative***
a) $y = (2x^4 - 4x + 3)^5$	$y = z^5$ where $z = 2x^4 - 4x + 3$	$dy/dz = 5z^4$ $dz/dx = 8x^3 - 4$	$\frac{dy}{dx} = \frac{dy}{dz} \cdot \frac{dz}{dx} = 5z^4 \frac{d}{dx} z$ $= 5(2x^4 - 4x + 3)^4 \cdot (8x^3 - 4)$
b) $t = 5^{(3x-2)}$	$t = 5^z$ where $z = 3x - 2$	$dt/dz = (\ln 5) 5^z$ $dz/dx = ?$	$\frac{dt}{dx} = \frac{dt}{dz} \cdot \frac{dz}{dx} = (\ln 5) 5^z \cdot \frac{d}{dx} z$ $= (\ln 5) 5^{(\ ?\ )} \cdot (3)$ $= (3 \ln 5)(5^{(3x-2)})$
c) $g = e^{-t^3}$	$g = e^?$ where $z = -t^3$	$dg/dz = e^z$ $dz/dx = ?$	$\frac{dg}{dt} = \frac{dg}{dz} = \frac{dz}{dt} = e^z \cdot \frac{d}{dt} z$ $= e^{(\ ?\ )} \cdot (-3t^2)$ $= -3t^2 e^{-t^3}$

2. *Can you use the chain rule with the product and quotient rules?*

✍ ____

____ *(See text p. 124-125.)*

The chain rule can also be used with the product and quotient rules and can often become quite challenging. With practice you will be able to take the derivative using the chain rule without the substitution technique.

✎ Your Turn

Differentiate the following by
a) using the quotient rule and
b) using the product rule. You fill in the missing parts.

$$f(w) = \frac{\sqrt[3]{3w}}{e^{2w}} = \frac{(3w)^{\frac{1}{3}}}{e^{2w}}$$

Solution:

a) Use the quotient rule:

$$f'(w) = \frac{\frac{d}{dw}(3w)^{\frac{1}{3}} \cdot e^{2w} - (3w)^{\frac{1}{3}} \cdot \frac{d}{dw}(\qquad)}{(e^{2w})^2}$$

$$f'(w) = \frac{\frac{1}{3}(3w)^{\frac{1}{3}-1} \frac{d}{dw}(3w) \cdot (\qquad) - (3w)^{\frac{1}{3}} \cdot e^{2w} \frac{d}{dw}(2w)}{e^{(\qquad)}}$$

$$f'(w) = \frac{\frac{1}{3}(3w)^{(\qquad)}(3) \cdot e^{2w} - (\qquad)^{(\qquad)} \cdot e^{2w} \cdot 2}{e^{4w}} = \frac{e^{2w}[(\qquad)^{-\frac{2}{3}} - 2(\qquad)^{\frac{1}{3}}]}{e^{4w}}$$

$$f'(w) = \frac{(3w)^{-\frac{2}{3}} - 2(3w)^{\frac{1}{3}}}{e^{2w}}$$

b) Use the product rule

$$f(w) = \frac{\sqrt[3]{3w}}{e^{2w}} = \frac{(3w)^{\frac{1}{3}}}{e^{2w}} = (3w)^{\frac{1}{3}} \cdot (e^{-2w})$$

$$f'(w) = \frac{d}{dw}(\qquad)^{\frac{1}{3}} \cdot (e^{-2w}) + (3w)^{\frac{1}{3}} \cdot \frac{d}{dw}(\qquad)$$

$$f'(w) = \frac{1}{3}(3w)^{(\qquad)} \cdot \frac{d}{dw}(3w) \cdot e^{-2w} + (3w)^{(\qquad)} \cdot e^{-2w} \cdot \frac{d}{dw}(-2w)$$

$$f'(w) = \frac{(3w)^{-\frac{2}{3}}}{e^{2w}} + \frac{-2(3w)^{\frac{1}{3}}}{e^{2w}} = \frac{(3w)^{-\frac{2}{3}} - 2(3w)^{\frac{1}{3}}}{e^{2w}}$$

The final answer is the same whether we use the product or quotient rule. However to show that the expressions are the same, the algebra simplification may be tricky.

3.5 The Trigonometric Functions

3. *What are the derivatives of the a) sine, b) cosine and c) tangent functions?*

✍ Write the derivative for the trigonometric functions. *(See text p. 129-130)*

a)____

b)____

c)____

You should be able to look at the graph of the *sine* function and recognize that when the sine function is increasing its derivative is positive and when the sine function is decreasing its derivative is negative. The derivative of the *sine* will be the *cosine* function. The derivative of the *cosine* will be the *negative sine* function. The derivative of the *tangent* will be the *1/(cosine)* ^{2}= *secant* ^{2} function. You should look at the graph of each trigonometric function and its derivative to see if the derivative makes sense.

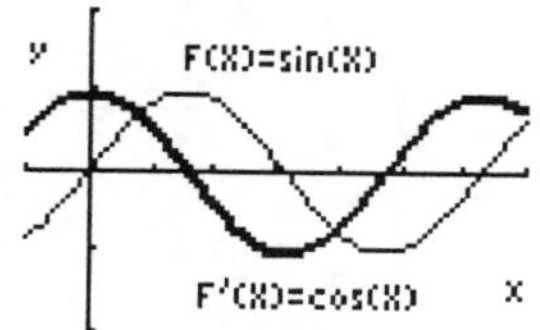	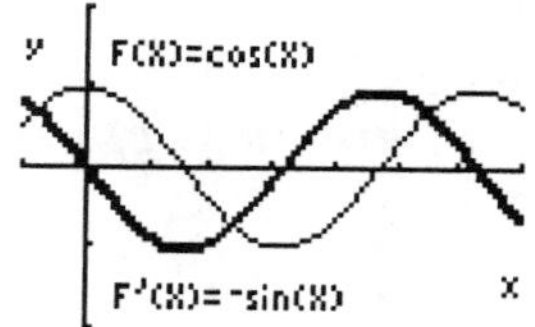	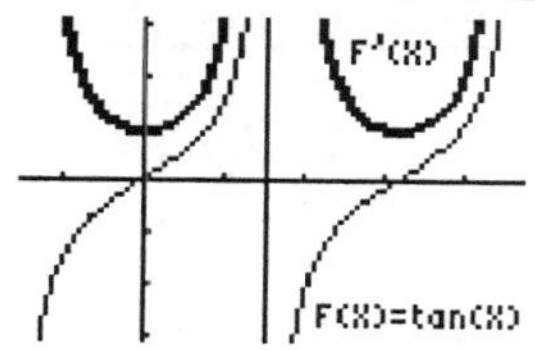
F'(x) is positive where the *sin(x)* is increasing and negative where the *sin(x)* is decreasing.	F'(x) is positive where the *cos(x)* is increasing and negative where the *cos(x)* is decreasing.	F'(x) is always positive where since *tan(x)* is always increasing .

Your Turn

Determine the derivatives of a) $y = cot\ x$ and b) $y = sec\ 2x$, using only the *sine* or *cosine* functions. You fill in the missing parts.

a) Rewrite the function: $cotx = \dfrac{cosx}{sinx}$

$$y' = \frac{d}{dx}(\cot x) = \frac{d}{dx}\left(\frac{\cos x}{\sin x}\right) = \frac{\frac{d}{dx}(\cos x)\cdot(\quad\quad) - (\quad\quad)\frac{d}{dx}(\sin x)}{(\sin x)^2}$$

$$y' = \frac{(\quad\quad)(\sin x) - (\cos x)(\quad\quad)}{(\sin x)^2} = \frac{-\sin^2 x - \cos^2 x}{(\quad\quad)} = \frac{-1}{\sin^2 x}$$

Note: The last step requires the identity $sin^2\ x + cos^2\ x = 1$

b) To find the derivative rewrite the function $\sec(2x) = \dfrac{1}{\cos(2x)}$.

$$y' = \frac{d}{dx}(\sec 2x) = \frac{d}{dx}\left(\frac{1}{\cos 2x}\right) = \frac{\frac{d}{dx}(1)\cdot(\quad\quad) - (\quad\quad)\frac{d}{dx}(\cos 2x)}{(\quad\quad)^2}$$

$$y' = \frac{(\quad)(\cos 2x) - (1)(-\sin(2x)\frac{d}{dx}(\quad)}{\cos^2 2x} = \frac{2\sin 2x}{\cos^2 2x}$$

Note: There are many ways to simplify trigonometric derivatives using trigonometric identities. Check with your instructor to see what he/she expects.

On Your Own 1

Chapter Four Review Quiz 3.1-3.5.

Find the derivatives of the following functions.

1. $y(x) = \cos^4(3x-1)$
2. $h(s) = \cos(\tan 2s)$
3. $t(x) = x^k + k^x - kx + \sin kx - e^k$
4. $z(t) = \dfrac{\cos t}{\sqrt{2-\cos 2t}}$
5. $p(r) = -2\sin((3r-\pi)^2)$

(The answers appear at the end of section 3.6.)

3.6 Applications of the Chain Rule

4. *What are the derivatives of a) ln x, b) arctan x, c) arcsin x?*

✍ Write the derivatives. *(See text p. 133-135.)*

a)____

b)____

c)____

The chain rule was used in each case to derive the rule for *ln x, arctan x, arcsin x*. If the argument used is not *x* but a function *z*, use the chain rule to find *dz/dx*. In some cases you will need to apply the chain rule more than one time.

Example 3.6.1

Find the derivative of the given function

a) $y(x) = \ln(e^{-x} - 2x)$

Let $z = (e^{-x} - 2x)$, so $y = \ln z$ and $y'(x) = (\ln z)'$.

$$\frac{dy}{dx} = \frac{dy}{dz}\cdot\frac{dz}{dx} = \frac{d}{dz}(\ln z)\cdot\frac{d}{dx}(e^{-x} - 2x) = \frac{1}{z}\cdot[e^{-x}\cdot\frac{d}{dx}(-x) - 2]$$

$$y'(x) = \frac{1}{(e^{-x} - 2x)}(-e^{-x} - 2)$$

b) $f(t) = \tan(\arcsin(3t - \pi))$

Let $z = \arcsin(3t - \pi)$, so $f = \tan z$ and $f'(t) = (\tan z)'$

$$\frac{df}{dt} = \frac{df}{dz}\cdot\frac{dz}{dt} = \frac{d}{dz}(\tan z)\cdot\frac{d}{dt}(\arcsin(3t-\pi)) = \frac{1}{\cos^2 z}\cdot\frac{dz}{dt} = \frac{1}{\cos^2(\arcsin(3t-\pi))}\cdot\frac{d}{dt}(\arcsin(3t-\pi))$$

$$f'(t) = \frac{1}{\cos^2(\arcsin(3t-\pi))}\cdot\frac{1}{\sqrt{1-(3t-\pi)^2}}\cdot\frac{d}{dt}(3t-\pi)$$

$$f'(t) = \frac{1}{\cos^2(\arcsin(3t-\pi))}\cdot\frac{1}{\sqrt{1-(3t-\pi)^2}}\cdot(3)$$

Let $w = \arcsin(3t-\pi)$ and $\sin(\arcsin(3t-\pi)) = 3t-\pi$, so $\sin w = 3t - \pi$

by trig identiy: $\cos^2 w = 1 - \sin^2 w$ and by substitution

$$f'(t) = \frac{1}{1-(3t-\pi)^2}\cdot\frac{3}{\sqrt{1-(3t-\pi)^2}}$$

c) Derive the derivative of *arccos x*, where $-1 < x < 1$.

Note: Recall that $\arccos x = w$ if $\cos w = x$, with $0 \le w \le \pi$ (see problem 42, p.36 of text). If $-1 < x < 1$, then $0 < w < \pi$ and also recall that $\cos(\arccos x) = x$, since they are inverses.

Take the derivative of both sides of the equation.

$$\frac{d}{dx}(\cos(\arccos x)) = \frac{d}{dx}(x)$$

$$-\sin(\arccos x)\cdot\frac{d}{dx}(\arccos x) = 1$$

$$\frac{d}{dx}(\arccos x) = \frac{1}{-\sin(\arccos x)} = \frac{-1}{\sin(w)}$$

If $w = \arccos x$ where $0 < w < \pi$, by trig identity $\sin^2 w + \cos^2 w = 1$

we have $\sin w = \sqrt{1-\cos^2 w}$,

thus we can simplify the answer.

$$\frac{d}{dx}(\arccos x) = \frac{-1}{\sqrt{1-\cos^2 w}} = \frac{-1}{\sqrt{1-\cos^2(\arccos x)}} = \frac{-1}{\sqrt{1-x^2}}$$

We take the positive square root because $\sin w > 0$ on the interval $0 < w < \pi$.

On Your Own 2

1. Find r' for $r(t) = \arccos(4t^2-5)$
2. Find the derivative of $p(v) = \ln(2 - \cos 7v)^3$.
3. Find the equation of the line tangent to $y = \ln(\sin(x/2) + 2)$, at $x = 5$, then graph both the function and the tangent.

(The answer appears at the end of this section)

Key Notation

- $(f(x))' = (g(x))'$; means "take the derivative of the function on both sides of the equation".
- $\frac{d}{dx}(f(g(x)) = f'(g(x))g'(x)$ The chain rule means, "take the derivative of the outside function without changing the inside function and multiply this result by the derivative of the inside function".

Key Study Skills

- Choose at least two problems that represent each section: 3.4, 3.5, and 3.6; put each on a 3 x 5 note card with the solution on the back. Review all the note card problems 3.1- 3.6.
- Use the ✓ ? ★ system on homework problems and get answers to all ★ problems.
- Use a computer algebra system to find the simplified answer to each derivative problem in sections 3.4-3.6. Practice algebraic simplification.
- Work with a friend each giving the other derivative problems and checking each other's answers.

Answer to On Your Own 1

1. $y(x) = \cos^4(3x-1) = (\cos(3x-1))^4$

   $y'(x) = 4(\cos(3x-1))^3(-\sin(3x-1))(3) = -12\cos^3(3x-1)(\sin(3x-1))$
2. $h(s) = \cos(\tan(2s))$

   $$h'(s) = -\sin(\tan(2s))(\frac{1}{\cos^2(2s)}(2) = \frac{-2\sin(\tan(2s))}{\cos^2(2s)}$$
3. $t(x) = x^k + k^x - kx + \sin kx - e^k$ assume k is a constant

   $t'(x) = kx^{k-1} + (\ln k)(k^x) - k + k\cos kx$

4. $z(t) = \dfrac{\cos t}{\sqrt{2-\cos 2t}} = (\cos t)(2-\cos 2t)^{-\frac{1}{2}}$

$$z'(t) = (-\sin t)(2-\cos 2t)^{-\frac{1}{2}} + (\cos t)(-\frac{1}{2}(2-\cos 2t)^{-\frac{3}{2}})(2\sin 2t)$$

5. $p(r) = -2\sin((3r-\pi)^2)$

$$p'(r) = -2\cos((3r-\pi)^2)(2(3r-\pi)(3)) = -12(3r-\pi)\cos((3r-\pi)^2)$$

Answer to On Your Own 2

1. *$r(t) = \arccos(4t^2-5)$*

Let $z = 4t^2 - 5$, so that $r = \arccos z$ and $dz/dt = 8t$

$$r'(t) = \frac{dr}{dz}\cdot\frac{dz}{dt} = \frac{-1}{\sqrt{1-z^2}}\cdot\frac{dz}{dt} = \frac{-1}{\sqrt{1-(4t^2-5)^2}}\cdot(8t) = \frac{-8t}{\sqrt{1-(4t-5)^2}}$$

2. *$p(v) = \ln(2 - \cos 7v)^3 = 3\ln(2 - \cos 7v)$*

Let $z = 2 - \cos 7v$, so that $p = 3\ln z$, and $dz/dv = -(-\sin 7v)(7)$

$$p'(v) = \frac{dp}{dz}\cdot\frac{dz}{dv} = 3\cdot\frac{1}{z}\cdot\frac{dz}{dv} = \frac{3}{(2-\cos 7v)}\cdot(7\sin 7v) = \frac{21\sin 7v}{2-\cos 7v}$$

3. $y = \ln(\sin(\frac{x}{2}) + 2)$

$$y' = \frac{1}{\sin(\frac{x}{2})+2}\cdot\frac{d}{dx}(\sin(\tfrac{x}{2})+2)$$

$$y' = \frac{1}{\sin(\frac{x}{2})+2}\cdot\cos(\tfrac{x}{2})\cdot\frac{1}{2} = \frac{\cos(\frac{x}{2})\cdot}{2(\sin(\frac{x}{2})+2)}$$

$$y'(5) = -0.154157$$

$$y_{\tan} = -0.154157(x-5) + y(5) = -0.154157(x-5) + 0.95424$$

$$y_{\tan} = -0.154157x + 1.7257$$

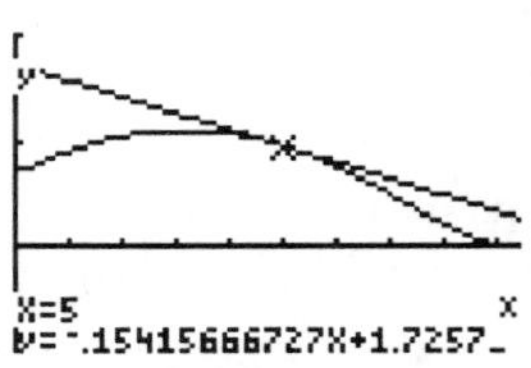

The graph of $y = \ln(\sin(x/2) + 2)$ and the tangent line : $y = -0.154157x + 1.7257$

3.7 Implicit Functions

Objective: To use the chain rule for finding the derivative of implicit functions.

Preparations: Read and take notes on section 3.9.
Suggested practice problems: 1 – 19 odd, 20, 21, 23, 27.

✠ Key Ideas of the Section

You should be able to:

- Recognize that an implicit function as one which is not in the form $y = f(x)$.
- Use the chain rule on an implicit function to find the expression for dy/dx.
- Find the equation of the tangent line for an implicit function.

Key Questions to Ask and Answer

1. How can you find a formula for dy/dx when y satisfies an equation but you do not have a formula for y?

✍ ____

____ *(See text p. 138-140.)*

When an equation that is not explicitly solved for y, that is, not in the function form $y = f(x)$, we say y is an *implicit* function of x. Here are some examples of implicit functions. Some can be put into explicit form, $y = f(x)$, and some can not.

$a)\quad x - 3y + 2 = 0$

$b)\quad x - 3y^2 = 5$

$c)\quad y^2 = 2x - 5y$

$d)\quad y^2(xy - 4) = x^2$

We think of y as a function of x, therefore we use the chain rule to find the derivative of y with respect to x. The derivative of y is $1\cdot\left(\frac{dy}{dx}\right)$, the derivative of y^2 is $2y\cdot\left(\frac{dy}{dx}\right)$, the derivative of $\ln y$ is $\frac{1}{y}\cdot(\frac{dy}{dx})$. You must always "chain" the $\frac{dy}{dx}$ to the derivative of y.

Example 3.7.1

Find the derivative of the implicit functions above.

Equation	Steps
$a)\quad x - 3y + 2 = 0$	
$\frac{d}{dx}(x) - 3\frac{d}{dx}(y) + \frac{d}{dx}(2) = \frac{d}{dx}(0)$	• Take the derivatives of all terms in the equation.
$1 - 3(1)\frac{dy}{dx} + 0 = 0$	• Use the chain rule on the y terms.
$-3\frac{dy}{dx} = -1 \Rightarrow \frac{dy}{dx} = \frac{1}{3}$	• Solve for *dy/dx*.

Of course in this equation we could have solved for y first and then taken the derivative explicitly. Solving for y gives: $y = \frac{1}{3}x + \frac{2}{3}$ and $y' = \frac{1}{3}$, the same result as above.

Equation	Steps
$b)\quad x - 3y^2 = 5$	
$\frac{d}{dx}(x) - 3\frac{d}{dx}(y^2) = \frac{d}{dx}(5)$	• Take the derivatives of all terms in the equation.
$1 - 3(2y)\frac{dy}{dx} = 0$	• Use the chain rule on the y terms.
$-6y\frac{dy}{dx} = -1 \Rightarrow \frac{dy}{dx} = \frac{1}{6y}$	• Solve for *dy/dx*.

To find the derivative of $x - 3y^2 = 5$ at $x = 8$, you also need to find the value of y. Since $y = \pm\sqrt{\dfrac{x-5}{3}}$. when $x = 8$, $y = \pm 1$., giving two points (8,1) and (8,-1) on the graph.. If $y' = \dfrac{1}{6y}$ then for $y = 1$, $dy/dx = 1/6$, for $y = -1$, $dy/dx = -1/6$. Look at the tangent lines to the graphs below.

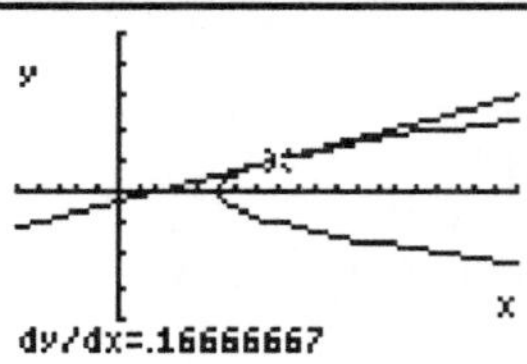 The graph of the equation and the line tangent at (8,1).	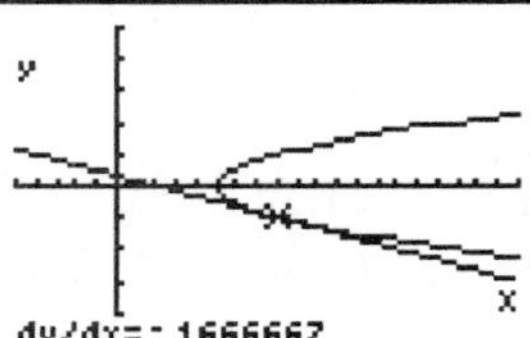 The graph of the equation and the line tangent at (8,-1).

When the derivative is positive the tangent line has an increasing slope and when the derivative is negative the tangent line has a decreasing slope. Note that y is not a function of x.

Equation	***Steps***
c) $y^2 = 2x - 5y$	
$\dfrac{d}{dx}(y^2) = \dfrac{d}{dx}(2x) - 5\dfrac{d}{dx}(y)$	• Take the derivatives of all terms in the equation. Look for products and quotients.
$2y\dfrac{dy}{dx} = 2 - 5(1)\dfrac{dy}{dx}$	• Use the chain rule on the y terms.
$2y\dfrac{dy}{dx} + 5\dfrac{dy}{dx} = 2$	• Put dy/dx on one side of the equation.
$\dfrac{dy}{dx}(2y+5) = 2 \Rightarrow \dfrac{dy}{dx} = \dfrac{2}{2y+5}$	• Factor out dy/dx. • Solve for dy/dx.
d) $y^2(xy-4) = x^2$	• Expand.
$xy^3 - 4y^2 = x^2$	
$(x)'(y^3) + x(y^3)' - (4y^2)' = (x^2)'$	• Take the derivative of all terms in the equation. Use the y' notation.
$1(y^3) + x(3y^2)y' - 8yy' = 2x$	
$y'(3xy^2 - 8y) = 2x - y^3$	• Use the product rule on xy^3 and use the chain rule on the y terms. • Factor out y'.
$y' = \dfrac{2x - y^3}{3xy^2 - 8y}$	• Solve for y'.

Key Notation

- $y' = dy/dx = (f(x))'$, several notations for *"the derivative of y with respect to x"*.

🗁 Key Study Skills

- Choose at least two problems that represent each section and put each on a 3 x 5 note card with the solution on the back. Review all the note card problems 3.1- 3.7.
- Use the ✓ ? ★ system on homework problems and get answers to all ★ problems.
- Students find this section particularly challenging. Ask your instructor for additional practice problems.

3.8 PARAMETRIC EQUATIONS

Objective: To find the parametric representation for curves in the plane.

Preparations: Read and take notes on section 3.8.
Suggested practice problems: 2, 3, 6, 14, 15, 17, 20, 24, 29, 30.

✠ Key Ideas of the Section

You should be able to:

- Describe the motion of an object with parametric equations.
- Find the slope of parametric equations.
- Find the velocity of parametric curves.

Key Questions to Ask and Answer

1. *How do you represent motion in a plane?*

✍ ____

____ *(See text p. 155.)*

To represent motion you must let both x and y be functions of time, t. As time changes both x and y will change, this causes the point (x,y) to move based upon changing time. Let $x = f(t)$ and $y = g(t)$, be parametric equations, thus at time, t, the particle is at $(f(t),g(t))$. The following are true about parametric equations:

- $x = f(t)$ and $y = g(t)$, are called parametric equations, because both depend upon the parameter t.
- t regulates the speed at which the curve is traced out. To change the speed replace t with ωt.
- Motion on a circle $x^2 + y^2 = 1$ can be parametized by $\cos^2 t + \sin^2 t = 1$, so $x = \cos t$ and $y = \sin t$.
- A parametric equation for a straight line through the point (x_0, y_0) is $x = x_0 + at$, and $y = y_0 + bt$, with the rate of change for x, $\frac{dx}{dt} = a$ and the rate of change of y, $\frac{dy}{dx} = b$. The slope of the line is $m = \frac{b}{a} = \frac{dy/dt}{dx/dt}$.
- The instantaneous velocity in an x direction is $v_x = \frac{dx}{dt}$. The instantaneous velocity in a y direction is $v_y = \frac{dy}{dt}$ and these are called components of the velocity in an x and y direction. Instantaneous speed is $v = \sqrt{\left(\frac{dx}{dt}\right)^2 + \left(\frac{dy}{dt}\right)^2}$
- The slope of a parametric curve is $\frac{dy}{dx} = \frac{dy/dt}{dx/dt}$.
- To find the tangent line at point (x_0,y_0) to a given curve parametrically, find the straight line motion through (x_0,y_0), $x = x_0 + \frac{dx}{dt}t$, and $y = y_0 + \frac{dy}{dt}t$.
- The graph of any function $y = f(x)$. can be parametized by $x = t$ and $y = f(t)$.

Example 3.8.1

Find the equation of the line tangent to the ellipse $\frac{x^2}{4}+\frac{y^2}{9}=1$ in parametric form at $t=\frac{7\pi}{4}$.

Solution:

a) Find the parametric form of the equation:

Rewrite the ellipse as $\left(\frac{x}{2}\right)^2+\left(\frac{y}{3}\right)^2=1$, adapting to the parametric form of a circle: $\frac{x}{2}=\cos t$ and $\frac{y}{3}=\sin t$. A parameterization of the ellipse is:

$$x=2\cos t,\quad y=3\sin t,\quad 0\le t\le 2\pi.$$

b) Find the slope of the curve at $t=\frac{7\pi}{4}$:

$\frac{dx}{dt}=-2\sin t,\quad \frac{dy}{dt}=3\cos t,$ at $t=\frac{7\pi}{4}$ the slope of the tangent line is

$$\frac{dy}{dx}=\frac{dy/dt}{dx/dt}=\frac{3\cos\left(\frac{7\pi}{4}\right)}{-2\sin\left(\frac{7\pi}{4}\right)}=\frac{3\left(\frac{1}{\sqrt{2}}\right)}{-2\left(-\frac{1}{\sqrt{2}}\right)}=\frac{3}{2}=1.5$$

c) Find the parameterization of the tangent line:

At $t=\frac{7\pi}{4}$, $x_0=2\cos\frac{7\pi}{4}=\frac{2}{\sqrt{2}}$, $y_0=3\sin\frac{7\pi}{4}=\frac{-3}{\sqrt{2}}$

$x=x_0+\frac{dx}{dt}t,\quad y=y_0+\frac{dy}{dt}t$, so $x=\frac{2}{\sqrt{2}}+\frac{2}{\sqrt{2}}t$ and $\frac{-3}{\sqrt{2}}+\frac{3}{\sqrt{2}}t$

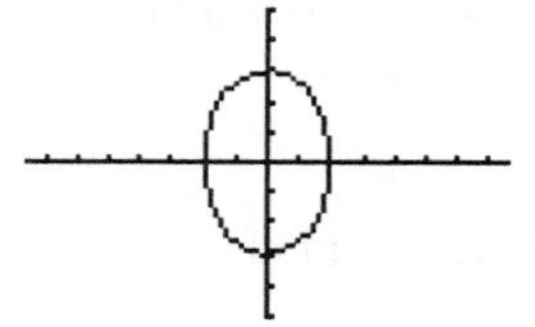

The graph of an ellipse $x=2\cos t,\quad y=3\sin t$ for $0\le t\le 2\pi$

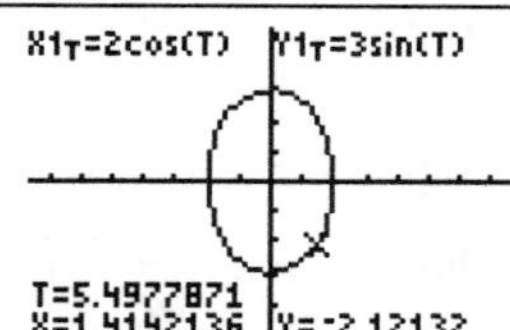

The point on the ellipse when $t=7\pi/4$.

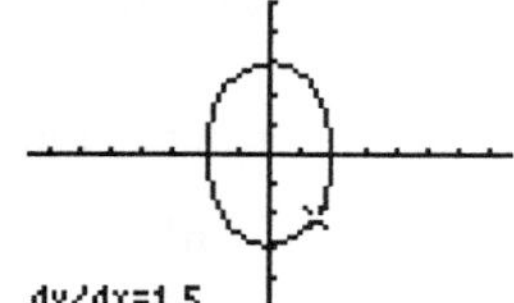

The slope of the curve at $t=7\pi/4$.

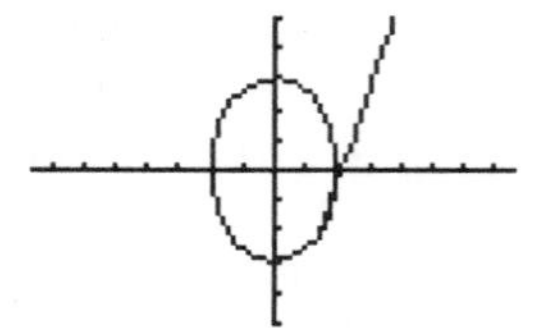

The line tangent to the curve at $t=7\pi/4$. for $0\le t\le 2\pi$

Key Notation

$x=f(t)$ "x is a function of the parameter t."

$y=f(t)$ "y is a function of the parameter t."

Key Algebra

Use unit circle trigonometry to change the equation of a circle into parametric form. Allow $r=1$.

$\sin t=\frac{y}{1}=y,\quad \cos t=\frac{x}{1}=x$

$\cos^2 t+\sin^2 t=1$

$x^2+y^2=1$

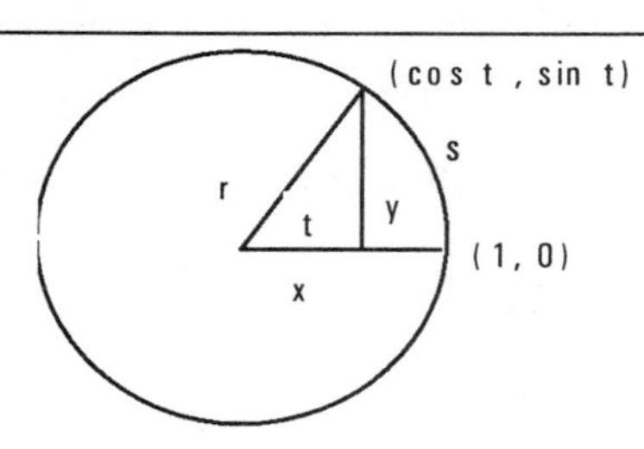

🗁 Key Study Skills

- Choose at least two problems that represent each section; put each on a 3 x 5 note card with the solution on the back. Review all the note card problems 3.1- 3.8.
- Use the ✓ ? ★ system on homework problems and get answers to all ★ problems.

3.9 Linear Approximations and The Derivative

Objective: To use the equation of a tangent line to approximate a function near a point and to estimate the error of the approximation.

Preparations: Read and take notes on section 3.9.
Suggested practice problems: 3, 4, 5, 10, 11, 15.

✠ Key Ideas of the Section

You should be able to:

- Write the equation of the tangent line to the graph of a function *f* at a point $(a,f(a))$ on the graph.
- Approximate the value of $f(x)$ for x near a using the tangent line, $f(x) \approx f(a)+f'(a)(x-a)$
-

Key Questions to Ask and Answer

1. ***When will the local linearization of f near x = a be an over approximation and will it be an under approximation?***

✍ ____

____ *(See text p. 152,153)*

We use the tangent line to approximate the value of the function $f(x)$, for x near a. Since the graph of *f* is nearly linear close to the point $(a, f(a))$, the graph of *f* and the graph of the line tangent to *f* at $(a, f(a))$ are nearly the same. To approximate the value of $f(x)$ near the point $x = a$ use the equation of the tangent line:

$$f(x)\approx f(a)+f'(a)(x-a)$$

- If $f(x)$ is concave up near $x = a$ the approximation will be an under approximation because the tangent line lies under the graph of *f*.
- If $f(x)$ is concave down near $x = a$ the approximation will be an over approximation because the tangent line lies above the graph of *f*.

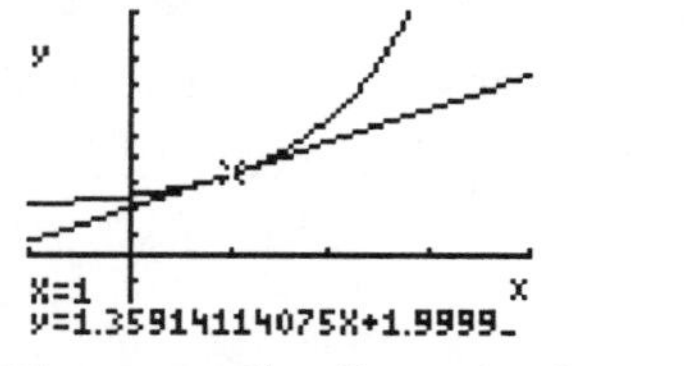

The tangent line lies under $f(x) = .5e^x + 2$

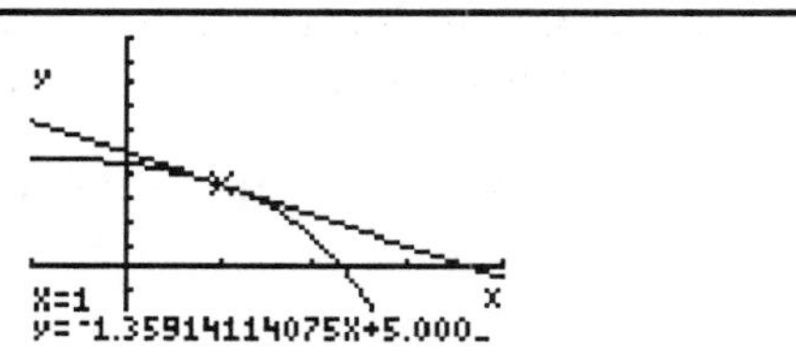

The tangent line lies above $g(x)= -.5e^x +5$

Example 3.9.1

a) Use the tangent line at the point *(1, f(1))* to find the approximate value of *f(1.5)* for the function $f(x) = .5e^{x} + 2$.

$$f(x) \approx f(a) + f'(a)(x-a)$$

$$Since\, f(1) = .5e + 2 \; and \, f'(x) = .5e^{x} \; and \, f'(1) = .5e$$

$$f(x) \approx f(1) + f'(1)(x-1) = (.5e+2) + .5e(x-1) = .5ex + 2 = 1.35914x + 2$$

$$f(1.5) \approx 1.35914(1.5) + 2 \approx 4.0387$$

This is an under approximation because the graph is concave up. See the graph below.

b) Use the tangent line at the point *(1,f(1))* to find the approximate value of *g(1.5)* for the function $g(x) = -.5e^{x} + 5$.

$$g(x) \approx g(a) + g'(a)(x-a)$$

$$Since\; g(1) = -.5e + 5 \; and \, g'(x) = -.5e^{x} \; and \, g'(1) = -.5e$$

$$g(x) \approx g(1) + g'(1)(x-1) = (-.5e+5) + (-.5e)(x-1) = -.5ex + 5 = -1.35914x + 5$$

$$g(1.5) \approx -1.35914(1.5) + 5 \approx 2.9613$$

This is an over approximation because the graph is concave down. See the graph below.

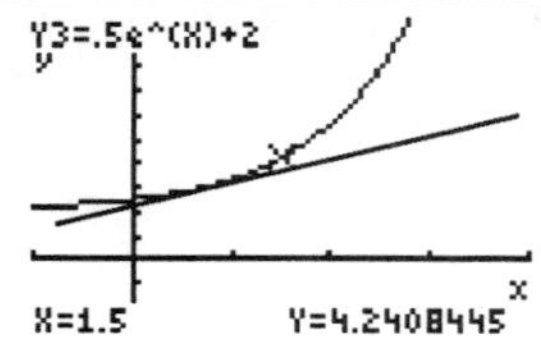

The tangent line at *f(1)*, shows the approximation of *f*(1.5) was an under approximation.

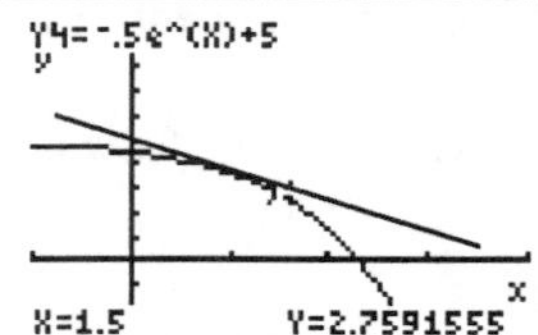

The tangent line at *f(1)*, shows the approximation for *g*(1.5) was an over approximation.

Note: ***The farther away you are from x = a, the less accurate your approximation of f(x).***

To calculate the error, *E(x)*, for your approximation we find the difference between the function, *f(x)* and the linearization: $E(x) = f(x) - (f(a) + f'(a)(x-a))$. As we zoom in on *f(x)*, the error *E(x)* is small and *(x-a)* is small, so $\lim_{x \to a} \frac{E(x)}{x-a} = 0$

Example 3.9.2

Find the linear approximation for $f(x) = x^2$ at $a = 1$ and state information about the error.

Since $f(1) = 1$, $f'(x) = 2x$ and $f'(1) = 2$

The linear approximation of *f(x)* near $x=1$ is:

$$L(x) = f(1) + f'(1)(x-1) = 1 + 2(x-1) = 2x - 1$$

The error would be $E(x) = f(x) - L(x)$ or $E(x) = x^2 - (2x-1) = (x-1)^2$.

Let $h = x - 1$, or $x = h + 1$, then $E(x) = h^2$. Numerically this means when h = 1.1, or you are .1 unit from 1, the error will be $.1^2 = .01$. So the error is small relative to $|x - a|$: $\frac{E(x)}{x-a} = \frac{.01}{.1} = .1$

Key Notation

≈ "Is approximately equal to."

🗁 Key Study Skills

- Choose at least two problems that represent each section; put each on a 3 x 5 note card with the solution on the back. Review all the note card problems 3.1- 3.9.
- Use the ✓ ? ★ system on homework problems and get answers to all ★ problems.

3.10 USING LOCAL LINEARITY TO FIND LIMITS

Objective: To use L'Hopital's rule to determine special limits.

Preparations: Read and take notes on section 3.10.
Suggested practice problems: 3, 4, 6, 11, 13, 14, 19.

✠ Key Ideas of the Section

You should be able to:

- Recognize that $\lim\limits_{x\to a}\frac{f(x)}{g(x)}$ where $f(a) = g(a) = 0$, gives $\frac{0}{0}$, which is undefined and requires a special technique called L'Hopital's rule.
- Use local linearity to show L'Hopital's rule.
- Recognize the appropriate use of L'Hopital's rule for $\lim\limits_{x\to\pm\infty}\frac{f(x)}{g(x)}$, or when f *and* g tend to $\pm\infty$.

Key Questions to Ask and Answer

1. What is L'Hopital's rule?

✍ ____

____ *(See text p. 154-158.)*

We use L'Hopital's rule to find limits of a ratio, which appear to be of the indeterminate form $\frac{0}{0}$ or $\frac{\pm\infty}{\pm\infty}$.

- If f and g are differentiable and $f(a) = g(a) = 0$, and $g'(a) \neq 0$, then:

$$\lim_{x\to a}\frac{f(x)}{g(x)} = \lim_{x\to a}\frac{f'(x)}{g'(x)}$$

- If f and g are differentiable and when $\lim\limits_{x\to a} f(x) = \pm\infty$ and $\lim\limits_{x\to a} g(x) = \pm\infty$ or when $a = \pm\infty$ or $\lim\limits_{x\to\infty} f(x) = \lim\limits_{x\to\infty} g(x) = \pm\infty$, or $\lim\limits_{x\to\infty} f(x) = \lim\limits_{x\to\infty} g(x) = 0$ then:

$$\lim_{x\to a}\frac{f(x)}{g(x)} = \lim_{x\to a}\frac{f'(x)}{g'(x)}$$

Example 3.10.1

Determine if it is appropriate to use L'Hopital's rule on the following and if possible find the limit:

a) $\lim_{x\to\pi}\frac{sin(2x)}{x}$, cannot use L'Hopital's rule since we find the limit using conventional means:

$\frac{sin(2\pi)}{\pi}=\frac{0}{\pi}=0$ is not of the form $\frac{0}{0}$ or $\frac{\pm\infty}{\pm\infty}$.

b) $\lim_{x\to\pi}\frac{sin(2x)}{3x-3\pi}$, use L'Hopital's rule since $\frac{sin(2\pi)}{3\pi-3\pi}=\frac{0}{0}$, therefore

$$\lim_{x\to\pi}\frac{sin(2x)}{3x-3\pi}=\lim_{x\to\pi}\frac{\frac{d}{dx}(sin(2x))}{\frac{d}{dx}(3x-3\pi)}=\lim_{x\to\pi}\frac{2cos(2x)}{3}=\frac{2}{3}$$

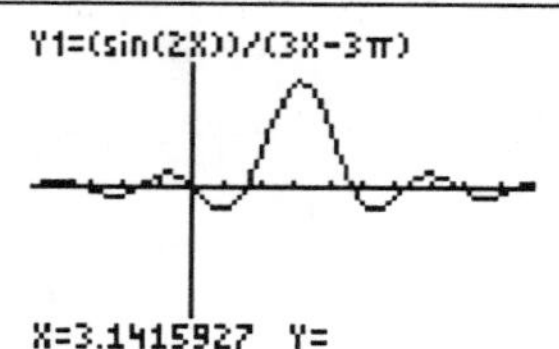 The function is undefined at $x=\pi$.	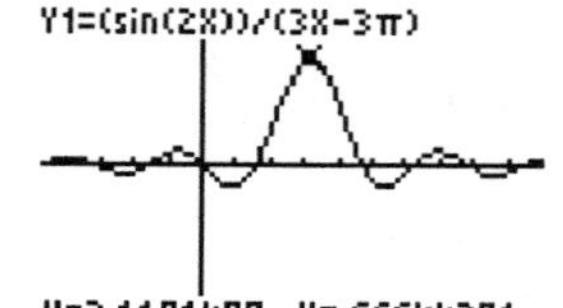 The function approaches 2/3 from the left of π.	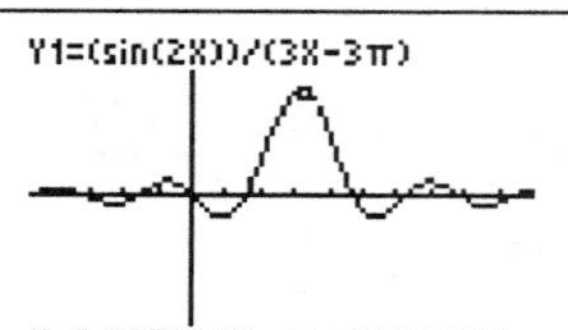 The function approaches 2/3 from the right of π.

c) $\lim_{x\to\infty}\frac{2}{x}$, cannot use L'Hopital's rule since $\lim_{x\to\infty}\frac{2}{x}=\frac{2}{\infty}=0$ is not of the form $\frac{0}{0}$ or $\frac{\pm\infty}{\pm\infty}$.

Example 3.10.2

Determine if it is appropriate to use L'Hopital's rule on the following and if possible find the limit, $\lim_{x\to\infty}\frac{t^2}{e^t}$.

Since $\lim_{x\to\infty}\frac{t^2}{e^t}\to\frac{\infty}{\infty}$ use L'Hopital's rule:

$$\lim_{x\to\infty}\frac{t^2}{e^t}=\lim_{x\to\infty}\frac{(t^2)'}{(e^t)'}=\lim_{x\to\infty}\frac{2t}{e^t}\to\frac{\infty}{\infty}$$, use L'Hopital's rule again.

$$\lim_{x\to\infty}\frac{(2t)'}{(e^t)'}=\lim_{x\to\infty}\frac{2}{e^t}=0 .$$

Key Notation

- Use the point slope form for finding the equation of the tangent line: $y - y_1 = m(x - x_1)$; solving for y gives $y = y_1 + m(x - x_1)$.
- If $x = a$ then $y = f(a)$ and $f'(a)$ is the slope of the graph of f at the point $(a, f(a))$; then the equation of the tangent line to the point $(a, f(a))$ becomes $y = f(a) + f'(a)(x-a)$. This is the same as the point slope form as above.
- $\approx$ "Approximately equal to"
- $\lim_{x\to a}$ "The limit as x approaches a ".

Key Study Skills

- Choose at least two problems that represent each section; put each on a 3 x 5 note card with the solution on the back. Review all the note card problems 3.1- 3.10.
- Use the ✓ ? ★ system on homework problems and get answers to all ★ problems.

CUMULATIVE REVIEW CHAPTER THREE

Key Study Skills

- Review all of the **Key Ideas** from 3.1 - 3.10.
- Randomly choose practice problems from each section. Mark beside each problem which derivative rule is necessary; i.e. product, quotient, power, exponential, chain, trig, log, implicit or a combination of the rules.
- Use a computer algebra system to get the answer to the derivative and practice algebraic simplification.
- Confirm with a graph that the derivative you derived and the graph of the derivative given by a calculator are the same.
- From the text do the *Chapter Review* problems pages 159-161 and the *Check Your Understanding* p. 162.
- Make up your own test from the problems on your 3x5 note cards.
- Do the Chapter 3 Practice Test that follows.

Chapter Three Practice Test

1. Find the derivative of the following:

   a. $f(x) = \dfrac{7x^2 + \sqrt{x} + 1}{x^2 + 8}$

   b. $y(x) = (a^x)\sqrt[3]{5x}$

   c. $P(t) = (6t^2 - 14t + 100)^3$

   d. $s(w) = \sin^4(5w - \pi)$

   e. $D(x) = (2x)e^{\tan x}$

2. Find the tangent line approximation to $f(x) = ln(x+2)$ for x near $x=0$. Will this be an overestimate or an underestimate? Use this to approximate $f(0.1)$, $f(0.2)$, $f(0.5)$.

3. Sketch the graph of a continuous function , $y(x)$, given the following information about y' and y'' over the indicated intervals of x:

$x < x_1$	$x = x_1$	$x_1 < x < x_2$	$x = x_2$	$x > x_2$
$y'>0$	$y'= 2$	$y' = 2$	$y' =2$	$y' > 0$
$y'' < 0$	$y'' = 0$	$y'' = 0$	$y'' = 0$	$y'' > 0$

<------------------------x_1--x_2-->

4. Find the dimensions of an aluminum can that is able to hold 40 in^3 of juice and that uses the least material (i.e. aluminum). Assume that the can is cylindrical and capped at both ends.

5. Determine the limit of the following.

   a) $\lim_{x \to o} \frac{\sin(5x)}{3x}$

   b) $\lim_{t \to \infty} \frac{t^2}{e^t}$

   c) $\lim_{x \to o^+} 3x \cdot \ln x$

6. a) Use technology to sketch the curve given by: $x = \cos(3t)$, $y = \sin t$ for $0 \le t \le 2\pi$.
   b) What is the slope of the curve at $t = \pi/4$?
   c) Find the equation of the tangent line at $t = \pi/4$?

(Answers appear at the end of this Study Guide.)

CHAPTER FOUR

USING THE DERIVATIVE

We will learn to use the first and second derivatives to analyze the behavior of functions and to solve optimization problems.

4.1 USING FIRST AND SECOND DERIVATIVES

Objective:	To use the first and second derivative to aid in the analysis of the behavior of a function.
Preparations:	Read and take notes on section 4.1. Suggested problems: 2, 4, 9, 10, 18, 19, 24, 25, 26, 32, 34, 36, 40, 42.

✠ Key Ideas of the Section

You should be able to:

- Use the derivative function to find when a function is increasing or decreasing.
- Use the second derivative to find the concavity.
- Use *critical points* to find local extrema and inflection points.

Key Questions to Ask and Answer

1. ***What do the first and second derivatives tell you about a function and its graph?***

✍

When $f' > 0$, then f is ________________________________

When $f' < 0$ then f is ________________________________

When $f'' > 0$, then f is ________________________________

When $f'' < 0$, then f is ________________________________

(See text p. 166.)

✍ If p is in the domain of f, $f'(p) = 0$ represents:

____ *(See text p. 167.)*

We use the first derivative to tell us where the graph of a function is increasing or decreasing and the second derivative to tell where the graph is concave up or concave down. We use *critical points* to help find *local minima* or *local maxima*.

For any function f, a point p in the domain of f where $f\,'(p) = 0$ or $f\,'(p)$ is undefined, is called a *critical point.*

To analyze a graph of a continuous function $f(x)$ do the following:

- Find the derivative $f\,'(x)$.
- Set the derivative equal to zero to find the critical point(s) p, where $f\,'(p) = 0$.
- Find any critical point(s) where $f\,'(p)$ is undefined.
- Find the intervals to the left and right of the critical point(s) and determine where $f\,'(x) > 0$ and where $f\,'(x) < 0$.
- A critical point can help determine possible local minima or maxima (also called *local extrema or turning points).* If the derivative changes sign there is a turning point on the graph of f at that critical point. Find the extrema point(s) on the graph of the function $(p, f(p))$.
- Find the *y-intercept.*
- Find the second derivative $f\,''(x)$.
- Set the second derivative equal to zero and find the point(s), where $f\,''(p) = 0$.
- Find any point(s) where $f\,''(p)$ is undefined.
- Find the intervals where $f\,''(x) > 0$ *or* $f\,''(x) < 0$ to determine the concavity of the graph of the function f.
- If the second derivative changes sign at a point there will be a change in concavity at that point (also called an *inflection point).* Find the inflection points on the graph $(p, f(p))$.
- Sketch the graph and use a grapher to confirm your results.

Example 4.1.1

Sketch the graph of $f(x)$ satisfying all of the following information about $f\,'(x)$ *and* $f\,''(x)$:

$f\,'(-3) = 0$ *and* $f\,'(4) = 0$

$f\,'(x) > 0$ *for* $x < -3$ or $x > 4$

$f\,'(x) < 0$ *for* $-3 < x < 4$

$f\,''(1/2) = 0$

$f\,''(x) > 0$ *for* $x > 1/2$

$f\,''(x) < 0$ *for* $x < 1/2$

Put this information all together on a number line, making note of the critical points:

	$x = -3$		$x = 1/2$		$x = 4$	
$f\,'(x) > 0$	$f\,' = 0$		$f\,'(x) < 0$		$f\,' = 0$	$f\,'(x) > 0$
		$f\,''(x) < 0$	$f\,''(x) = 0$	$f\,''(x) > 0$		

Add information about when f is increasing or decreasing and concave up or concave down.

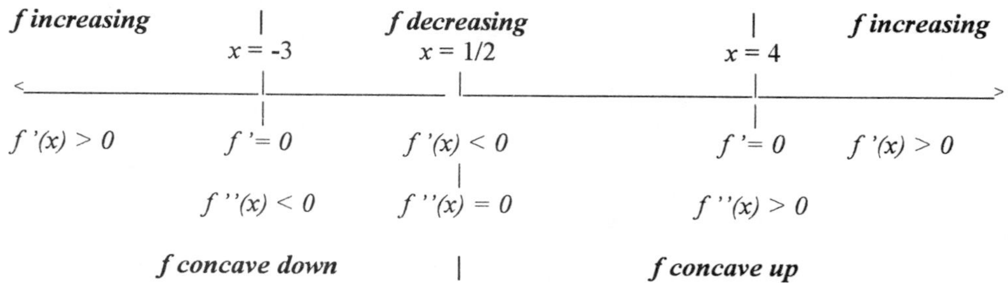

Steps in sketching the graph.	*Sketch of the graph in progress.*
• Sketch in the turning points. • *Local Max.* at $x = -3$, since f is increasing then decreasing and $f'(-3) = 0$. • *Local Min.* at $x = 4$, since f is decreasing then increasing and $f'(4) = 0$.	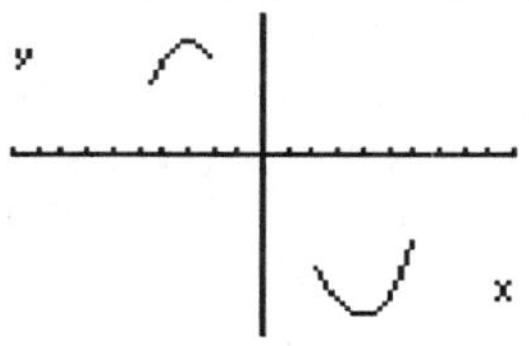 y x
• Sketch in where the graph is increasing. • f increases over $x < -3$ *and* $x > 4$.	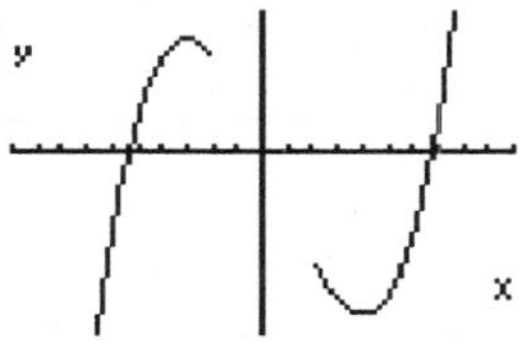 y x
• Sketch in where the graph is decreasing. • f decreases over $-3 < x < 4$. • Confirm the concavity: concave down over $x < 1/2$, concave up over $x > 1/2$. • Mark the inflection point on the graph.	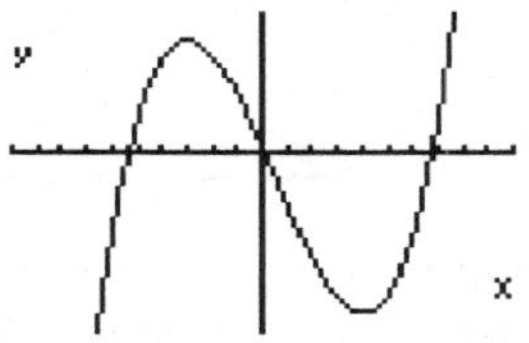 y x

Special notes about critical points:

- Geometrically, a critical point occurs when the tangent line to the graph of f at p is a horizontal line and where $f'(p) = 0$.
- A critical point of f at p occurs where $f'(p)$ is undefined. At $f(p)$ there is a either a vertical tangent or no tangent at all.
- Suppose p is a critical point of a continuous function and reading left to right:

If f' changes from negative to positive at p then f ________________________

If f' changes from positive to negative at p then f ________________________

(See text p. 168.)

2. *What is the second derivative test and when would you use it?*

If $f'(p) = 0$ and $f''(p) > 0$ then ____

If $f'(p) = 0$ and $f''(p) < 0$ then ____

If $f'(p) = 0$ and $f''(p) = 0$ then ____

(See text p. 170.)

Example 4.1.2

Determine if the function $f(x) = e^{-(x-3)^2}$ has any local extrema or inflection points.

Solution:

- Find $f'(x)$.

$$f'(x) = e^{-(x-3)^2} \cdot (-2(x-3)) = \frac{-2(x-3)}{e^{(x-3)^2}}$$

- Set $f'(x) = 0$ and find x. (For $f'(x) = 0$ only the numerator can be zero)

$$-2(x-3) = 0 \Rightarrow x = 3 \Rightarrow f'(3) = 0$$

For $x < 3,\ f'(x) > 0$ and for $x > 3,\ f'(x) < 0,$ therefore $f(3)$ is a local maximum of f.

- Confirm with the second derivative test. (Use the product rule on the first derivative.)

$f'(x) = e^{-(x-3)^2}(-2(x-3))$

$f''(x) = [e^{-(x-3)^2}(-2(x-3))][(-2(x-3))] + e^{-(x-3)^2}(-2)$

Factor out $e^{-(x-3)^2}$.

$f''(x) = e^{-(x-3)^2} \cdot [(4(x-3)^2 - 2]$

$f''(3) = 1(-2) = -2 < 0$. Since $f'' < 0$ the f is concave down at $x = 3$ and

$f(3)$ is a local max.

Observe the graph of *f* and *f '* to confirm your finding.

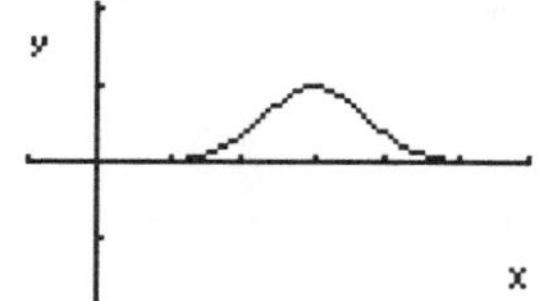

The graph of *f* with local *max.* at *x = 3.*

The graph of *f '* with a zero at *x = 3.*

It appears that *f* changes concavity twice, once to the left and once to the right of *x = 3*. The graph of *f '* confirms this because *f '* has a local *max.* and *min.* which would be inflection points on *f.*

Set *f ''(x) = 0* .

Since $e^{-(x-3)^2} \neq 0$ let $4(x-3)^2 - 2 = 0$.

Use the quadratic formula to find that *x = 3.70711 or x = 2.29289* . In summary,

$f(x) = e^{-(x-3)^2}$ has a local *max.* at *(3 , 1)* and inflection points at (3.71, .607) and (2.29,.607).

Special note on inflection points.

A function *f* with a continuous derivative has an inflection point at *p* if either of the following conditions hold:

- *f '* has ____
- *f ''* changes ____

(See text p. 171.)

Note: *f ''(p) = 0* is not a sufficient condition for an inflection point.

- Inflection points for continuous functions occur when the function *f '* has a local max or local min.

Key Notation

Using the correct notation is especially important here; as you know *f(p) = 0, f '(p) = 0* and *f ''(p) = 0* mean very different things.

- *f(p) = 0 "When x = p the value of the function is zero".*
- *f '(p) = 0 "When x =p the value of the derivative is zero".*
- *f ''(p) = 0 "When x = p the value of the second derivative is zero."*

🗀 Key Study Skills

- Choose at least two problems that represent this section; put each on a 3 x 5 note card with the solution on the back.
- Use the ✓ ? ★ system on homework problems and get answers to all ★ problems.

4.2 FAMILIES OF CURVES

Objective:	To use calculus to study the graphs of families of functions and to see what effect parameters have on the graphs.
Preparations:	Read and take notes on section 4.2. Suggested practice problems: 6, 9, 12, 13, 16, 17, 18, 19, 30. Problems requiring substantial use of algebra: 12-20, 21-27, 31.

✠ Key Ideas of the Section

You should be able to:

- Recognize that the graphs of all functions in a family of functions have the same general shape but parameters cause stretches or shifts.
- Use the derivative to find when a function is increasing or decreasing.
- Recognize the shape of the various families of functions given in the chapter.

Key Questions to Ask and Answer

1. ***When given a function $y = a f(x + b) + c$, where a, b, and c are constants and are called parameters, how do you analyze this family of functions?***

✍ ____

____ *(See text p. 176-178.)*

Recall from our study of Chapter One that we could determine horizontal or vertical shifts and stretches of a function by observing the effect of constants in the function. We now use calculus to determine the general shape of a family of functions and if functions in the family will have local extrema and inflection points.

Example 4.2.1

If $a > 0$ and $b > 0$, analyze the graph of $y = axe^{-bx}$ (a and b are considered parameters).

Solution:

You might want to begin by holding a constant, since it is a multiplier of the parent function xe^{-bx}. Let $a = 1$, and concentrate on what b is doing to f. Graph several graphs for $b = 1, 2, 3, 4, \ldots$ and $b = 0.9, 0.8, 0.7, 0.8, \ldots$

- For $b > 1$, as b increases there appears to be a shrinking of the graph of f, with the maximum moving to the left.

- For $0 < b < 1$, as b decreases there appears to be a stretching of the graph of f with the maximum moving to the right.

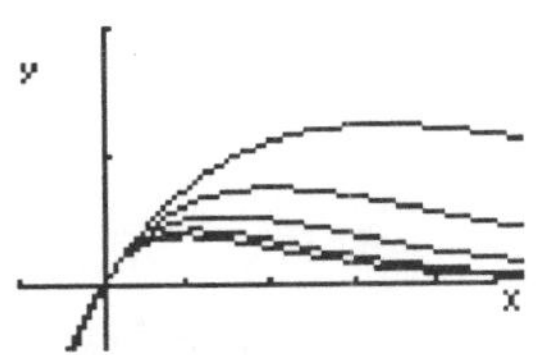

Analyze the curve using the first derivative.

$$f(x) = axe^{-bx}$$
$$f'(x) = (ax)'(e^{-bx}) + ax(e^{-bx})'$$
$$f'(x) = a[e^{-bx}] + ax[e^{-bx}(-b)]$$
$$f'(x) = e^{-bx}(-abx + a)$$

Set $f'(x) = 0$. Since $e^{-bx} = \frac{1}{e^{bx}} \neq 0$ we must have $-abx + a = 0 \Rightarrow x = \frac{1}{b}$

There is a critical point at $(1/b, f(1/b))$. The critical point is a local max., because of the following argument.

Observe that the sign of the derivative is determined by $(-abx + a)$, a linear expression with negative slope, which decreases as x increase and has a value of 0 at $x = 1/b$. If $x < 1/b$ then $f'(x) > 0$, so f is increasing over the interval $(-\infty, 1/b]$. If $x > 1/b$, $f'(x) < 0$, so f is decreasing over the interval $[1/b, \infty)$. This confirms the observation that if b increases, the local max. moves to the left.

Find the second derivative.

$$f'(x) = e^{-bx}(-abx + a)$$
$$f''(x) = (e^{-bx})'(-abx + a) + e^{-bx}(-abx + a)'$$
$$f''(x) = e^{-bx}(-b)(-abx + a) + e^{-bx}(-ab)$$
$$f''(x) = abe^{-bx}(bx - 2)$$

Set $f''(x) = 0$. Since $abe^{-bx} = \frac{ab}{e^{bx}} \neq 0$ we must have $bx - 2 = 0 \Rightarrow x = \frac{2}{b}$.

There is a critical point for f'' at $(2/b, f(2/b))$. This critical point is an inflection point, because of the following argument.

Observe that the sign of the second derivative is determined by the expression $(bx - 2)$, a linear function with positive slope, which increases as x as increases and has value 0 at $x=2/b$. If $x < 2/b$, $f''(x) < 0$ so the curve is concave down over the interval $(-\infty, 2/b]$. When $x > 2/b$, $f''(x) > 0$ so the curve is concave up over the interval $[2/b, \infty)$. As with the local max, as b increases the inflection point moves left.

The second derivative test confirms that f has a max at $x = 1/b$.

$$f''(x) = \frac{ab}{e^{bx}} \Rightarrow f''(1/b) = \frac{ab}{e^{(b)(\frac{1}{b})}} = \frac{ab}{e} > 0, \text{ since } a > 0, b > 0 \text{ and } e > 0.$$

When the critcal value of the first derivative is substituted into the second derivative, $f''(1/b) > 0$ indicating that $f(1/b)$ is a local maximum value for f.

{ EMBED Word.Picture.6 }

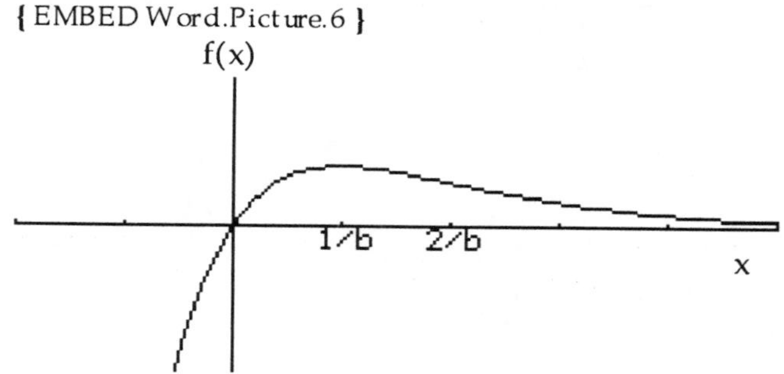

The graph of $f(x) = ae^{-bx}$ with max. at 1/b and inflection at 2/b.

What caused the stretches and shrinks that we saw in the graphs at the beginning of the example? Calculate the maximum point for several values of b.

$$f(x) = axe^{-bx} = \frac{ax}{e^{bx}}$$

$$f(1/b) = \frac{a(1/b)}{e^{b(1/b)}} = \frac{a}{be}$$

If $b = 2$	$f(1/2) = a/(2e)$
If $b = 5$	$f(1/5) = a/(5e)$
If $b = 1/2$	$f(2) = 2a/e$
If $b = 1/7$	$f(7) = 7a/e$

Summary of analysis:

- As b gets larger the value of $x = 1/b$ for the max. gets closer to zero and the value of y gets smaller(x shifts left and y shrinks).
- As b gets smaller the value of $x = 1/b$ for the max. gets larger and the value of y gets larger(x shifts right and y stretches).
- Since the value of a in the function $f(x)$ is a multiplier it will either stretch or shrink the whole graph but not change its shape. Only b will change the shape of the graph.

Key Notation

- In a given function pay attention to variables which are considered constant.
- Use $f'(x)$ and $f''(x)$ carefully to note which function is the first derivative and which function is the second derivative.

Key Algebra

- When analyzing a function find the *x- intercepts,* if possible, use limits as $x \to \pm\infty$ to find the end behavior, and find any points where f is undefined to determine vertical asymptotes.

Key Study Skills

- Choose at least two problems that represent this section; put each on a 3 x 5 note card with the solution on the back.
- Use the ✓ ? ★ system on homework problems and get answers to all ★ problems.
- The problems in this section can be very challenging. Work slowly and methodically. Be careful to use precise notation. Don't expect to get answers immediately. Sometimes you may have to work on a problem for a few days before you really understand it.
- Work with a partner on homework problems and get help from your instructor, which by the way, is always a good idea.

4.3 OPTIMIZATION

Objective: To use derivatives and critical points to find points where optimization of a problem situation can occur and to decide if the critical points represent a global optimum.

Preparations: Read and take notes on section 4.3.
Suggested practice problems 2, 4, 6, 12, 15, 25, 31,32.

✠ Key Ideas of the Section

You should be able to:

- Find critical points by looking at points where the derivative is zero.
- Check end points and critical points to find the global maximum or global minimum on an interval.

- Define *optimization* of a problem situation as finding the largest or smallest value of a function over a specified domain.
- Define a *closed interval* $a \le x \le b$, as an interval that contains the end points.
- Distinguish local extrema from global extrema.

Key Questions to Ask and Answer

1. *Are the local extrema also the global extrema of a function over a specified interval?*

✍ ____

____ *(See text p. 181.)*

The single greatest (or least) value of a function *f* over a specified domain is called the *global maximum* (or *minimum*). This is also called the *absolute maximum (or minimum).*

If *f* is a continuous function on a closed interval $a \le x \le b$ then:

- *f* has a *global minimum* at *p* if *f(p)* is ______________________________
- *f* has a *global maximum* at *p* if *f(p)* is ______________________________

(See text p. 181)

Continuous functions on a closed interval must have a global maximum and minimum. To find the global maximum or minimum of a continuous function over a closed interval $a \le x \le b$:

- Take the first derivative , then find the value of *f* at the critical points.
- Compare the value of the function at the critical points to the value of *f* at the end points.
- The largest (or smallest) values on the interval are the absolute extrema.

Example 4.3.1

Find all global and local extrema for $f(x) = -3x^4 + 4x^3 + 72x^2 - 12$, over the interval $-5 \le x \le 6$.

- Since *f* is a polynomial it is continuous over the interval $-5 \le x \le 6$. We first look at the graph of *f* and *f '* .

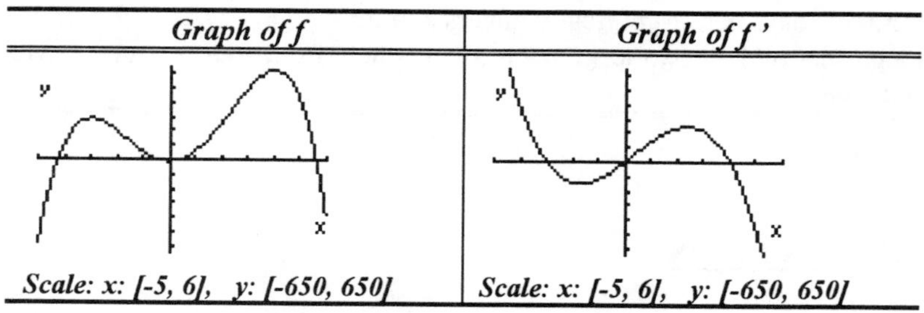

- Find the first derivative , set $f'(x) = 0$ and determine if any critical points exist.

$$f'(x) = -12x^3 + 12x^2 + 144x = 0$$
$$= -12x(x^2 - x - 12) = 0$$
$$= -12x(x+3)(x-4) = 0$$

The critical points are at $x = 0$, $x = -3$ *and* $x = 4$ and by looking at the sign changes of *f '*, the function *f* has local maximums at $x = -3$ and $x = 4$ and a local minimum at $x = 0$.

- Find the values of *f* at the critical points $x = -3$, 0 and 4 and also at the end points $x = -5$ and $x = 6$.

Graphs and values of f	*Local and Global Extrema*
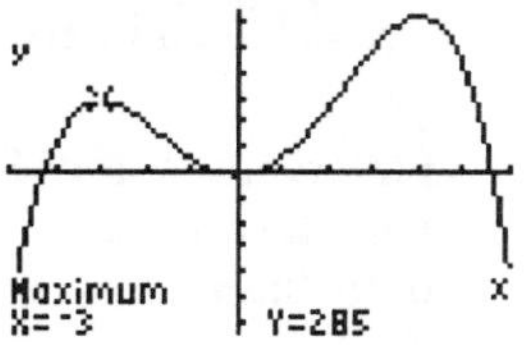 Maximum X=-3 Y=285	• The critical point $f(-3) = 285$ is a local max.
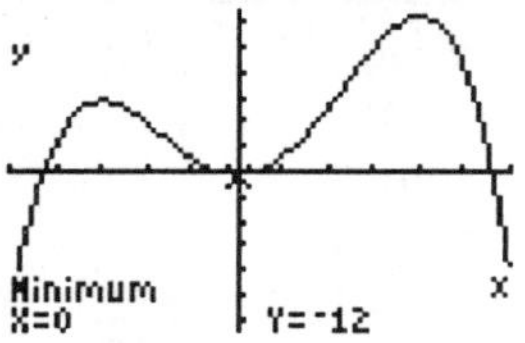 Minimum X=0 Y=-12	• The critical point $f(0) = -12$ is a local min.
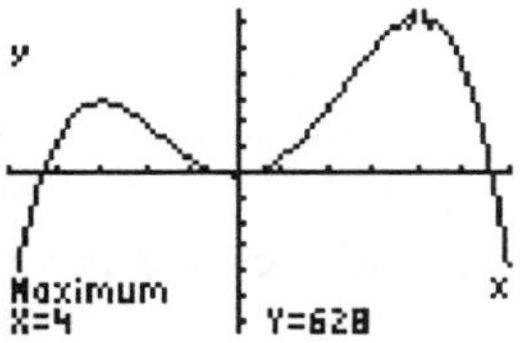 Maximum X=4 Y=628	• The critical point $f(4) = 628$ is a **global max.** (also a local max.).
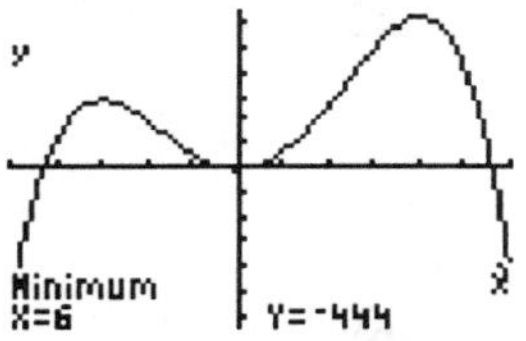 Minimum X=6 Y=-444	• The end point $f(6) = -444$ is a local min.
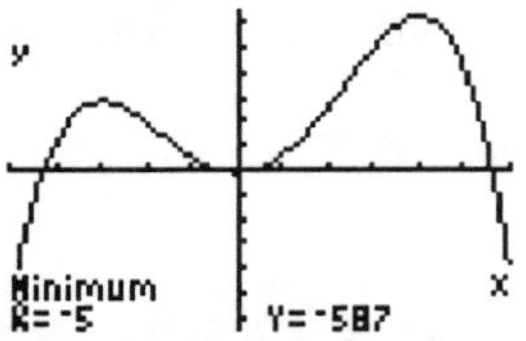 Minimum X=-5 Y=-587	• The end point $f(-5) = -587$ is a **global min.** (also a local min.).

A global maximum or minimum of a continuous function f, on a closed interval $[a,b]$, will occur at either a local maximum or minimum or at the end points $x = a$, $x = b$.

2. *How do you find a global extrema on an open interval $a < x < b$?*

✍ ____

____ *(See text p. 182.)*

Any continuous function on a closed interval will have a global max. and a global min. on that interval. This is not necessarily true if the interval is not closed.

If a function is defined and continuous on an open interval $a < x < b$ (end points not included), then:

- Find the value of the function at all critical points.
- Sketch the graph.
- Look at the function values where x approaches the end points of the interval or as $x \to \infty$

Example 4.3.2

Find all local and global extrema for $f(x) = -3x^4 + 4x^3 + 72x^2 - 12$ over the open interval $-5 < x < 6$.

- On a grapher you <u>cannot</u> tell the difference between the closed interval graph in **Example 4.3.1** and the one on the open interval below. Sometimes on hand drawn graphs, circles will be drawn at the endpoints to indicate that the end points are not included.

Graph of f over -5 < x < 6	*Local and Global Extrema*
	• Global max. (and local max.)at $(4,f(4))$. • Local min. at $(0,f(0))$. • Local max. at $(-3,f(-3))$. • No global min. exists since $x = -5$ is not in the domain of f.

✎ On Your Own

An electric current I in amps, is given by $I = \cos(wt) + \sqrt{3}\sin(wt)$ where $w \neq 0$ and is a constant. What are the max. and min. values of I for $t > 0$?

The answer appears at the end of this section.

Key Notation

- In a given function pay attention to variables that are considered constant.
- $a \le x \le b$, a closed interval including the end points a and b or [a, b].
- $a < x < b$, an open interval not including the end points a and b or (a, b).
- When working with an equation also draw a graph and look for key words that tell you to optimize the situation like " find the greatest, smallest, least, most,... etc."
- Use the units of the problem to decide how to find the optimum requested.

🗁 Key Study Skills

- Choose at least two problems that represent this section; put each on a 3 x 5 note card with the solution on the back.
- Use the ✓ ? ★ system on homework problems and get answers to all ★ problems.
- The problems in this section can be very challenging. Work slowly and methodically. Sometimes you may have to work on a problem for a few days before you really understand it.
- Always check the endpoints on a closed interval when finding global extrema.
- Work with a partner on homework problems and get help from your instructor.

Answer to On Your Own

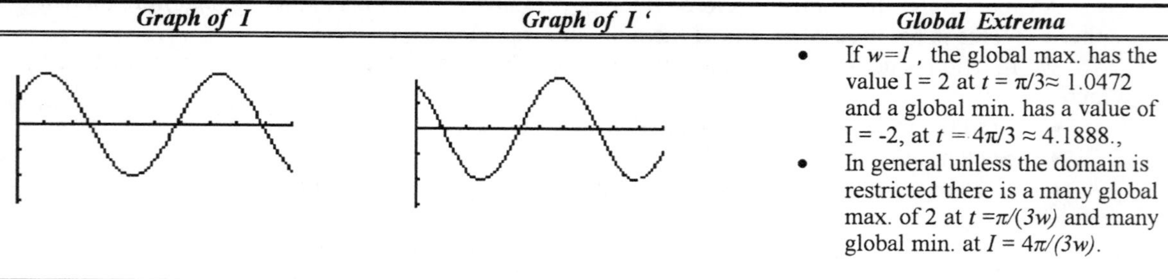

Graph of I	*Graph of I '*	*Global Extrema*
		• If $w=1$, the global max. has the value I = 2 at $t = \pi/3 \approx 1.0472$ and a global min. has a value of I = -2, at $t = 4\pi/3 \approx 4.1888$., • In general unless the domain is restricted there is a many global max. of 2 at $t = \pi/(3w)$ and many global min. at $I = 4\pi/(3w)$.

4.4 APPLICATIONS TO MARGINALITY

Objective: To use derivatives and critical points to find marginal cost, revenue and profit.

Preparations: Read and take notes on section 4.4.
Suggested practice problems 1, 6, 9, 15, 16.

✠ Key Ideas of the Section

You should be able to:

- Define the economic terms cost, revenue and profit.
- Recognize that the derivatives of the cost, revenue and profit functions is called the marginal cost, marginal revenue and marginal profit.
- Find maximum profit when marginal cost equals marginal revenue.

Key Questions to Ask and Answer

1. How are total cost and total revenue related?

✍ ____

____ *(See text p. 189.)*

To analyze economic ideas we use the cost, revenue and profit functions which depend upon the quantity q, bought or sold.

- **Cost** = $C(q)$ = (cost per item)(quantity produced) + fixed costs
- **Revenue** = $R(q)$ = (price per item) (quantity sold)
- **Profit** = $\pi(q)$ = Revenue - Cost = $R(q) - C(q)$

To maximize profit or loss examine the derivative of the cost and revenue functions, also known as *marginal cost and marginal revenue.*

The maximum (or minimum) profit occurs when

Marginal Cost = Marginal Revenue or $C'(q) = R'(q)$.

Example 4.4.1

The total cost C, in thousand of dollars, of producing q hundred goods is given by: $C(q) = .01q^3 - .5q^2 + 10q$ over the interval [0, 50]. What is the maximum profit if a lot of one hundred items is sold for \$6.50?

- Look at the graph of $C(q)$ and $C'(q)$

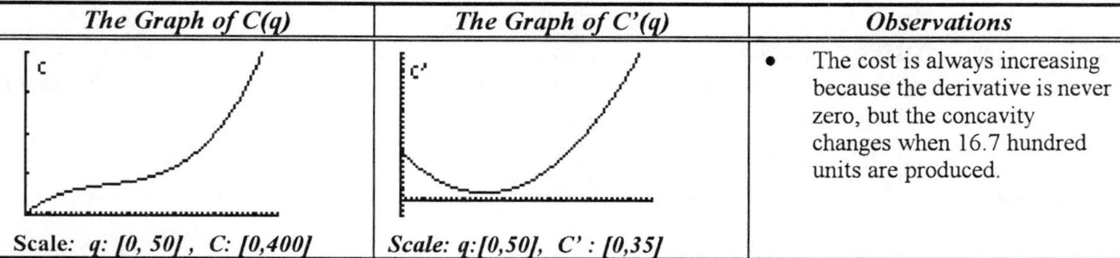

The Graph of C(q)	*The Graph of C'(q)*	***Observations***
C	C'	• The cost is always increasing because the derivative is never zero, but the concavity changes when 16.7 hundred units are produced.
Scale: *q: [0, 50], C: [0,400]*	*Scale: q:[0,50], C' : [0,35]*	

- Find the revenue equation and graph both the total cost and total revenue equations. The revenue equation is $R(q) = 6.5q$, since the price is \$6.50 per hundred units.

Graphs of Cost and Revenue	*Observations*
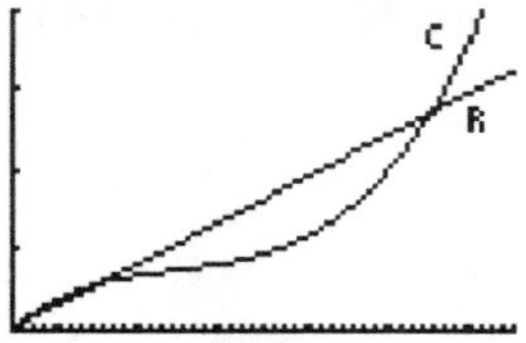 C R	• The revenue is below the cost equation over the intervals (0,8.4) and (41.5,50), thus these are the intervals where a loss occurs. • The cost is above the revenue over the interval (8.4, 41.5), thus a profit will occur in this interval.
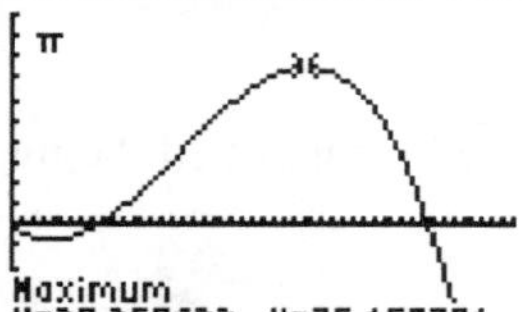 π Maximum X=29.359622 Y=75.158781 *Profit graph $\pi(q)$*	• Profit is $\pi(q) = R(q) - C(q)$. • The maximum profit of 74.16 thousand dollars is made when 29.35 hundred items are produced. This is a local max., however it is clear from the graph that it is also a global max. Checking the end points confirms this, $\pi(0) = 0$ and $\pi(50) = -174$.
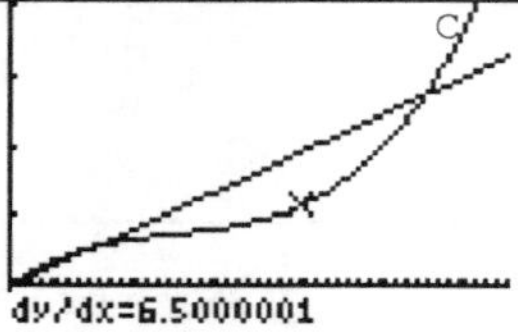 C dy/dx=6.5000001	• The derivative of the cost $C'(29.35) = 6.5$ thousand dollars/hundred units, exactly the same as the derivative of the revenue function, $R'(29.35) = 6.5$ thousand dollars/hundred units.
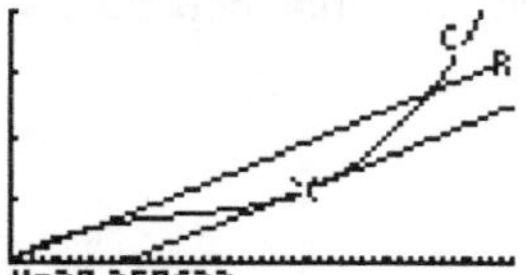 C R X=29.359622 Y=6.500000125X+-75.1587_	• At the point where the maximum profit occurs, the tangent line to C is parallel to the tangent line at R, since $C' = R'$.
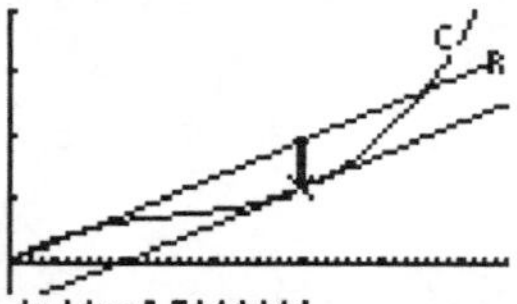 C R dy/dx=6.5000001	• The vertical distance between Revenue and Cost is largest where the slopes of R and C are parallel. This represents the difference $R - C =$ *profit.*

Key Notation

- $\pi(q)$: "The total profit for q units produced." In economics π is the preferred variable for profit since p is used for price.

Key Study Skills

- Choose at least two problems that represent this section; put each on a 3 x 5 note card with the solution on the back. Review all problems from sections 4.1-4.4.
- Use the ✓ ? ★ system on homework problems and get answers to all ★ problems.
- Drawing graphs is very helpful in this section.

4.5 MORE OPTIMIZATION: INTRODUCTION TO MODELING

Objective: To use mathematical modeling to find a function for a problem situation, and then use derivatives to find global extrema.

Preparations: Read and take notes on section 4.4.
Suggested practice problems 1, 2, 4, 5, 7, 9, 14, 17.

✠ Key Ideas of the Section

You should be able to:

- Translate a problem into a formula, called a mathematical model.
- Use geometry ideas of distance, perimeter, area and volume to model problem situations.
- Draw a sketch of the problem situation.
- Write the modeling formula in terms of one variable.

Key Questions to Ask and Answer

1. How do you set up a modeling optimization problem when no formulas are given?

✍ ____

____ *(See text p. 198-200)*

Modeling a problem situation and analyzing it mathematically are what most scientists try to do. Most students find this the most difficult of tasks. You may think, "If I only knew the formula I could solve this problem." In this section, we concentrate on building the formula to optimize.

Many situations rely on geometrical relationships to build an understanding of the problem. Since Babylonian times, math students have struggled with the idea of maximizing the area of rectangular objects. You must learn to systematize your problem solving strategies. Here are some practical tips:

- Read and re-read the problem, identify all the quantities listed, attach variables to the quantities and identify the quantity to be optimized. Write clear statements in words about the meaning of each variable, the units, and the relationship among variables.
- Draw a sketch of the problem situation.
- List all formulas that may be helpful.
- Try to obtain your optimization formula first in words, and then in symbols.
- Put your formula in terms of one variable, (designate other variables as constants if necessary).
- Identify the domain of the problem situation.
- Find all of the critical points and evaluate the function at these points and at the end points. Remember, in most applications you are most likely wanting to find the global maximum or minimum.

✎ Your Turn

One of the authors of this study guide has a daughter who raised a Jersey calf for her 4-H project. This question came up: "What size rectangular pen could be made from 1000 ft. of fencing to maximize Buttercup's grazing area?".

✍ (Fill in the missing information.)

- What quantities are given, what is being asked?

?________________________
?________________________
?________________________

Answers
• 1000 ft of fencing (length)
• Rectangular pen
• Grazing area ?

- Draw a sketch and label the sketch. Let $x = length$ and $y = width$.

Area	y
x	

- List helpful formulas.

Distance around = ?________ Area of a rectangle = ?________ Fence length = ?________	**Answers** $P = 2x + 2y \quad (ft)$ $A = xy \quad (ft^2)$ $P = 1000 \; ft$

- What is to be maximized? How do the formulas relate to the problem?

Fill in the missing quantities	***Answers***
Maximize the Area:	Domain: $0 < x < 500$
$A = xy$ $1000 = 2x + 2y$ $y = \quad ? \quad = 500 - x$	$y = \dfrac{1000 - 2x}{2}$
Rewrite the area in terms of x only. $A = x(\quad ? \quad) = 500x - x^2$	$A = xy = x(500 - x)$
Find the critical points: $A' = \quad ? \quad = 0$ $x = ?$	$A' = 500 - 2x = 0$ $x = 250 \; ft,$ where A has a local max.
$y = ?$	$y = 500 - 250 =$ $= 250 \; ft$

- Answer the question.

  The dimension of a rectangle that maximizes the grazing area is *250 ft x 250 ft* or a square. The total grazing area would be 62,500 ft^2. Check to see if this is a local max using the second derivative test or looking at the graph of the function.

- Verify with a graph.

The Graph of A	***The graph of A'***	
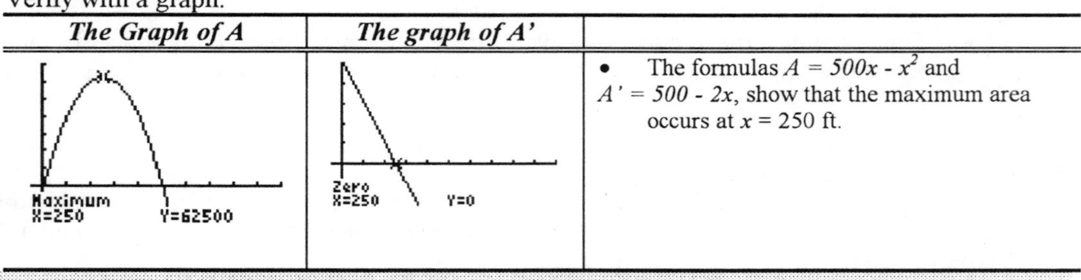		• The formulas $A = 500x - x^2$ and $A' = 500 - 2x$, show that the maximum area occurs at $x = 250$ ft.

✎ On Your Own

Extend the idea of maximizing the area of a fenced-in field, using p *ft* of fencing. Two equally sized rectangular fenced-in areas, (with a fence down the middle) are to be made so that cows can be switched from pasture to pasture to ensure that the grass has a chance to grow. The fields are set up as in the drawing below. What dimensions maximize the area?

Area 1 y	Area 2 y
x	x

(The answer appears at the end of this section.)

Key Notation

- Set the derivative of the quantity to be maximized equal to zero to find critical points. Use the prime notation for the quantity: Volume change = V', Area change = A', Distance change = D', Population change = P' ...etc.

Key Algebra

- Use formulas for area and volume that relate to regular geometric figures.

Figure	*Perimeter*	*Area*	*Surface Area*	*Volume*
Rectangle	$P = 2l + 2w$	$A = lw$		
Circle	$C = 2\pi r$	$A = \pi r^2$		
Box			$SA = 2wl + 2wh + 2wl$	$V = lwh$
Cylinder			$SA = 2\pi r^2 + 2\pi rh$	$V = \pi r^2 h$

- Distance between two points = $\sqrt{(x_2 - x_1)^2 + (y_2 - y_1)^2}$
- Review the definitions of trigonometric functions.

Key Study Skills

- Choose at least two problems that represent this section; put each on a 3 x 5 note card with the solution on the back. Review all problems from sections 4.1-4.4.
- Use the ✓ ? ★ system on homework problems and get answers to all ★ problems.
- Drawing sketches of the problem situation is very helpful in this section.
- Do problem 23 in the text p. 203, it is a classic and makes use of the arctangent function.

Answer to On Your Own

Maximizing Area for P ft of fencing	*Example*
Length of fence = $P = 3y + 4x$ Solve for y $\Rightarrow y = \frac{P-4x}{3}$ $Area_{total} = Area_1 + Area_2 = xy + xy = 2xy$ $Area_{total} = A = 2x(\frac{P-4x}{3}) = \frac{2Px}{3} - \frac{8x^2}{3}$ $A' = \frac{2P}{3} - \frac{16x}{3} = 0 \Rightarrow x = \frac{1}{8}P \Rightarrow y = \frac{P - 4(\frac{1}{8}P)}{3} = \frac{P}{6}$ To Maximize the area choose the length of each pasture to be 1/8 of the fencing and the width to be 1/6 of the fencing. This will work for any size fence of length *p*.	*Let P = 1/2 mile of fence.* *P = 5280/2 = 2640* *x=2640/8=330 ft.* *y=2640/6=440 ft.* *A=2xy=2(330)(440) = 290,400 ft².* Maximum X=329.99995 Y=290400 The graph of $A = 2x((2640-4x)/3)$

4.6 HYPERBOLIC FUNCTIONS

Objective: To define and determine properties of hyperbolic functions.

Preparations: Read and take notes on section 4.6.
Read,
Suggested practice problems: 1, 2, 11 13, 18.
Problems requiring substantial algebra: 1-11, 17, 18, 20.

✠ Key Ideas of the Section

You should be able to:

- Define the *hyperbolic cosine (cosh)* , the *hyperbolic sine (sinh) and hyperbolic tangent (tanh)* .
- Discover properties of hyperbolic functions by looking at their graphs.
- Find the derivative of the hyperbolic functions.

Key Questions to Ask and Answer

1. *What are the properties of the hyperbolic functions?*

✍ ____

____ *(See text p. 203-205)*

The hyperbolic functions are defined using e^x *and* e^{-x}. They have applications to various engineering problems, such as suspension cables. They are defined as follows:

- *hyperbolic cosine* = *cosh(x)* = $\dfrac{e^x + e^{-x}}{2}$
- *hyperbolic sine* = *sinh(x)* = $\dfrac{e^x - e^{-x}}{2}$
- *hyperbolic tangent* = *tanh(x)* = $\dfrac{\sinh(x)}{\cosh(x)} = \dfrac{\frac{e^x - e^{-x}}{2}}{\frac{e^x + e^{-x}}{2}} = \dfrac{e^x - e^{-x}}{e^x + e^{-x}}$.

Looking at the graphs we begin to see some of the properties of the hyperbolic functions.

Graphs of e^x and e^{-x}	***f(x) = cosh(x)***	***g(x) = sinh(x)***	***h(x) = tanh(x)***

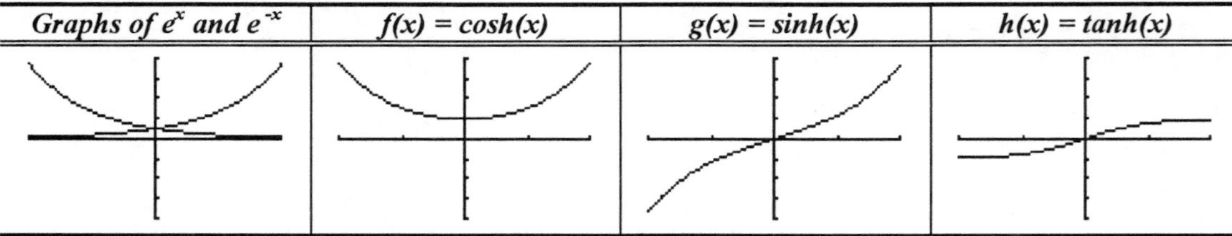

Since the *cosh (x)* function is the sum of the functions $\frac{e^x}{2}$ and $\frac{e^{-x}}{2}$ and *sinh(x)* is the difference of the functions $\frac{e^x}{2}$ and $\frac{e^{-x}}{2}$ the shapes of the graphs become evident when you compare them. In addition, looking at the graphs will help understand the properties listed below.

Properties	*f(x) = cosh(x)*	*g(x) = sinh(x)*	*h(x) = tanh(x)*
$x = 0$	$cosh(0) = 1$	$sinh(0) = 0$	$tanh(0) = 0$
symmetry	*even*	*odd*	*odd*
$x \to \infty$	$cosh(x) \to \frac{e^x}{2}$	$sinh(x) \to \frac{e^x}{2}$	$tanh(x) \to 1$
$x \to -\infty$	$cosh(x) \to \frac{e^{-x}}{2}$	$sinh(x) \to -\frac{e^{-x}}{2}$	$tanh(x) \to -1$

Example 4.6.1

Use the definition of the *tanh(x)* to determine the end behavior of the function as $x \to \infty$.

$$tanh(x) = \frac{sinh(x)}{cosh(x)} = \frac{\frac{e^x - e^{-x}}{2}}{\frac{e^x + e^{-x}}{2}} = \frac{e^x - e^{-x}}{e^x + e^{-x}} \Rightarrow \lim_{x\to\infty} \frac{e^x - e^{-x}}{e^x + e^{-x}} = \lim_{x\to\infty} \frac{e^x - \frac{1}{e^x}}{e^x + \frac{1}{e^x}} = \lim_{x\to\infty} \frac{e^x}{e^x} = 1$$

2. *What are the derivatives of the hyperbolic functions?*

✍ ____

____ *(See text p. 205.)*

You can derive the derivatives of the hyperbolic functions from their definitions and simplify them by using the identity $cosh^2(x) - sinh^2(x) = 1$.

$$\frac{d}{dx}(cosh(x)) = \frac{d}{dx}\left(\frac{e^x + e^{-x}}{2}\right) = \frac{e^x}{2} + \frac{-e^{-x}}{2} = sinh(x)$$

$$\frac{d}{dx}(sinh(x)) = \frac{d}{dx}\left(\frac{e^x - e^{-x}}{2}\right) = \frac{e^x}{2} - \frac{-e^{-x}}{2} = cosh(x)$$

$$\frac{d}{dx}(tanh(x)) = \frac{d}{dx}\left(\frac{sinh(x)}{cosh(x)}\right) = \frac{cosh(x)\cdot cosh(x) - sinh(x)\cdot sinh(x)}{(cosh(x))^2} = \frac{1}{cosh^2(x)}$$

Example 4.6.2

The hyperbolic cosine is also called a *catenary*. The curve describes a cable suspended from two points. Find the lowest point on a suspended cable given by the formula $y = cosh(0.2x - 3) + 4$ where x is in feet from the left end point and y is feet above the ground.

The Derivative	*The Graph of y and y'*
$\frac{d}{dx}(\cosh(0.2x-3)+4)$ $= 0.2\sinh(0.2x-3)$ From the graph of y' we see that when $x = 15$, $y' = 0$ and $y(15) = 5$ *ft* above the ground.	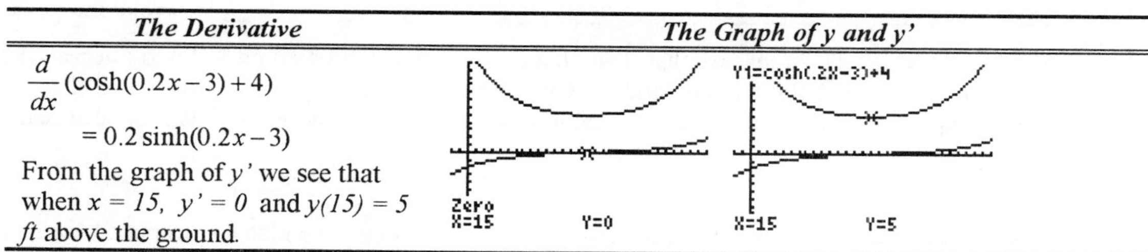

Key Notation

- $(cosh(x))^2 = cosh^2(x)$; $(sinh(x))^2 = sinh^2(x)$; $(tanh(x))^2 = tanh^2(x)$.
- $tanh(x) = sinh(x)/cosh(x)$.
- $cosh^2(x) - sinh^2(x) = 1$.

🗁 Key Study Skills

- Choose at least two problems that represent this section; put each on a 3 x 5 note card with the solution on the back. review all problems from sections 4.1-4.6.
- Use the ✓ ? ★ system on homework problems and get answers to all ★ problems.
- Use the graph of the hyperbolic function and the graph of the derivative to check that problems are done correctly.

4.7 THEOREMS ABOUT CONTINUOUS AND DIFFERENTIABLE FUNCTIONS

Objective: To obtain results about functions which are continuous on a closed interval $[a.b]$ and differentiable on the open interval (a,b).

Preparations: Read and take notes on section 4.7.
Suggested practice problems: 1-8, 11, 15, 23, 29.

✠ Key Ideas of the Section

You should be able to:

- Use the Extreme Value Theorem to determine the existence of global extrema.
- Use the Mean Value Theorem to find the connection between the average and the instantaneous rate of change.
- Use the Racetrack Principle to show the relation between one function and another given either their common starting points or ending points and their rates of change.

Key Questions to Ask and Answer

1. What is the Extreme Value Theorem?

✍ ____

____ *(See text p. 207)*

The Extreme Value Theorem says that if you have a continuous function on a closed interval, the function must have a global maximum and a global minimum on that interval. Notice that the hypothesis requires that the function be continuous on that interval and that the interval be closed.

Example 4.7.1

Consider the function $f(x) = \dfrac{1}{x}$ on the following intervals. Does f have global extrema on the given interval and, if not, explain why this isn't a contradiction to the Extreme Value Theorem.

Interval		*Explanation*
a) $[1,\infty)$	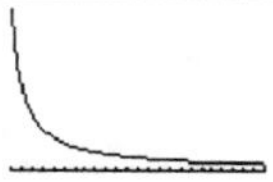 window: $x:[1,25]$, $y:[0,1]$	f is continuous on this interval, but has no global minimum. This does not contradict the Extreme Value Theorem because the interval is not finite.
b) $(0,1]$	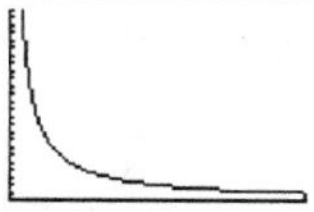 window: $x:[1,25]$, $y:[0,1]$	f has no global maximum on this interval because the interval is not closed. This does not contradict the Extreme Value Theorem because the interval is not closed.
c) $[0.5,1]$	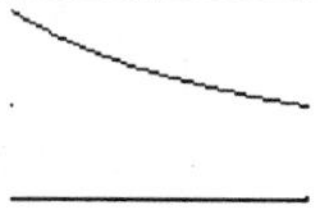 window: $x:[0.5,1]$, $y:[0,2]$	f has a global maximum at x = 0.5 and a global minimum at x = 1. The function is continuous and the interval is closed and the Extreme Value Theorem applies.

2. *What is the Mean Value Theorem?*

✍ ____

____ *(See text p. 208.)*

The Mean Value Theorem says that if you have a "nice" function (i.e. continuous on $[a,b]$ and differentiable on (a,b)), then there is a number c in that interval such that the slope of the curve at c is the same as the slope of the secant line joining the two endpoints. Again, the assumptions are critical here.

Example 4.7.2

Consider the following functions. Explain if the hypotheses of the Mean Value Theorem are met, and if so, apply it.

Function	*Graph*	*Explanation*
a) $f(x)=x^2$ on x: [1,2] and y:[1,4]	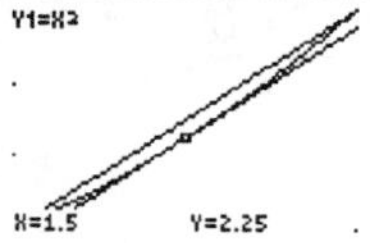 Y1=X2 X=1.5 Y=2.25	The Mean Value Theorem applies since the function is continuous on the closed interval and differentiable on the open interval. $\frac{f(2)-f(1)}{2-1}=\frac{4-1}{1}=3$ so we need a value c such that $f'(c)=3$. Since $f'(x)=2x$, then c =1.5
b) $f(x)=x^{\frac{2}{3}}$ on x: [-8,8] and y:[-2,4]	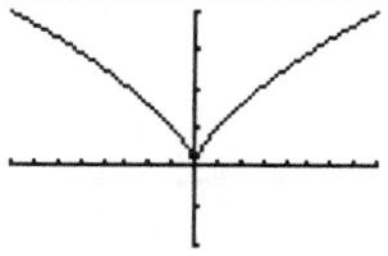	f is continuous on the closed interval, but is not differentiable at x = 0. Therefore the Mean Value Theorem does not apply.

On Your Own 1

Explain if the hypotheses of the Mean Value Theorem applies, and if so, apply it for $f(x) = x^3 - 4x$ on [-2,3].

(The answer appears at the end of this section.)

3. *What is Racetrack Principle?*

____

____ (See text p. 209.)

The Racetrack Principle says that if you have two "nice" functions f and g (i.e. continuous on $[a,b]$ and differentiable on (a,b)), and the rate of change of g is always less than or equal to the rate of change of f, then: if they start in the same place (i.e. $g(a)=f(a)$) g will always be less than or equal to f on that interval, or, if they end at the same place (i.e. $g(b)=f(b)$) g will always be greater than or equal to f on that interval.

Example 4.7.3

Use the Racetrack Principle to show that $\cos x \le x - \frac{\pi}{2}$ for $x \ge \pi/2$.

Let $g(x) = \cos x$ and $f(x) = x - \frac{\pi}{2}$. $g(\frac{\pi}{2}) = 0$ and $f(\frac{\pi}{2}) = 0$. Furthermore, $-1 \le g'(x) = -\sin x \le 1 = f'(x)$. Hence, the Racetrack Principle applies and says that $g(x) \le f(x)$ for $x \ge \pi/2$

window: $x:[\pi/2,10]$ and $y:[-1,10]$

Key Study Skills

- Choose at least two problems that represent this section; put each on a 3 x 5 note card with the solution on the back. review all problems from sections 4.1-4.7.
- Use the ✓ ? ★ system on homework problems and get answers to all ★ problems.
- Make sure that the hypotheses of each theorem are satisfied, that is, that the functions are continuous on a closed interval and differentiable on the open interval.

Answer to On Your Own 1

The Mean Value Theorem applies as the function is continuous on the closed interval and differentiable on the open interval. $\frac{f(3) - f(-2)}{3-(-2)} = \frac{15-0}{5} = 3$ so we need a value c such that $f'(c) = 3$. $f'(c) = 3c^2 - 4 = 3$ or $c = \pm\sqrt{\frac{7}{3}} \approx \pm 1.527$ which are both in the interval [-2,3]

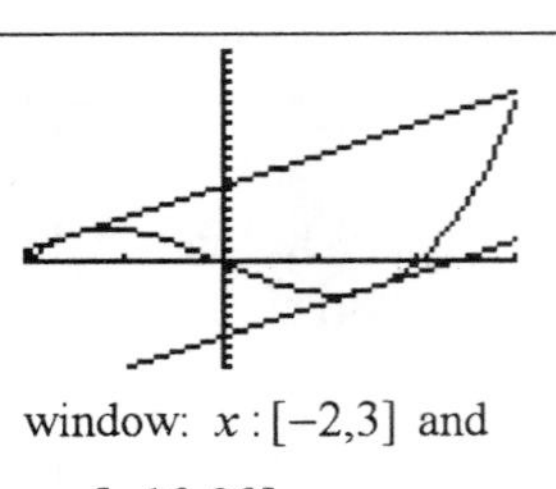

window: $x:[-2,3]$ and $y:[-10,20]$

CUMULATIVE REVIEW CHAPTER FOUR

Key Study Skills

- Do the review problems at the end of Chapter Five in the text.
- Make up your own test from the 3 x 5 note cards you have made.
- Review all of the **Key Ideas** from 4.1 - 4.7 and make sure you can answer all the Key Questions in the chapter.
- From the text do the Review Exercies p. 212-215 and Check Your Understanding p. 216-217.
- Do the Chapter 4 Practice test.

Chapter 4 Practice Test

1. The graph below is the graph of the derivative of a function *f*. Use this graph to answer the following questions about *f* on the interval from 0 to 10.

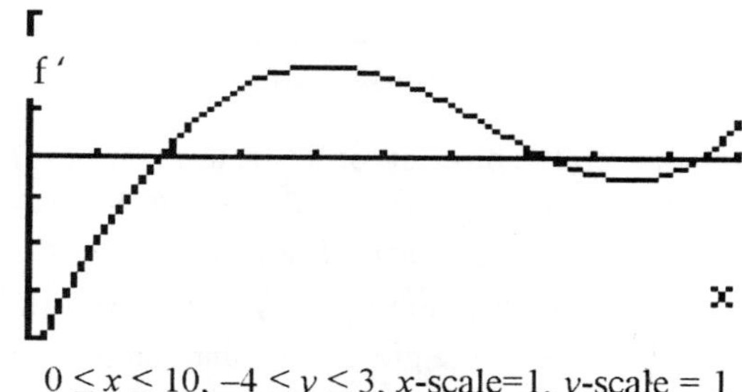

$0 \le x \le 10$, $-4 \le y \le 3$, x-scale=1, y-scale = 1

   a. On which intervals is *f* increasing and on which is *f* decreasing?
   b. On which intervals is *f* concave up and on which is *f* concave down?
   c. Find where *f* has a local maxima or a local minima.
   d. Where on the interval [0,10] would *f* have a global maximum?

2. On the axes below, sketch a smooth, continuous curve (i.e. no sharp corners, no breaks) which passes through the point $P(3,4)$, and which clearly satisfies all of the following conditions.
   - concave up to the left of P
   - concave down to the right of P
   - increasing for $x > 0$
   - decreasing for $x < 0$
   - Does *not* pass through the origin

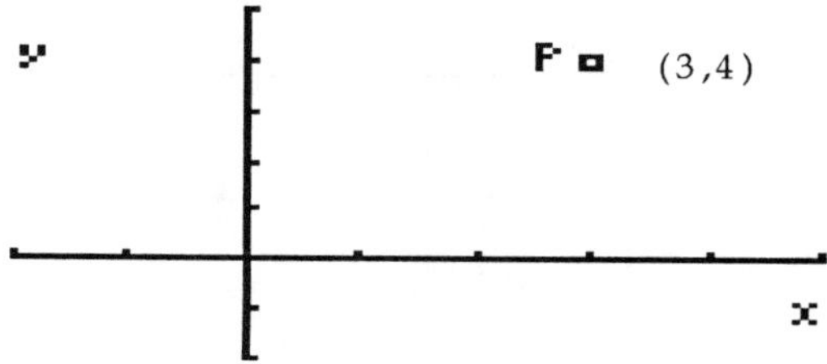

3. Sketch the graph of a continuous function that satisfies all the following conditions:

   $f'(x) > 0$ *for all real numbers* $x \neq 3$

   $f'(3)$ *does not exist*

   $f''(x) > 0$ *for all* $x < 3$

   $f''(x) < 0$ *for all* $x > 3$

4. The cost $C(q)$ (in dollars) of producing a quantity q of a certain product is shown in the graph below.

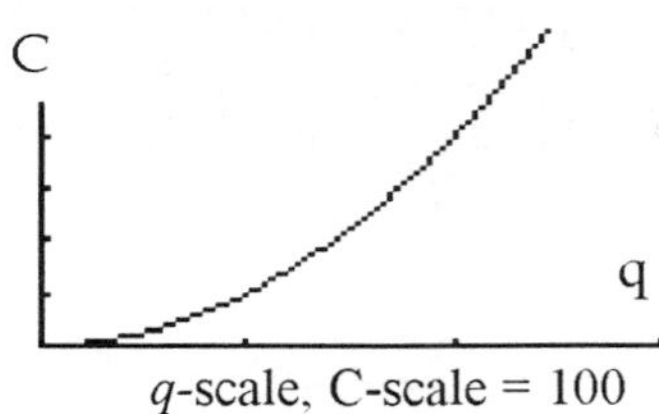

q-scale, C-scale = 100

Suppose the manufacturer can sell the product for \$2 each (regardless of how many are sold), so that the total revenue from selling a quantity q is $R(q) = 2q$. The difference $P(q) = R(q) - C(q)$ is total profit. Let q_0 be the quantity that will produce the maximum profit. You are told that q_0 can be found by finding the point on the graph where $C'(q_0) = 2$. Draw the graph of R on the figure above and then *explain* why this rule makes sense graphically and mathematically.

5. You have a manufacturing plant which produces televisions. The hourly cost, H, of operating the production line is proportional to the square of the rate, r, at which the televisions are produced. You know that this cost is \$400 per hour when you produce 10 televisions per hour. Other fixed costs are \$1600 per hour. You receive an order for T televisions. Find the rate at which you should produce the televisions to minimize the total production cost , C, of the order.

6. For the following function: $f(x) = \dfrac{1 + \cosh x}{1 - \cosh x}$,

   a) Give the domain
   b) Find the derivative
   c) Graph both the function and its derivative.

7. Explain if the hypotheses of the Mean Value Theorem are met, and if so, apply it to the function $f(x) = 2^x - x^2$ on [0,4].

CHAPTER FIVE

KEY CONCEPT: THE DEFINITE INTEGRAL

The definite integral computes the total change in a function from its rate of change. The definite integral is used to calculate area under a curve and the average value of a function. Calculating derivatives and calculating definite integrals are, in a sense, reverse processes.

5.1 HOW DO WE MEASURE DISTANCE TRAVELED?

Objective: To calculate total distanced traveled, given the velocity function.

Preparations: Read and take notes on section 5.1.
Suggested practice problems: 2, 3, 4, 6, 7, 11, 14.

✠ Key Ideas of the Section

You should be able to:

- Recognize that distance is the product of velocity and time.
- Add estimates of distance traveled over small intervals, using left and right rectangular sums.
- Find the difference between left and right sums using : $|f(b) - f(a)|\Delta x$.
- Visualize that area under the velocity curve represents distance.

Key Questions to Ask and Answer

1. How do calculate distance traveled given the velocity?

✍ ____

____ *(See text p. 222-226.)*

We know that *distance = (velocity)(time)*. If velocity is positive, the total distance can be calculated by summing the product of velocity and time over individual time intervals.

Example 5.1.1

a) The velocity data below represents a car that comes to a stop six seconds after the brakes are applied. Use this data to calculate a lower and upper estimate for the distance traveled by the car.

Time (sec)	0	1	2	3	4	5	6
Velocity (ft/sec)	45	40	35	23	16	8	0

Calculate an estimate of the distance over one second intervals (Δt) using $d = v \cdot t$

Upper sum =

$$v_0 \Delta t + v_1 \Delta t + v_2 \Delta t + v_3 \Delta t + v_4 \Delta t + v_5 \Delta t =$$
$$45(1) + 40(1) + 35(1) + 23(1) + 16(1) + 8(1) = 167\, ft.$$

Lower sum =

$$v_1 \Delta t + v_2 \Delta t + v_3 \Delta t + v_4 \Delta t + v_5 \Delta t + v_6 \Delta t =$$
$$40(1) + 35(1) + 23(1) + 16(1) + 8(1) + 0(1) = 122\, ft.$$

b) Draw the graph of the data and the graphs of the rectangular upper and lower sums.

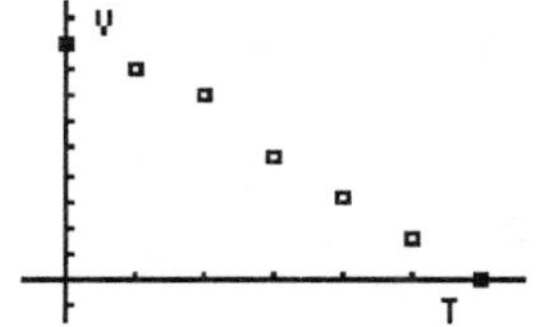

Graph of velocity data against time.

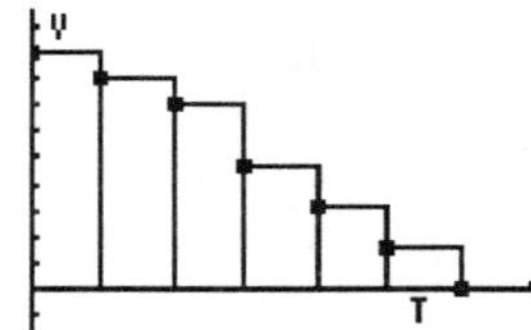

Upper sum for distance = 167 ft.
Use the left data point.

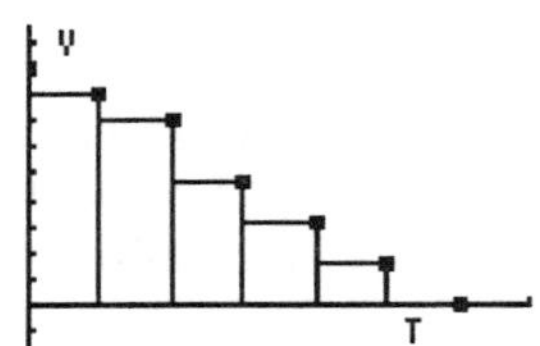

Lower sum for distance = 122 ft.
Use the right data point.

Note: The area of each of the rectangles corresponds to one of the terms in the upper sum and lower sum calculation from part (a) above.

2. *How do you make your estimate for distance more precise?*

✍ ____

____ *(See text p. 225-226.)*

By making your time increments smaller you can improve your precision. You need to:

- Identify your interval for t, where $a \le t \le b$.
- Divide the time interval into n equally spaced subdivisions: $t_0, t_1, t_2, t_3, \ldots t_n$. The larger you make n the more subdivisions you have.
- Calculate the width of each subdivision, $\Delta t = \dfrac{b-a}{n}$. The larger you make n the smaller Δt becomes.
- For each increment the distance traveled is approximately $f(t_i)\Delta t$, where f is the velocity and Δt is the time.
- The left sum is approximately equal to:
  Total distance traveled between $t = a$ and $t = b$
  $\approx f(t_0)\Delta t + f(t_1)\Delta t + f(t_2)\Delta t + \ldots + f(t_{n-1})\Delta t$.
- The right sum is approximately equal to:
  Total distance traveled between $t = a$ and $t = b$
  $\approx f(t_1)\Delta t + f(t_2)\Delta t + f(t_3)\Delta t + \ldots + f(t_n)\Delta t$.
- The difference between the upper and lower estimates $= |f(b) - f(a)| \cdot \Delta t$. This is the maximum error of the estimate, if the function is always increasing or always decreasing.
- The smaller the difference the more precise your estimate.

✎ On Your Own

Velocity is a rate. Suppose oil is leaking out of a container at a decreasing rate. The rate is measured in half-hour intervals as given by the table below:

Time (hours)	0	.5	1.0	1.5	2.0
Rate (liters/hours)	35	33	27	24	22

Estimate the total amount of oil that has leaked out for the time shown.

The answer appears at the end of this section.

Key Notation:

- Δt : "A small change in t" .
- $\Delta t = \dfrac{b-a}{n}$; the method used to find the change in t, or the width of the subdivisions, Δt .
- Δt = (end point - begin point) /(number of subdivisions)
- $\lim\limits_{n\to\infty}$: "The limit as n goes to infinity", i.e. , let n get very large.
- t_0 : the beginning of the subdivisions, $t_0 = a$.
- t_n : the end of the subdivisions, $t_n = b$.
- $\lim\limits_{n\to\infty}$ (Left-hand sum) = $\lim\limits_{n\to\infty}$ $[f(t_0)\Delta t + f(t_1)\Delta t + f(t_2)\Delta t + \ldots + f(t_{n-1})\Delta t]$ =

area under the graph of $f(t)$ from a to b, if $f(t) \geq 0$ on $[a,b]$.

- $\lim\limits_{n\to\infty}$ (Right-hand sum) = $\lim\limits_{n\to\infty}$ $[f(t_1)\Delta t + f(t_2)\Delta t + f(t_3)\Delta t + \ldots + f(t_n)\Delta t\ ]$ =
  area under the graph of $f(t)$ from a to b, if $f(t) \geq 0$ on $[a,b]$.

🗁 Key Study Skills

- Choose at least two problems that represent this section; put each on a 3 x 5 note card with the solution on the back.
- Use the ✓ ? ★ system on homework problems and get answers to all ★ problems.

Answer to On Your Own

Δt = (2-0)/4 = .5

Left Sum = 35(.5) + 33(.5) + 27(.5) + 24(.5) = 59.5 liters.

Right sum = 33(.5) + 27(.5) + 24(.5) + 22(.5) = 53 liters.

53 < Total liters spilled < 59.5.

Estimate of the total liters spilled ≈ (53 + 59.5)/2 = 56.25 liters of oil spilled.

5.2 THE DEFINITE INTEGRAL

Objective: The definite integral of a function can be calculated using the limit of Riemann Sums and can be visualized as the *signed* area trapped between the x-axis and the graph of the function.

Preparations: Read and take notes on section 5.2.
Suggested practice problems: 1, 2, 4, 8, 9, 13, 16, 20, 23, 24, 27,30.

✠ Key Ideas of the Section

You should be able to:

- Calculate the left-hand and right-hand rectangular sums (Riemann Sum) of a function f to approximate the definite integral.
- Identify the summation symbols: $\sum_{i=0}^{n-1} f(t_i)\Delta t$ and $\sum_{i=1}^{n} f(t_i)\Delta t$ as the left and right Riemann sums respectively.
- Recognize that the definite integral is a number, which is the limit of the left-hand and right-hand sums.
- Use the notation $\int_a^b f(x)dx$ for the definite integral.
- Demonstrate that the definite integral represents the area between the x-axis and the graph of the function provided $f(x) \geq 0$ on the interval. The definite integral represents the negative of the area between the x-axis and the graph of the function provided $f(x) < 0$ on the interval.
- Find the definite integral for non-monotonic functions by breaking the graph of the function up into increasing and decreasing portions and summing the areas separately.
- Recognize that the general Riemann sum allows for subdivisions of different lengths and to choose any point in the subdivision.

Key Questions to Ask and Answer

1. *How do you use the left and right Riemann sums to approximate the definite integral?*

(See text p. 229-231.)

The definite integral of a function f, $\int_a^b f(t)dt$, is a number approximated by the left and right rectangular sums (also know as the Riemann Sum). To approximate the definite integral over an interval $[a, b]$, for any continuous function $f(t)$ for $a \leq t \leq b$:

- divide the interval from a to b into n equal subdivisions,
- find the width Δt of each subdivision so that $\Delta t = \frac{b-a}{n}$
- let the endpoints of each subdivision be: $t_0, t_1, t_2, t_3, .. t_{n-1}, t_n$, with $t_0 = a$, $t_1 = a + \Delta t$, $t_2 = t_1 + \Delta t$, $t_3 = t_2 + \Delta t$, etc. and $t_n = t_{n-1} + \Delta t = b$
- calculate the product of $f(t_i)\,\Delta t$, then sum all the rectangular products.

The following figure helps you to visualize one of the rectangles, Δt units wide and $f(t_i)$ units high.

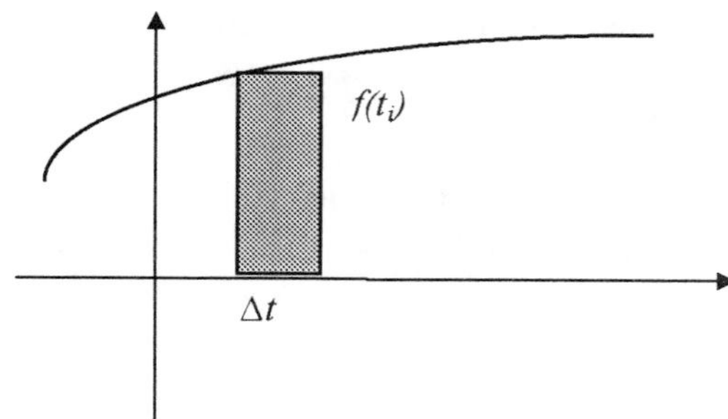

We use the $\sum$ (sigma) notation for the summation process.

The *left -hand Riemann sum* begins at t_0 and ends at t_{n-1} :

$$\sum_{i=0}^{n-1} f(t_i)\Delta t = f(t_0)\Delta t + f(t_1)\Delta t + f(t_2)\Delta t + \ldots f(t_{n-1})\Delta t$$

The *right -hand Riemann sum* begins at t_1 and ends at t_n :

$$\sum_{i=1}^{n} f(t_i)\Delta t = f(t_1)\Delta t + f(t_2)\Delta t + f(t_3)\Delta t + \ldots f(t_n)\Delta t$$

The *definite integral* is a number approximated by the left and right-hand sums as the number of subdivisions n gets very large (thus the width of the rectangles is small).

- For a monotonic increasing[1] function, the left-hand sum is smaller than the right-hand sum and the definite integral is trapped in-between:

$$\sum_{i=0}^{n-1} f(t_i)\Delta t \le \int_a^b f(t)dt \le \sum_{i=1}^{n} f(t_i)\Delta t$$

Increasing and Concave Up   ***Increasing and Concave Down***

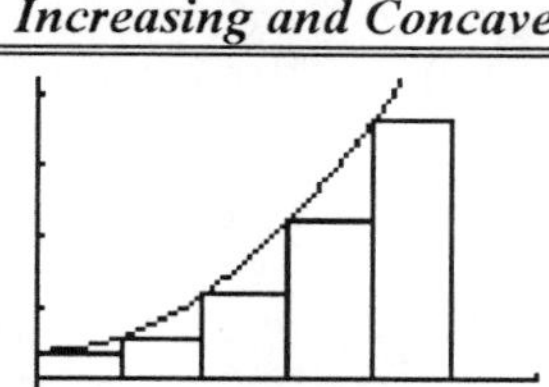

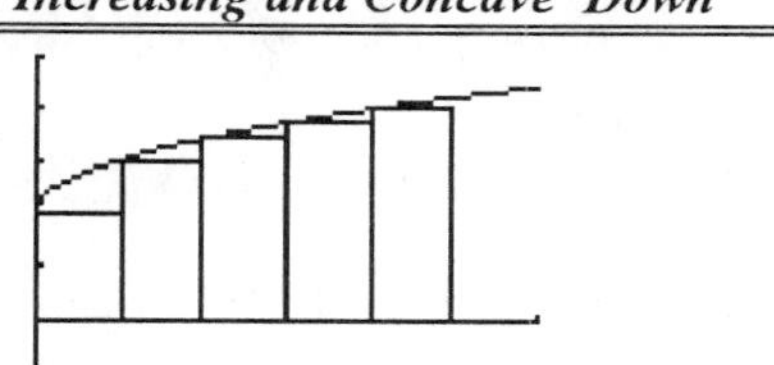

For both types of increasing functions the left-hand sum lies under the curve and will always be less than the right-hand sum.

- For a monotonic decreasing[2] function, the left-hand sum is greater than the right-hand sum and the definite integral is trapped in-between:

$$\sum_{i=1}^{n} f(t_i)\Delta t \le \int_a^b f(t)dt \le \sum_{i=0}^{n-1} f(t_i)\Delta t$$

Decreasing and Concave Up   ***Decreasing and Concave Down***

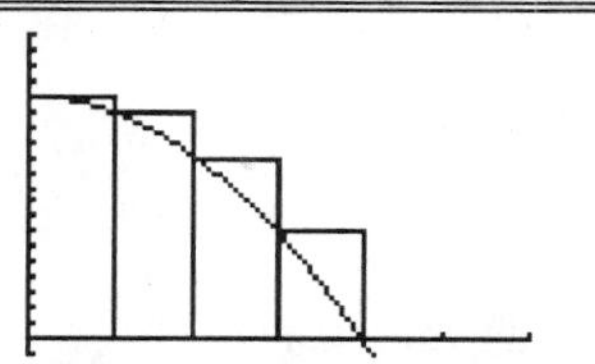

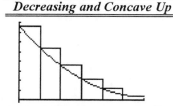

For both types of decreasing functions the left-hand sum lies above the curve and will always be greater than the right-hand sum.

- $\int_a^b f(t)dt = \lim_{n\to\infty}(\text{Left-hand sum})$ and $\int_a^b f(t)dt = \lim_{n\to\infty}(\text{Right-hand sum})$.

[1] monotonic increasing: always increasing on the interval.
[2] monotonic decreasing: always decreasing on the interval.

Example 5.2.1

Approximate the definite integral using first 5, then 50 subdivisions of the intervals.

$$f(x) = e^{-x^2}\text{, over the interval } a = 1\text{ , } b = 2\text{: } \int_1^2 e^{-x^2}dx\text{ .}$$

a) For the left-hand sum:

$n = 5,\ \Delta x = (2 - 1)/5 = 1/5 = .2$, and

$t_0 = 1,\ t_1 = 1.2,\ t_2 = 1.4,\ t_3 = 1.6,\ t_4 = 1.8$

$$\sum_0^4 f(x)\Delta x = f(1)(.2) + f(1.2)(.2) + f(1.4)(.2) + f(1.6)(.2) + f(1.8)(.2)$$

$$= e^{-(1)^2}(.2) + e^{-(1.2)^2}(.2) + e^{-(1.4)^2}(.2) + e^{-(1.6)^2}(.2) + e^{-(1.8)^2}(.2)$$

$$\approx .17243$$

For the right-hand sum:

$n = 5,\ \Delta x = (2 - 1)/5 = 1/5 = .2$, and

$t_1 = 1.2,\ t_2 = 1.4,\ t_3 = 1.6,\ t_4 = 1.8,\ t_5 = 2$

$$\sum_1^5 f(x)\Delta x = f(1.2)(.2) + f(1.4)(.2) + f(1.6)(.2) + f(1.8)(.2) + f(2)(.2)$$

$$= e^{-(1.2)^2}(.2) + e^{-(1.4)^2}(.2) + e^{-(1.6)^2}(.2) + e^{-(1.8)^2}(.2) + e^{-(2)^2}(.2)$$

$$\approx .10251$$

Note: The right-hand sum is less than the left-hand sum since *f(x)* is a decreasing function.

$$.10251 \le \int_1^2 e^{-x^2}dx \le .17243$$

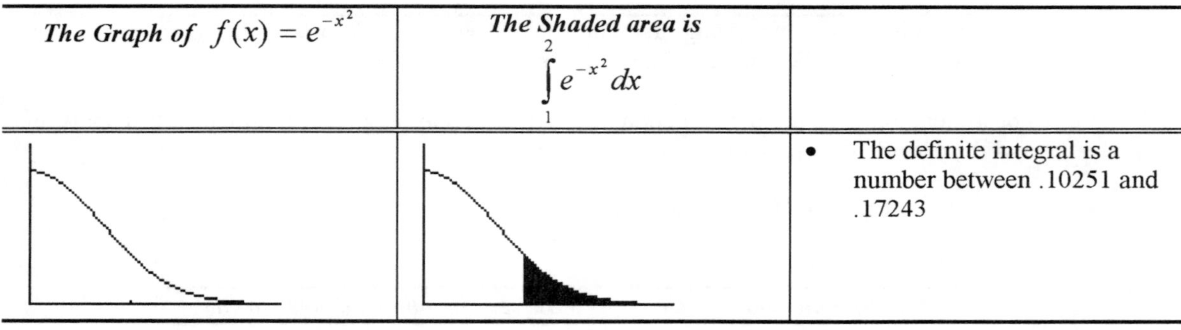

The Graph of $f(x) = e^{-x^2}$
The Shaded area is $\int_1^2 e^{-x^2}dx$
• The definite integral is a number between .10251 and .17243

b) For 50 subintervals:

$n = 50,\ \Delta x = (2 - 1)/50 = 1/50 = .02$

Using a program or calculator (ask your instructor for this program, it is in the instructor's manual):

Left hand sum = 0.13877

Right hand sum = 0.13178

$$0.13178 \le \int_1^2 e^{-x^2}dx \le 0.13877$$

The definite integral is a number approximately equal to 0.135 accurate to the one-hundredth or second decimal place. A calculator shows the value to be 0.13525726.

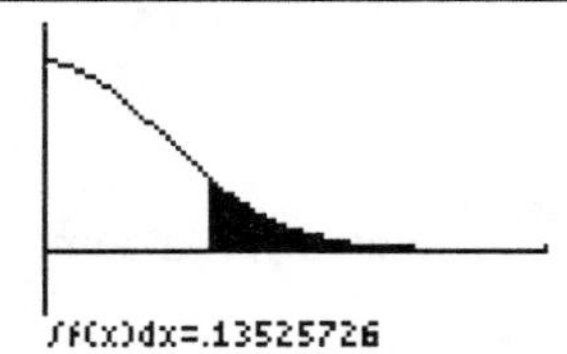

- The calculator has shaded the area under the curve from 1 to 2 and displayed the value of the definite integral to be 0.13525726.

2. *How is the definite integral related to area under the graph of f?*

✍ ____

____ *(See text p. 231-233)*

For *f(x)* ≥ 0 on the interval the left-hand or right-hand Riemann sum is the sum of the area of many rectangles Δx units wide. As the width Δx approaches zero the rectangles fit the area under the graph of *f* more exactly. In other words, you are summing the areas of thinner and thinner rectangles.

When *f(x)* is positive for some values and negative for other values, and $a < b$, then $\int_a^b f(x)dx$ is the area above the *x*-axis minus the area below the *x*-axis.

Remember that area is always positive, but for arithmetic purposes we say

- area above the *x-axis* is counted positively,
- area below the *x-axis* is counted negatively and
- the integral equals the total signed area or
- if all area is taken to be positive, then the integral equals the area above the *x*-axis minus the area below the *x*-axis.

Example 5.2.2

Find the signed area under the curve, $f(x) = \dfrac{\sin x}{x}$, between 1 and 6, or $\int_1^6 \dfrac{\sin x}{x} dx$.

The area between the curve and the *x*-axis is above the *x*-axis for $1 < x < \pi$, and is taken to be positive.

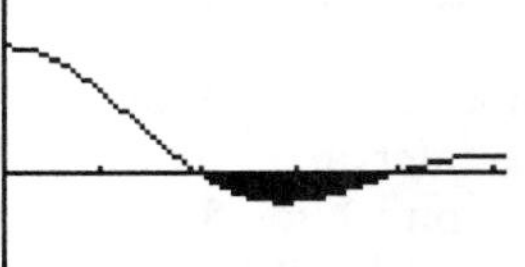

The area between the curve and the *x*-axis is below the *x*-axis for $\pi < x < 6$, and is taken to be negative.

$\int_1^6 \dfrac{\sin x}{x} dx = 0.47860448$,

The sum of the area counted as positive and negative over 1 to 6 is: 0.90585398 - 0.4272495) = 0.47860448.

3. *What is the important feature of the General Riemann Sum?*

✍ ____

____ *(See text p. 233-234.)*

The general Riemann Sum allows you to sum using subdivisions of different lengths and instead of using the endpoints we evaluate anywhere in the subdivision.

Many people prefer the midpoint of the subdivision. The area of a *i-th* subdivision is

A_i = **(value of *f* at some point in the *i*-th subdivision)(length of the *i*-th subdivision).**

The total sum of all subdivisions for *f* on the interval $[a,b]$ is:

$$\sum_{i=1}^{n} f(c_i)\cdot \Delta x_i \ ,$$

where $\Delta x_i = x_i - x_{i-1}$ or the length of the subinterval and $x_{i-1} \le c_i \le x_i$, that is c_i is a point within the subdivision.

Key Notation

- $\sum_{i=1}^{n} f(x_i)\Delta x$: "The sum of the product of $f(x_i)$ and Δx, beginning at x_1 and ending at x_n."
- $\int_a^b f(x)dx$: "The definite integral of $f(x)$ over the interval a to b" or "The sum of the area between *f* and the
  x-axis from a to b."
- Area between the graph of *f* and the *x*-axis is always positive, but the total area is the area above the *x*-axis minus the area below the x-axis. The value of an integral can be negative, which indicates there is more area below the *x*-axis than above the *x*-axis.
- $-\int_a^b f(x)dx$ means take the opposite of the value of the integral.

Key Study Skills

- Choose at least two problems that represent this section; put each on a 3 x 5 note card with the solution on the back.
- Use the ✓ ? ★ system on homework problems and get answers to all ★ problems.
- Review the **Key Ideas** from sections 5.1, 5.2.

5.3 INTERPRETATIONS OF THE DEFINITE INTEGRAL

Objective: The definite integral can be interpreted as area and as the average value of a function on a given interval.

Preparations: Read and take notes on section 5.3.
Suggested practice problems: 1, 3, 5, 6, 9, 11, 12, 14, 17, 21, 25.

✠ Key Ideas of the Section

You should be able to:

- Identify the three meanings of the definite integral:
  1). the limit of the Riemann Sums,
  2). the area between the graph of the function and the *x*-axis and
  3). the total change from a rate.
- Use the notation $f(x)dx$ to find the units of the definite integral.
- Define the average value of a function f from a to b as $\frac{1}{b-a}\int_a^b f(x)dx$.
- Define the definite integral as the product: (average value)$(b - a)$.
- Visualize the definite integral as a rectangular area whose height is the average value and whose width is $b - a$.

Key Questions to Ask and Answer

1. How can you find the units of the definite integral?

✍ ____

____ *(See text p. 238.)*

Since the definite integral is the limit of sums of rectangular areas, and each rectangle is *f(x)* units high and Δx units wide, the unit of the definite integral is the product:
(unit of the vertical axis)(unit of the horizontal axis).

Example 5.3.1

An oil spill and its subsequent clean up was monitored. It was determined that the rate of change of the area affected by the spill could be modeled by:

$$R(t) = 1000\,\text{Sin}\,((\pi/24)\,t)$$

Where R is measured in ft^2/hr and time, t, is measured in hours. The graph below is of the function $R(t)$.

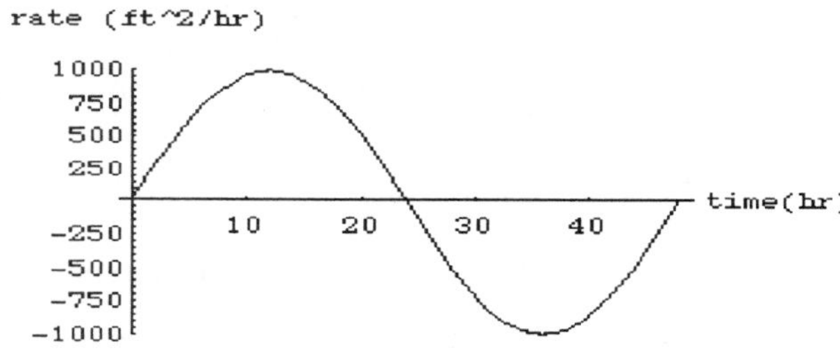

a) Interpret $\int_0^{10} R(t)dt$.

The definite integral $\int_0^{10} R(t)dt = 5662.2\ ft^2$; it is the total area affected by the oil spill after ten hours. The unit is ft^2 because you are multiplying small rectangles $R(t)$ by Δt : $(ft^2/hr)(hr) = ft^2$.

b) When will the oil spill be at a maximum?

Since the area under the graph represents the total area in ft^2 affected by the spill, the area between the graph of $R(t)$ and the t-axis, which is positive, shows the spill accumulating. At $t = 24$ hours the positive area is at a maximum $(1000 \sin((\pi/24)(24)) = 0)$.

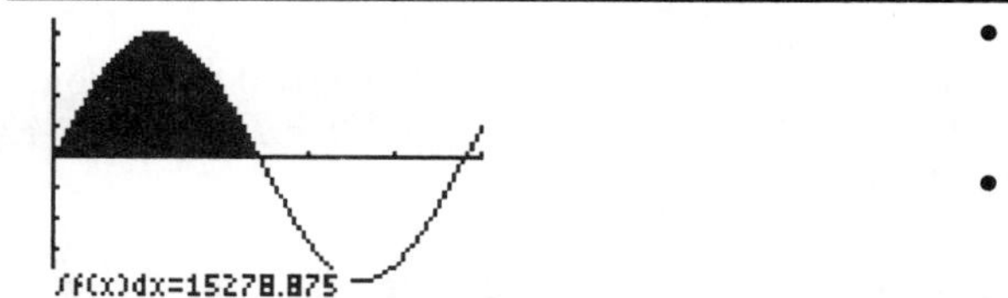

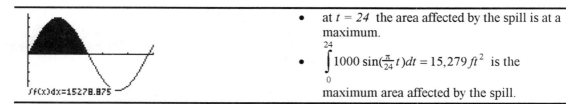

- at $t = 24$ the area affected by the spill is at a maximum.
- $\int_0^{24} 1000 \sin(\frac{\pi}{24}t)dt = 15{,}279\,ft^2$ is the maximum area affected by the spill.

c) When was the oil spill clean up making progress?

The area that represents more oil being taken away than oil spilling out is the area between the graph of $R(t)$ and below the t-axis. Therefore between the hours of 24 and 48, the spill was being cleaned up, because the total area affected is getting smaller. See the graphs below.

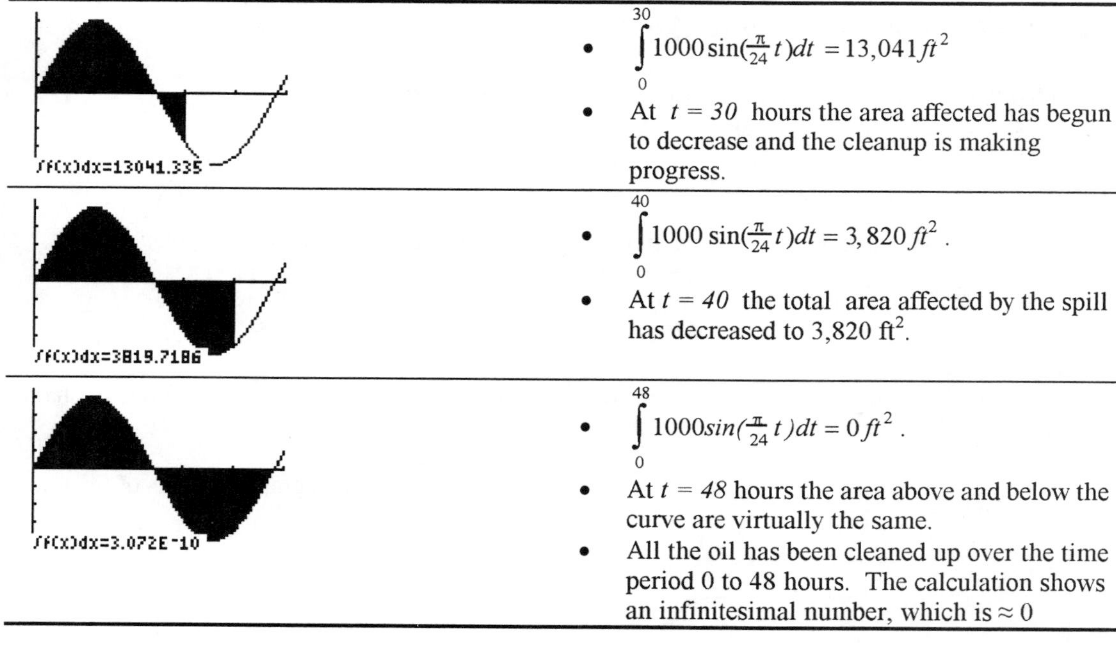

- $\int_0^{30} 1000\sin(\frac{\pi}{24}t)dt = 13{,}041 ft^2$
- At $t = 30$ hours the area affected has begun to decrease and the cleanup is making progress.

- $\int_0^{40} 1000\sin(\frac{\pi}{24}t)dt = 3{,}820 ft^2$.
- At $t = 40$ the total area affected by the spill has decreased to 3,820 ft^2.

- $\int_0^{48} 1000 sin(\frac{\pi}{24}t)dt = 0 ft^2$.
- At $t = 48$ hours the area above and below the curve are virtually the same.
- All the oil has been cleaned up over the time period 0 to 48 hours. The calculation shows an infinitesimal number, which is ≈ 0

On Your Own

In **Example 5.3.1**, at $t = 36$ hours, is there more or less area affected by the spill than at $t = 20$ hours?

(The answer appears at the end of the section)

2. *What is meant by the average value of the function over an interval?*

✍ ____

____ *(See text p. 238-239.)*

The average value of a function on an interval [*a, b*] can be calculated by finding the limit of rectangular sums and dividing by the interval length (*b-a*) or

- **Average value of *f* from *a* to *b*** $= \frac{1}{b-a}\int_a^b f(x)dx.$

Example 5.3.2

From **Example 5.3.1**, we know that the rate of the oil being spilled was defined by the function $R(t) = 1000 \, Sin\,((\pi/24)\,t)$. We know $R(0) = 0$ and $R(12) = 1000 \, Sin\,((\pi/24)(12) = 1000 \, ft^2/hr$. Each value represents the rate at which the oil is spilling out.

Find the average rate at which the oil is spilling out over the time period 0 to 12 hours.

$$\text{Average value of } R(t) \text{ from 0 to 12} = \frac{1}{12-0}\int_0^{12} 1000 sin(\frac{\pi}{24}t)dt = 636.6\, ft^2/hr$$

We can interpret this as "on average" over the first twelve hours the oil is spilling out at a rate of 636.6 ft^2/hr.

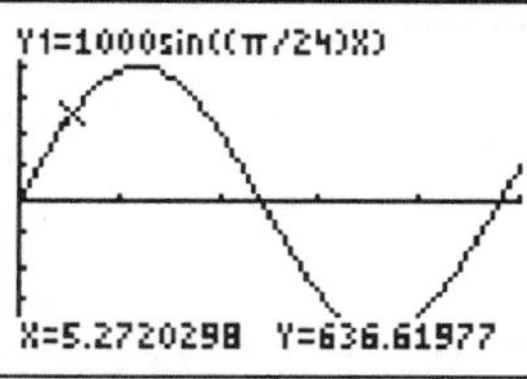

- We can visualize the average value of R from 0 to 12 as a point on the function: $R(5.272) = 636.6$ or *average value=height.*

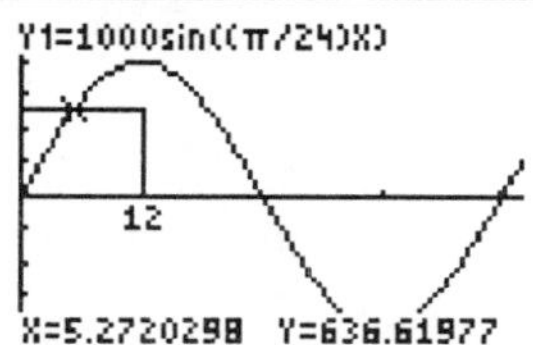

- The rectangular area (636.62) by (12) is equal to the definite integral because :

$$(\text{Avg. value})(b - a) = \int_a^b f(x)dx$$

$(636.62)(12) = 7636.44$ ft^2.

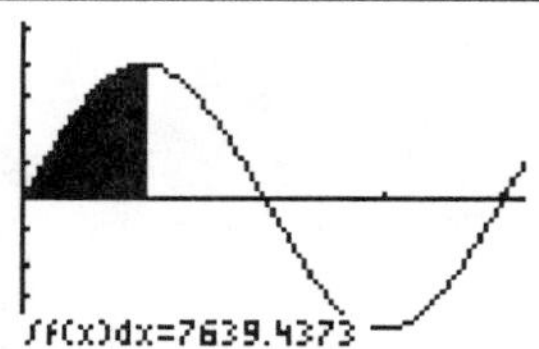

- The definite integral from 0 to 12 is $\int_0^{12} R(t)\,dt = 763944\,ft^2$.
- This is the same value as the average value times 12 or (636.62 ft^2/hr) (12 hr) = 7639.44 ft^2 .

Key Notation

- The average value of a function $f = \dfrac{1}{b-a}\displaystyle\int_a^b f(x)dx,$ from a to b.
- $\int_a^b f(x)dx = (\text{Average value of } f)(b - a).$

Key Study Skills

- Choose at least two problems that represent this section; put each on a 3 x 5 note card with the solution on the back.
- Use the ✓ ? ★ system on homework problems and get answers to all ★ problems.
- Review the **Key Ideas** from sections 5.1, 5.2, 5.3.
- List the three meanings of the definite integral.

Answer to On Your Own

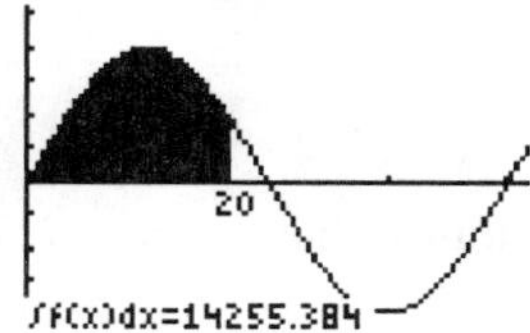

The area from 0 to 20 is all positive.

$$\int_0^{20} R(t)dt = 14,255 ft^2.$$

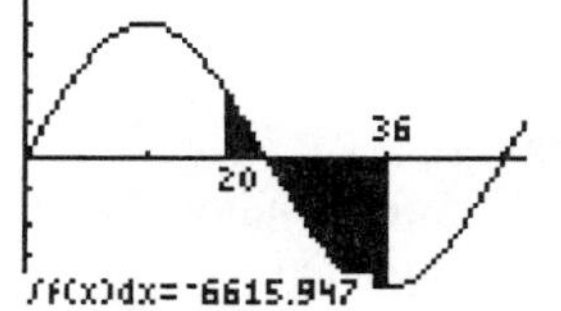

The signed area from 20 to 36 is counted negative.

$$\int_{20}^{30} R(t)dt = -6,616 ft^2.$$

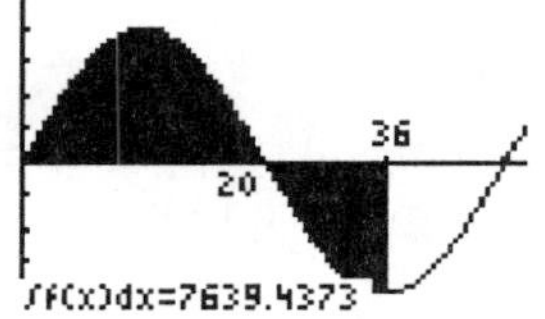

The total signed area from 0 to 36 is less than the area from 0 to 20

$$\int_0^{20} R(t)dt + \int_{20}^{36} R(t)dt =$$

$$14,255 + (-6,616) =$$

$$7,639 ft^2.$$

5.4 THEOREMS ABOUT DEFINITE INTEGRALS

Objective: Recognize that the definite integral of a rate of change gives total change: the Fundamental Theorem of Calculus.

Preparations: Read and take notes on section 5.4.
Suggested practice problems: 1, 4, 5, 8, 9, 15, 17, 19, 21.

✠ Key Ideas of the Section

You should be able to:

- Recognize that the definite integral of a rate of change makes the connection between the derivative and the total change.
- Find the total change in a function *F*, given its derivative $F'(x) = f(x)$.

  The Fundamental Theorem of Calculus:
  If $f(x)$ is continuous over $[a, b]$ and $F'(x)=f(x)$. then

  $$\int_a^b f(x)dx = F(b) - F(a)$$

- State the theorems and the properties of definite integrals for
  - Limits of Integration,
  - Sums and Constant multiples of the integrand and
  - The Comparison of definite integrals.

Key Questions to Ask and Answer

1. How does the Fundamental Theorem of Calculus link the derivative to its function?

✍ ____

____ *(See text p.244-247.)*

Given the definite integral $\int_a^b f(x)dx$, we want to think of the $f(x)$ part of the integrand as a rate of change function, therefore it is the derivative of some function $F(x)$, or $f(x) = F'(x)$. Calculating the value of the definite integral, $\int_a^b f(x)dx$, measures the total change that has occurred for $F(x)$, over the interval from *a* to *b*, $F(b) - F(a)$.

Example 5.4.1

Recall the oil spill problem from **Example 5.3.1**:
An oil spill and its subsequent clean up were monitored and it was determined that the rate of change of the area affected by the spill could be modeled by

$$R(t) = 1000 \, Sin\,((\pi/24)\, t)$$

where R is measured in ft^2/hr and time, *t*, is measured in hours.

✍

a) Fill in the table below given that $R(t) = F'(t)$ and $F(0) = 0$. Use the definite integral to find the total change in $F(x)$ over $a = 0$ and $b = 6$, 12, 18, 24, 30, 36, 42, and 48.

t	0	6	12	18	24	30	36	42	48
F(t)	0								

The Definite Integral	*Area Under Curve of R(t)*	*Total change in F(b) -F(a).*
$\int_0^6 R(t)\,dt = 2{,}237\,ft^2$	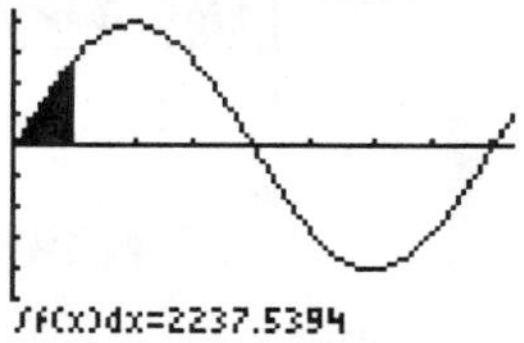 ∫f(x)dx=2237.5394	$F(6) - F(0) = 2{,}237\ ft^2$
$\int_0^{12} R(t)\,dt = 7{,}639\,ft^2$	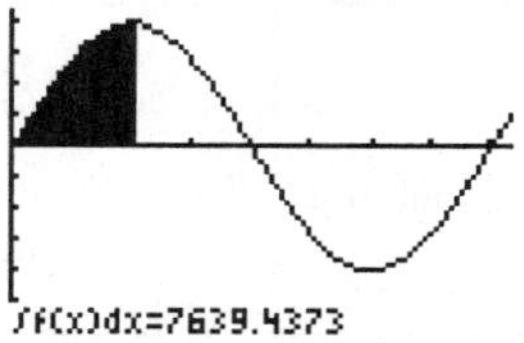 ∫f(x)dx=7639.4373	$F(12) - F(0) = 7{,}639\ ft^2$
$\int_0^{18} R(t)\,dt = 13{,}041\,ft^2$	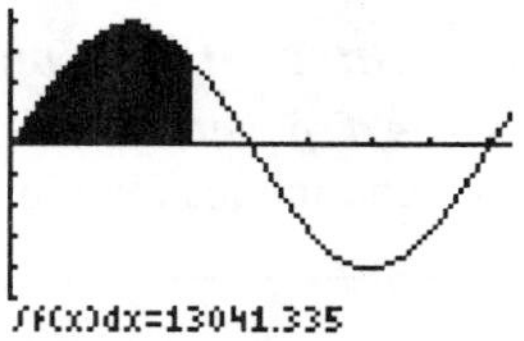 ∫f(x)dx=13041.335	$F(18) - F(0) = 13{,}041\ ft^2$

Continue the process of accumulating area...

The table filled in would be:

t	0	6	12	18	24	30	36	42	48
F(t)	0	2,237	7,639	13,041	15,279	13,041	7,639	2,237	0

b) Plot the points *(t, F(t))* then trace the graph of *F(t)* and compare it to the graph of *R(t)*.

Plot of the total change in F(t)	*The graph of F(t)*	*The graph of F'(t) = R(t)*
 24 48	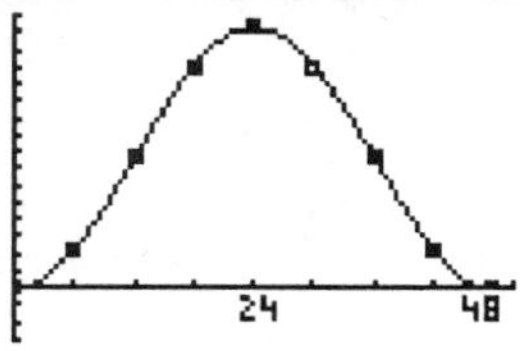 24 48	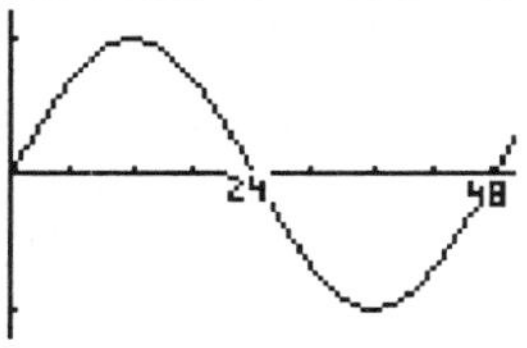 24 48

The points on the plot represent the accumulation of area under the curve *R(t)* . One can see that where the graph of *F(t)* is increasing, the rate of change graph *R(t)* is positive; at $t = 24$, *F(t)* is at a maximum when $t=24$ and $R(24) = 0$, and where *F(t)* is decreasing $R(t) < 0$. Thus we have connected *F(t)* and the derivative of *F(t)* , $F'(t) = R(t)$, with the area under the curve or the definite integral.

2. *How does symmetry help in evaluating integrals?*

✍ ____

____ *(See text p. 246.)*

Functions that are symmetric about the *y*-axis (even functions) and symmetric about the origin (odd functions) have the same area under the curve on either side of the axis.

Even functions have identical area over [-a,0] and [0,a].	***Odd functions have identical area but opposite sign over [-a,0] and [0,a]***
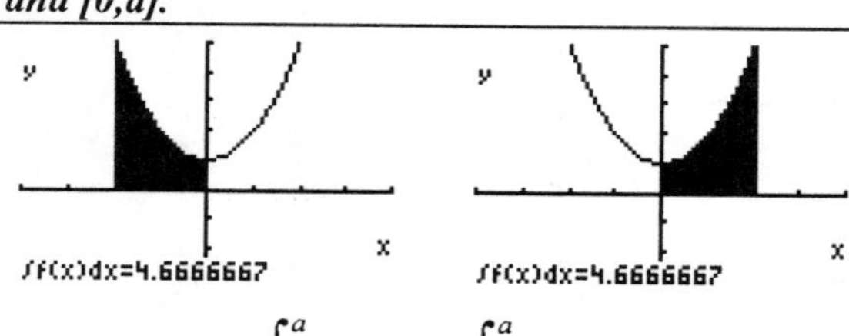	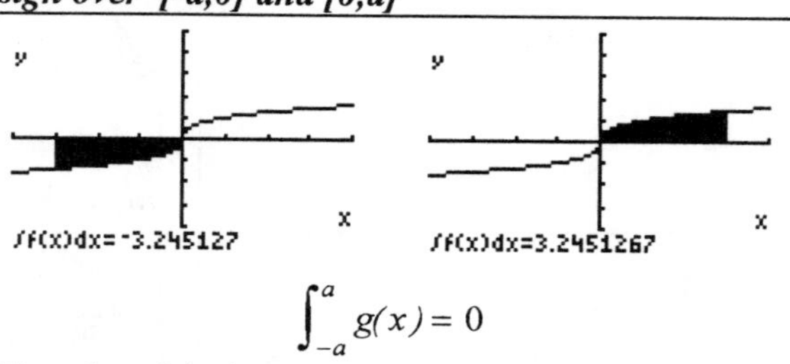
$\int_{-a}^{a} f(x) = 2\int_{0}^{a} f(x)$	$\int_{-a}^{a} g(x) = 0$
Total area under the *even* curve can be found by doubling the area from zero to the upper bound *a*.	The value of the integral is zero for the *odd* curve from [-*a*,*a*] because areas from [-*a*,0] and [0,*a*] cancel each other.

2. *List four properties of the definite integral in symbols and in words.*

✍ Write the properties in symbols. *(See text p. 246-248)*

If *a*, *b*, and *c* are any numbers and *f* is a continuous function, then

1. ____
2. ____
3. ____
4. ____

✍ Write the properties in words *(See text p. 246-248)*

1. ____
2. ____
3. ____
4. ____

Example 5.4.2

a) Examine $\int_{0}^{\pi/2} e^{\sin x^2} dx$ and $\int_{\pi/2}^{0} e^{\sin x^2} dx$.

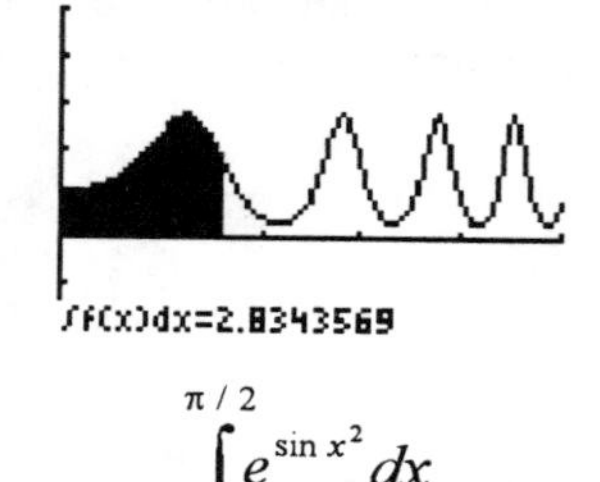	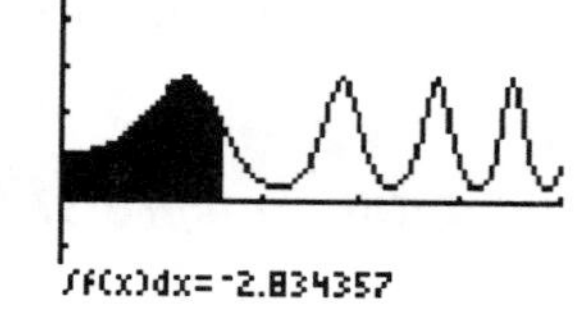	$\int_{\pi/2}^{0} e^{\sin x^2} dx = -\int_{0}^{\pi/2} e^{\sin x^2} dx =$ -0.2.834357
$\int_{0}^{\pi/2} e^{\sin x^2} dx$	$\int_{\pi/2}^{0} e^{\sin x^2} dx$	• The integral from $\pi/2$ to 0 is the negative of the integral from 0 to $\pi/2$. • $\int_{b}^{a} f(x)dx = -\int_{a}^{b} f(x)dx$.

b) If the definite integral of $f(x) = e^{\sin x^2}$ from 0 to $\pi/2$ is 2.834 and the integral from $\pi/2$ to π is 1.865, find the definite integral from 0 to π. Write the solution in symbols and demonstrate with a graph.

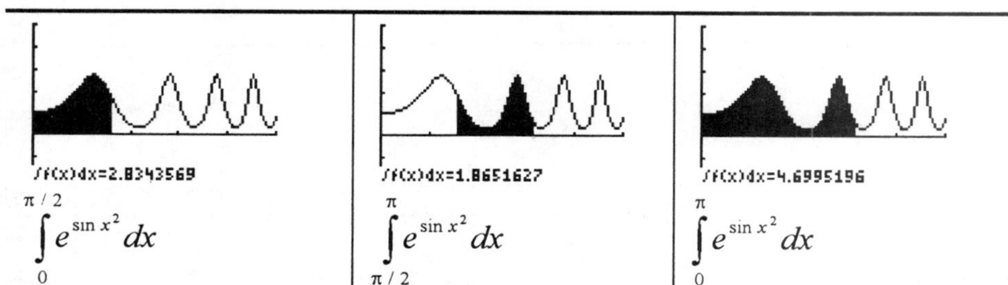

$$\int_0^{\pi/2} e^{\sin x^2}dx \qquad \int_{\pi/2}^{\pi} e^{\sin x^2}dx \qquad \int_0^{\pi} e^{\sin x^2}dx$$

$$\int_0^{\pi/2} e^{\sin x^2}dx + \int_{\pi/2}^{\pi} e^{\sin x^2}dx = \int_0^{\pi} e^{\sin x^2}dx \text{ or } \int_0^{\pi} e^{\sin x^2}dx = 2.834 + 1.865 = 4.699$$

c) Find the area described by $\int_0^2 (x+1)dx$.

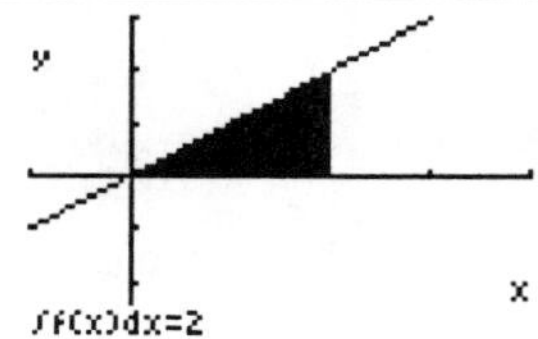	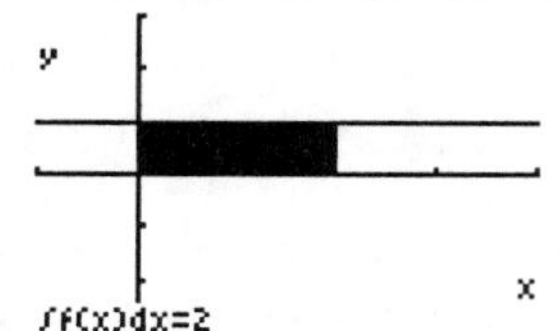	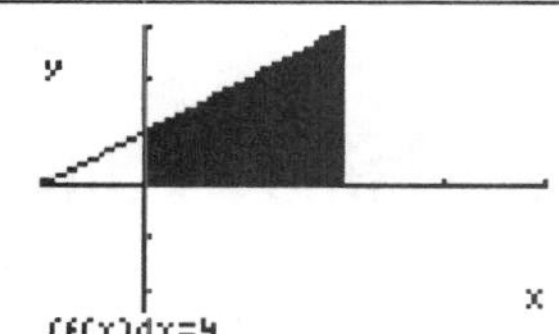
∫f(x)dx=2	∫f(x)dx=2	∫f(x)dx=4
The area formed by $y = x$ over [0,2] is a triangle with area (1/2)(2)(2)=2, so $\int_0^2 xdx = 2$.	The area formed by $y = 1$ over [0,2] is the area of a rectangle (2)(1)=2, so $\int_0^2 1dx = 2$.	The total area is the sum of the triangle and rectangle: $\int_0^2 (x+1)dx = \int_0^2 xdx + \int_0^2 1dx = 2 + 2 = 4$

d) Find the area of described by $\int_0^2 3xdx$.

The area for $y = x$ over [0,2] as seen from above was a triangle A= (1/2)(2)(2)=2, therefore the area of $y = 3x$ over [0,2] must be three times as large.

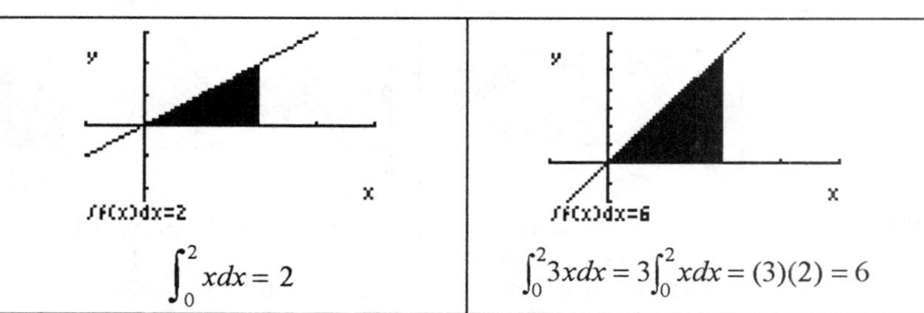

$\int_0^2 xdx = 2$	$\int_0^2 3xdx = 3\int_0^2 xdx = (3)(2) = 6$

3. *List the two theorems about the comparison of definite integrals.*

✍ Write the theorems *(See text p. 249.)*

1. ______________________________

2. ______________________________

On Your Own

Show that $6 \le \int_0^2 (3 + e^{-x})\,dx \le 8$.

(Answer appears at this end of this section.)

Key Notation

- If $m \le f(x) \le M$ for $a \le x \le b$, then $m(b-a) \le \int_a^b f(x)dx \le M(b-a)$. The meaning:

  "Let f be a continuous function, where m *and* M are lower and upper bounds for the function on the interval $[a,b]$. These bounds can be represented as straight horizontal lines beginning at a and ending at b. The area under the curve f must be more than the area of a rectangle $m(b-a)$, where m is the height and $(b-a)$ is the width of the rectangle and the area under the curve f must be less than the area of a rectangle $M(b-a)$, where M is the height and $(b-a)$ is the width of the rectangle." *(See diagram in text p. 249.)*

Key Study Skills

- Choose at least two problems that represent this section; put each on a 3 x 5 note card with the solution on the back. Review all the note card problems from Chapter Three.
- Use the ✓ ? ★ system on homework problems and get answers to all ★ problems.
- Review the **Key Ideas** from sections 5.1, 5.2, 5.3 and 5.4.
- Do the Chapter Three Review problems in the text.
- Make up your own test for Chapter Three from your note card problems and the Chapter Three Review problems.

Answer to On Your Own

If $f(x) = 3 + e^{-x}$, we know that $f(0) = 4$ and $f(x)$ is a decreasing function such that $\lim_{x \to \infty} f(x) = 3$, so $3 < f(x) \le 4$. The area under the graph of $f(x)$ is between the horizontal lines y = 3 and y=4 over [0,2]. See the graphs below.

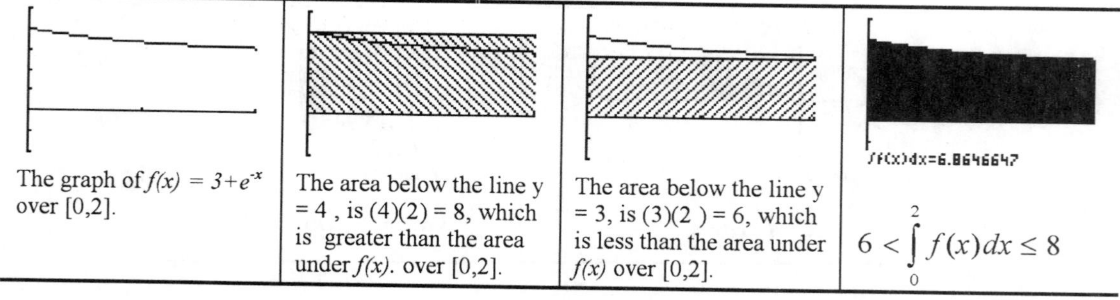

The graph of $f(x) = 3 + e^{-x}$ over [0,2].

The area below the line y = 4 , is (4)(2) = 8, which is greater than the area under $f(x)$. over [0,2].

The area below the line y = 3, is (3)(2) = 6, which is less than the area under $f(x)$ over [0,2].

$6 < \int_0^2 f(x)dx \le 8$

CUMULATIVE REVIEW CHAPTER FIVE

Key Study Skills

- Review all of the **Key Ideas** from 5.1 - 5.4.
- Do the Review Problems for Chapter Five in the text.
- Make up your own test from the 3 x 5 note card problems.
- Go back and redo all the homework problems that you put a ★ beside.
- Do the Chapter Five practice test.

Chapter Five Practice Test

1. The rate at which a river was rising during a flood was modeled by the function *r(t)* graphed below. *r(t)* is measured in inches per hour and time *t* is measured in hours.

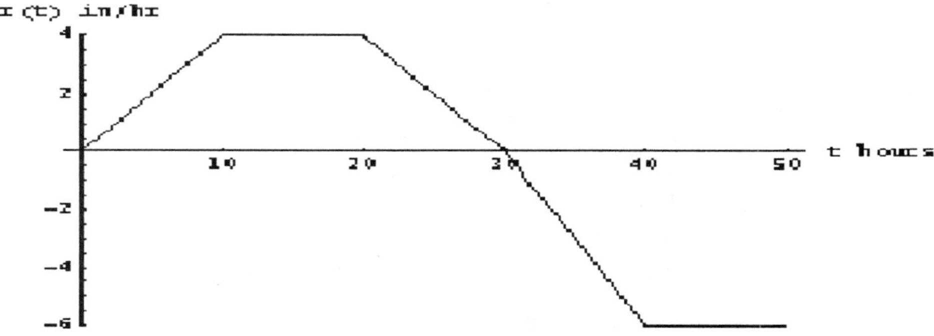

The rate of a river rising.

1a. If *h(t)* measures the height of the river above flood stage (in inches), and if at $t = 0$ the river was at flood stage, express *h (t)* in terms of an integral of *r(t)*.

1b. Determine the height of the river above flood stage at $t = 10, 20,30\ 40\ ,50$ hours.

1c. Sketch the graph of the height of the river above flood stage, *h(t)*, over the 50 hour time period.

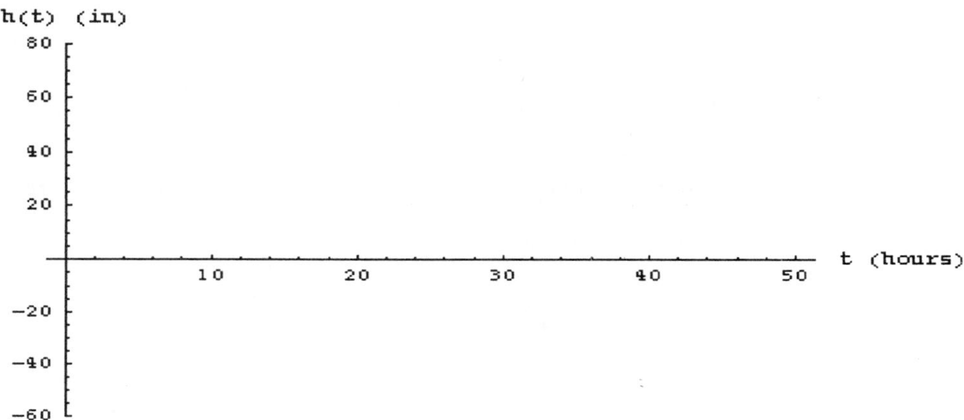

1d. Determine when the river was at a maximum height above flood stage and find the maximum height.

1e. At which of the time(s): 10,20,30, 40 or 50 hours is the river below flood stage? Explain.

2. Below is the graph of $f(x) = (5x)e^{-2x} + 5$ and a list of some values of $f(x)$ over the interval [-1,5], at 0.5 increments.

x	-1	-0.5	0	0.5	1	1.5	2	2.5	3	5.5	4	4.5	5
$f(x)$	-31.9452	-1.7957	5	5.9197	5.6767	5.3734	5.1832	5.0842	5.0372	5.016	5.0067	5.0028	5.0011

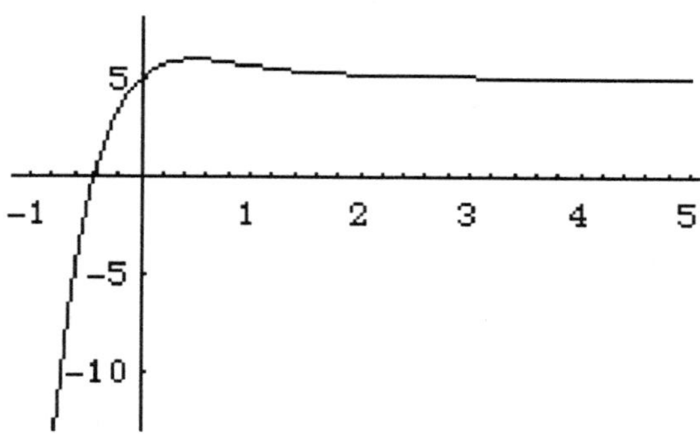

The graph of $f(x) = (5x)e^{-2x} + 5$

2a. Explain why the following is true or false.

$$\int_{-1}^{0} f(x)dx < \int_{0}^{1} f(x)dx$$

2b. Determine an approximation for $\int_{1}^{5} f(x)dx$ using left and right sums , with n= 4.

(Be sure to write your formula for calculating the sums.)

2c. How far from the true value of the definite integral could your estimate be? What is your error?

(The answers appear at the end of this Study Guide.)

CHAPTER SIX

CONSTRUCTING ANTIDERIVATIVES

We will learn to reconstruct a function from its derivative. This will be done graphically by analyzing the derivative function, numerically by calculating the area under the graph of the function and analytically by using the properties of antiderivatives.

6.1 ANTIDERIVATIVES GRAPHICALLY AND NUMERICALLY

Objective: To use the graph of the derivative to work backwards to reconstruct original function, $\pm$ a constant.

Preparations: Read and take notes on section 6.1.
Suggested practice problems: 1-8, 11, 13, 15, 17, 18, 23.

✠ Key Ideas of the Section

You should be able to:

- Use the derivative to find when the original function is increasing or decreasing.
- Graphically reconstruct the original function F, $\pm$ .a constant, from the graph of its derivative function f.
- Numerically reconstruct the original function F, $\pm$ .a constant, by calculating the area under the graph of f.
- Use the Fundamental Theorem of Calculus, $\int_a^b f(x)dx = F(b) - F(a)$, to find the graph of the antiderivative F, of a continuous function f.

Key Questions to Ask and Answer

1. ***What is an antiderivative and how do I find it?***

✍ ____

___________________________________ *(See text p. 262-264.)*

If the derivative of F is f, $F'(x) = f(x)$, then F is called an *antiderivative* of f.

Special note: There are many antiderivatives F, of the function f since adding a constant C to F does not change the shape of its graph. Adding C only shifts the graph of F up or down.

In this section we are given the graph of the derivative f and asked to find the graph of an original function F. The original function F is an antiderivative function.

We can make a sketch of an antiderivative function F if we know information about the derivative $f = F'$

- if $f > 0$ on an interval then F *is increasing* on the interval.
- if $f < 0$ on an interval then F *is decreasing* on the interval.
- if $f = 0$ at a point then F *has a critical point (possible local max. or min.).*

Example 6.1.1

Determine the shape of the graph of an antiderivative Y, using the graph of the derivative Y' below.

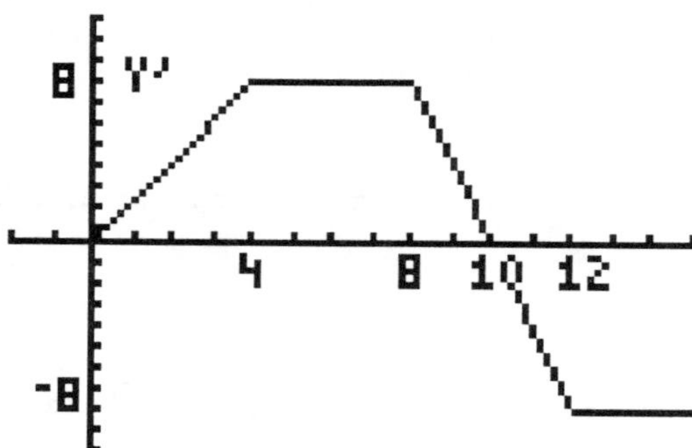

Note: This is the graph of the derivative.

Interval	*Behavior of Y'*	*Behavior of Y*
$0 < x < 4$	$Y' > 0$ and increasing	Increasing and concave up
$4 < x < 8$	$Y' > 0$ and constant	Increasing and a straight line
$8 < x < 10$	$Y' > 0$ and decreasing	Increasing and concave down
$x = 10$	$Y' = 0$	Local maximum
$10 < x < 12$	$Y' < 0$ and decreasing	Decreasing and concave down
$12 < x < 15$	$Y' < 0$ and constant	Decreasing and straight line

2. ***How do we use the Fundamental Theorem of Calculus to sketch the graph of an antiderivative?***

✍ ____

____ *(See text p. 263-264.)*

In chapter 5 we learned that the Fundamental Theorem of Calculus allows us to accumulate the area under the graph of the derivative curve f, counted either positive or negative, over an interval $[a,b]$ and the definite integral will represent the total change in the graph of the antiderivative F.

The *Fundamental Theorem of Calculus* :

If F is the antiderivative of a continuous function f, then

$$\int_a^b f(x)dx = F(b) - F(a)$$

Example 6.1.2

Sketch the graph of the antiderivative Y, from the graph of the derivative Y' in Example 6.1.1. If it is given that $Y(0) = 0$, sketch the graph of $Y(x)$ showing all critical points and inflection points and giving their coordinates.

a. From geometry we can determine the area under the graph of Y'. Please note that area below the x-axis is counted negative.

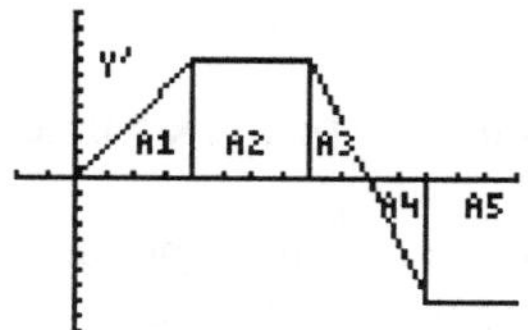

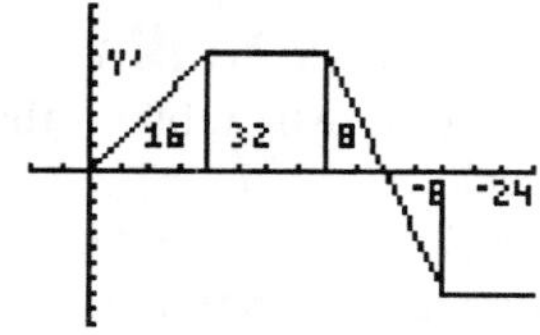

b. From the Fundamental theorem of Calculus we know that

$$\int_0^b Y'(x)dx = Y(b) - Y(0) \Rightarrow Y(b) = Y(0) + \int_0^b Y'(x)dx = \text{starting value of} Y + \text{signed area } [0, b]$$

The starting point is (0, 0), since the starting value is Y = 0. We can determine the values of the antiderivative by accumulating the <u>signed</u> areas over certain intervals.

$$Y(4) = 0 + \int_0^4 Y'(x)dx = 0 + A_1 = 0 + 16 = 16$$

$$Y(8) = 0 + \int_0^8 Y'(x)dx = 0 + A_1 + A_2 = 0 + 16 + 32 = 48$$

$$Y(10) = 0 + \int_0^{10} Y'(x)dx = 0 + A_1 + A_2 + A_3 = 0 + 16 + 32 + 8 = 56$$

$$Y(12) = 0 + \int_0^{12} Y'(x)dx = 0 + A_1 + A_2 + A_3 - A_4 = 0 + 16 + 32 + 8 - 8 = 48$$

$$Y(15) = 0 + \int_0^{15} Y'(x)dx = 0 + A_1 + A_2 + A_3 - A_4 - A_5 = 0 + 16 + 32 + 8 - 8 - 24 = 24$$

c. Plot the points on the antiderivative graph, sketch the graph using the behavior of the antiderivative from Example 1.

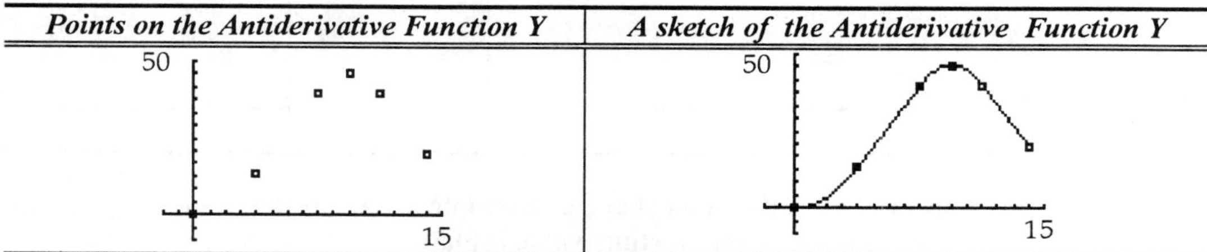

✎ On Your Own

Sketch the graph of the antiderivative as in Example 6.1.1, but let the starting point be (0,5).

(The answer appears at the end of this section)

Key Notation

- $F(b) = F(a) + \int_a^b f(x)dx$. This is an alternative form of the Fundamental Theorem of Calculus. It allows you to find a value of an antiderivative function.
- If F is an antiderivative of the function f, then $F' = f$ and $F(a)$ represents the starting value of an antiderivative function.

Key Study Skills

- Choose at least two problems that represent this section; put each on a 3 x 5 note card with the solution on the back.
- Use the ✓ ? ★ system on homework problems and get answers to all ★ problems.
- Problems 1-21 in section 6.1 of the text will give you good practice in sketching the graph of antiderivative functions.

Answer to On Your Own

Since the starting point is (0, 5) we can determine the points on the antiderivative by accumulating the signed areas over certain intervals.

$$Y(4) = 5 + \int_0^4 Y'(x)dx = 5 + A_1 = 5 + 16 = 21$$

$$Y(8) = 5 + \int_0^8 Y'(x)dx = 5 + A_1 + A_2 = 5 + 16 + 32 = 53$$

$$Y(10) = 5 + \int_0^{10} Y'(x)dx = 5 + A_1 + A_2 + A_3 = 5 + 16 + 32 + 8 = 61$$

$$Y(12) = 5 + \int_0^{12} Y'(x)dx = 5 + A_1 + A_2 + A_3 + A_4 = 5 + 16 + 32 + 8 + (-8) = 53$$

$$Y(15) = 5 + \int_0^{15} Y'(x)dx = 5 + A_1 + A_2 + A_3 + A_4 + A_5 = 5 + 16 + 32 + 8 + (-8) + (-24) = 29$$

Points on the Antiderivative Function beginning at (0,5)	***A Sketch of the Antiderivative function.***
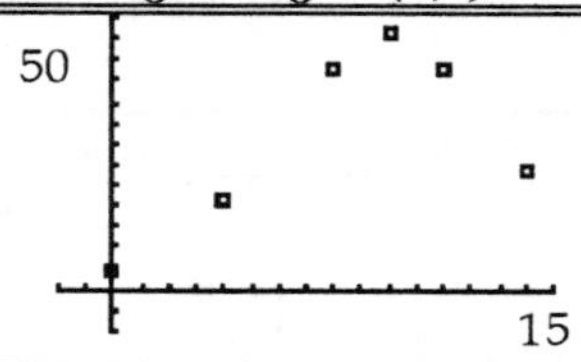 50 15	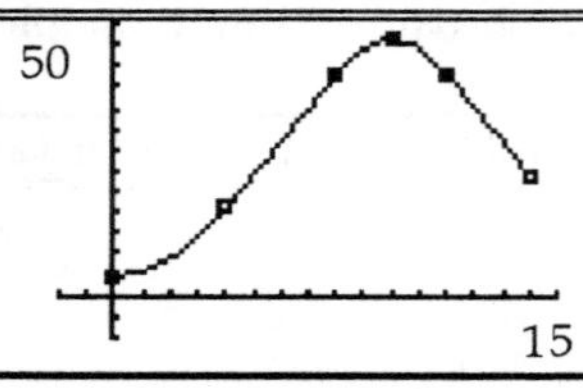 50 15

Note: **This graph has the exact shape as Example 6.1.2, but the whole graph is shifted up five units because the starting value is (0,5).**

6.2 CONSTRUCTING ANTIDERIVATIVES ANALYTICALLY

Objective: To identify the functions that have elementary antiderivatives and to find indefinite integrals of sums and differences.

Preparations: Read and take notes on section 6.2.
Suggested practice problems: 1-9, 13, 16, 18, 25, 31, 35, 37, 41, 53, 55, , 67, 72,77, 81, .

✠ Key Ideas of the Section

You should be able to:

- Identify functions that have elementary antiderivatives of powers of x, e^x, $\ln x$, $\sin x$, $\cos x$.
- Recognize that the antiderivative of an elementary function f is a family of antiderivatives called the *indefinite integral* : $\int f(x)dx = F(x) + C$
- Compare and contrast the definite integral, $\int_a^b f(x)dx$, which is a number and the indefinite integral, $\int f(x)dx$, which is a family of functions.
- Find the indefinite integrals of constants, sums, and differences.

Key Questions to Ask and Answer

1. *What is the indefinite integral and how do you find it?*

✍ ____

____________________________________ *(See text p. 268.)*

We have seen from the previous section that the definite integral is a number. We now want to find a formulas for antiderivative functions, if possible. The indefinite integral is a family of antiderivative functions.

antiderivative of f(x) = *indefinite integral* = $\int f(x)dx = F(x) + C$, where

f(x) is called the *integrand,* and $F'(x) = f(x)$.

If you have a function f as an integrand which is recognized as a derivative, then you can find the antiderivative easily.

✍ Your Turn

Consult the text p.268-270 and write the antiderivative rules for the following.

1. $\int k dx =$ ____________________

In words: The antiderivative of a constant is the constant times x, *plus a constant.*

2. $\int x^n dx =$ ____________________

In words: The antiderivative of a power function is the power increased by one, divided by the new power, *plus a constant.*

3. $\int \frac{1}{x} dx =$ ______________________________

In words: The antiderivative of *1/x* is *ln |x|*, *plus a constant.*

4. $\int e^x dx =$ ______________________________

In words: The antiderivative of e^x is itself, *plus a constant.*

5. $\int \cos x dx =$ ______________________________

In words: The antiderivative of *cos x* is *sin x*, *plus a constant.*

6. $\int \sin x dx =$ ______________________________

In words: The antiderivative of *sin x* is *-cos x, plus a constant.*

Example 6.2.1

Find the antiderivative of the following:

Derivative	*Antiderivative*
$f(x) = -7$	$F(x) = -7x + C$
$g(x) = x^4$	$G(x) = \frac{x^5}{5} + C$
$h(x) = \frac{1}{x^3}$	$H(x) = \frac{x^{-2}}{-2} + C = -\frac{1}{2x^2} + C$
$k(x) = 2\sin 2x$	$K(x) = -\cos 2x + C$

Note: **To check that the antiderivative is correct, take the derivative and see if it is the original integrand function.**

2. *What are the properties of antiderivatives that involve Sums and Constant Multiples?*

✎ Fill in the missing parts *(See text p. 270.)*

The antiderivative of sums:

$\int [f(x) \pm g(x)] dx =$ __________________________

In words: ____

The antiderivative of constant multiples:

$\int cf(x) dx =$ ______________________________

In words: ____

Example 6.2.2

Find the general antiderivative in the following problems.

Indefinite Integral	*Answers*
a. $\int\left(\frac{2}{x}-\frac{5}{x^3}\right)dx$	$=2\int\frac{1}{x}\,dx-5\int x^{-3}dx=2\,ln\,|\,x\,|+\frac{5}{2x^2}+C$
b. $\int \pi x^5 dx$	$=\pi\int x^5dx=\frac{\pi}{6}x^6+C$
c. $\int(\frac{sin\,x}{2}-3\,cos\,x)dx$	$=\frac{1}{2}\int sinxdx-3\int cosxdx=-\frac{1}{2}\,cosx-3\,sinx+C$
d. $\int\frac{e^x}{e}\,dx$	$=\frac{1}{e}\int e^xdx=\frac{e^x}{e}+C=e^{x-1}+C$

Note: Check each antiderivative by taking the derivative.

3. How can you use the fundamental theorem of calculus with elementary antiderivatives ?

✍ ____

____ *(See text p. 271)*

Example 6.2.3

Determine which is larger: $\int_0^1 x^2dx$ or $\int_0^1 x^3dx$.

a. Using a graphical method, we see that over $[0,1]$, x^3 lies under x^2,

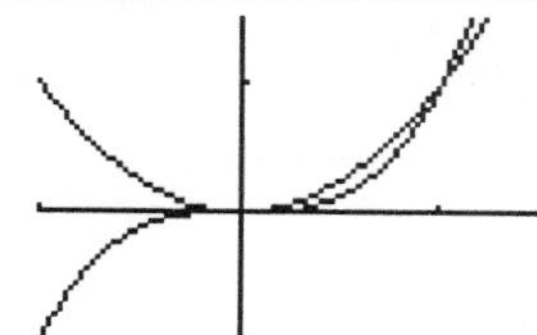

The graph of x^2 and x^3.

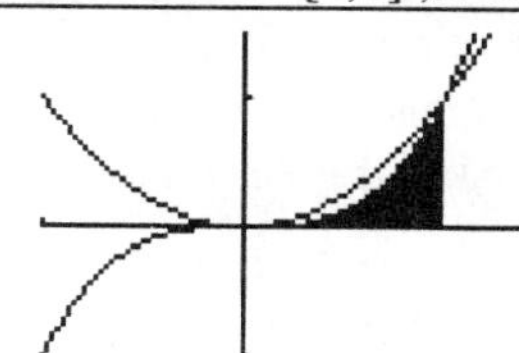

Area under x^3 for $[0,1]$.

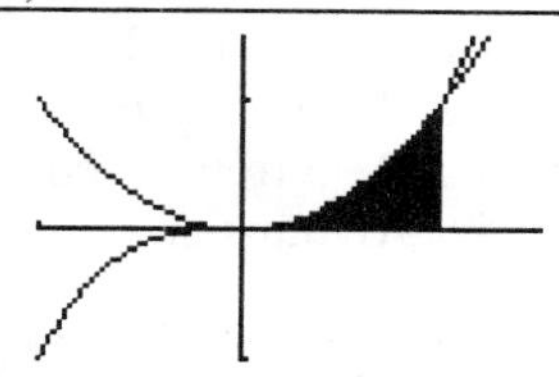

Area under x^2 for $[0,1]$.

therefore the area under x^3 would be smaller than the area under x^2, so $\int_0^1 x^2dx > \int_0^1 x^3dx$.

b. Using an analytic method, you find the elementary antiderivative and then calculate the definite integral using the Fundamental Theorem of Calculus, which would be the precise area.

$$\int_0^1 x^2dx=\frac{x^3}{3}\Big|_0^1=\frac{1^3}{3}-\frac{0^3}{3}=\frac{1}{3}$$

$$\int_0^1 x^3dx=\frac{x^4}{4}\Big|_0^1=\frac{1^4}{4}-\frac{0^4}{4}=\frac{1}{4}$$

therefore:

$$\int_0^1 x^2dx>\int_0^1 x^3dx$$

✎ On Your Own

Using the Fundamental Theorem of Calculus, evaluate $\int_1^3 \left(\frac{(1-t)^2}{t}\right)dt$ both exactly and numerically.

(The answer appears at the end of this section.)

Key Notation

- $\int_a^b f(x)dx = F(x)\Big|_a^b = F(b) - F(a)$. This notation is used when you can find a formula for the elementary antiderivative *F(x)*. It means that you should substitute the upper bound into the antiderivative, then substitute the lower bound and take the difference.
- $\int f(x)dx = F(x) + C$. The indefinite integral. The antiderivative is a family of functions.
- $\int_a^b f(x)dx = F(b) - F(a)$. The definite integral. The definite integral is a *number* that represents the total accumulated area between the *x*-axis and the curve of *f(x)* over [a,b], and it also represents the vertical distance between two points on the graph of the antiderivative function.
- $\int f(x)dx$ is a single notation. You never see the integrand without the *dx*. The *dx* tells you that the independent variable is *x*.
- Since $f = F'$, remember $\int_a^b f(x)dx \neq f(b) - f(a)$. Students sometimes make the mistake of substituting the bounds into the integrand rather than the antiderivative function.

Key Study Skills

- Choose at least two problems that represent this section; put each on a 3 x 5 note card with the solution on the back. Review all problems from sections 6.1-6.2.
- Use the ✓ ? ★ system on homework problems and get answers to all ★ problems.
- Problems 42-73 from section 6.2 in the text will give you excellent practice in finding the antiderivative function.

Answer To On Your Own

When you can not readily find an antiderivative, expand the function algebraically then simplify.

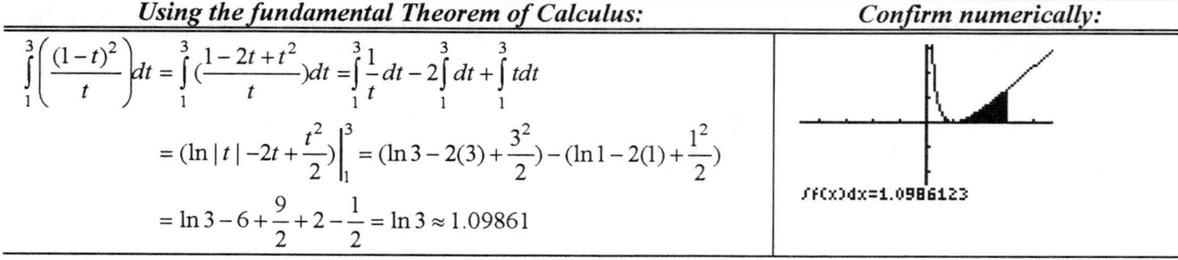

Using the fundamental Theorem of Calculus:	*Confirm numerically:*
$\int_1^3 \left(\frac{(1-t)^2}{t}\right)dt = \int_1^3 \left(\frac{1-2t+t^2}{t}\right)dt = \int_1^3 \frac{1}{t}dt - 2\int_1^3 dt + \int_1^3 tdt$ $= (\ln\lvert t\rvert - 2t + \frac{t^2}{2})\Big|_1^3 = (\ln 3 - 2(3) + \frac{3^2}{2}) - (\ln 1 - 2(1) + \frac{1^2}{2})$ $= \ln 3 - 6 + \frac{9}{2} + 2 - \frac{1}{2} = \ln 3 \approx 1.09861$	∫f(x)dx=1.0986123

6.3 DIFFERENTIAL EQUATIONS

Objective: To solve a differential equation using an initial condition in the antiderivative.

Preparations: Read and take notes on section 6.3.
Suggested practice problems: 3, 4, 5, 7, 12, 13, 14, 19, 23.
Do the **Project** problems on p. 288.

✠ Key Ideas of the Section

You should be able to:

- Identify that antidifferentiation is the process of "working backwards" from the derivative and that this process produces a family of antiderivatives.
- Find the solution to a differential equation that represents a unique antiderivative due to a given initial condition.
- Find a position equation using a differential equation for velocity.

Key Questions to Ask and Answer

1. What is a differential equation?

✍ ____

____ *(See text p. 273-276)*

A differential equation is a rate of change equation. To solve the differential equation of the type: $\frac{dy}{dx} = f(x)$:

- find a family of antiderivatives $y = F(x) + C$, then
- solve for the constant C if you know a unique value of the antiderivative $y(x_0) = y_0$.

This will give you a unique antiderivative $y = G(x)$.

Example 6.3.1

A person throws a ball straight up from a position 3 ft. above the ground. What initial velocity would allow the ball to reach 100 ft after 2 seconds? How long would it take to hit the ground?

a. We know that gravity causes a uniform downward acceleration of 32 ft/sec^2 and acceleration is the derivative of velocity. Working backwards we can find a velocity equation, so our differential equation is:

$$acceleration = \frac{dv}{dt} = -32$$

$$velocity = v = \int -32dt = -32t + v_0$$

where v_0 = *the initial velocity* , a constant.

b. Velocity is the derivative of position (distance above the ground). Working backwards we can find the position equation, so our differential equation is:

$$velocity = \frac{ds}{dt} = -32t + v_0$$

$$position = s = \int (-32t + v_0)dt$$

$$s = \frac{-32t^2}{2} + v_0 t + s_0 = -16t^2 + v_0 t + s_0$$

c. We know the initial position $s_0 = 3\ ft$ and the ball reaches a height $s = 100$ at time $t = 2$. Substituting into the position equation we can find the initial velocity.

$$100 = -16(2)^2 + v_0(2) + 3$$
$$v_0 = 80.5\,ft/\sec$$

d. When the ball hits the ground the position is $s = 0$. Now solve the position equation for t.

$$0 = -16t^2 + 80.5t + 3$$

Using the quadratic formula:

$t = 5.07$ sec, The solution $t = -.04$ sec. is not applicable.

e. A ball thrown straight up 3 ft off the ground needs an initial velocity of 80.5 ft/sec to reach a height of 100 ft after two seconds and will hit the ground after 5.07 sec.

f. Confirm graphically:

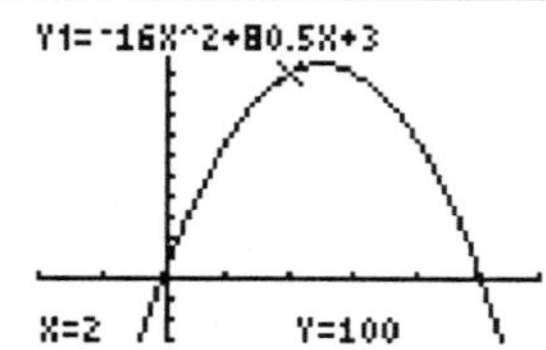 An initial velocity if 80.5 ft/sec allows the ball to reach a height of 100 ft at 2 sec.	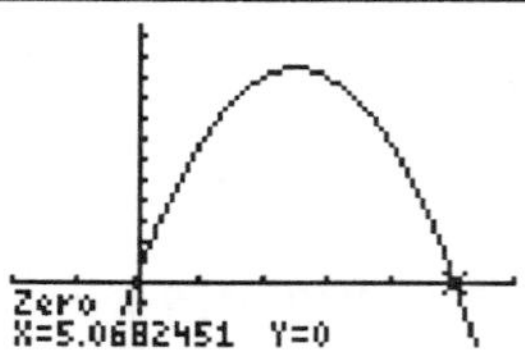 The ball hits the ground at 5.07 sec.

Key Notation

- $\dfrac{dy}{dx} = f(x)$ is a differential equation. Finding the general solution of the differential equation means finding the general antiderivative $y = \int f(x)dx = F(x) + C$.

🗁 Key Study Skills

- Choose at least two problems that represent this section; put each on a 3 x 5 note card with the solution on the back. Review all problems from sections 6.1-6.3.
- Use the ✓ ? ★ system on homework problems and get answers to all ★ problems.

6.4 SECOND FUNDAMENTAL THEOREM OF CALCULUS

Objective: To use the construction theorem for antiderivatives.

Preparations: Read and take notes on section 6.4.
Suggested practice problems: 1, 4, 6, 7, 12, 13, 16.

✠ Key Ideas of the Section

You should be able to:

- Recognize that many functions DO NOT have elementary antiderivatives.
- Use the construction theorem to define and analyze an antiderivative that is not elementary.

Key Questions to Ask and Answer

1. ***How can we find an antiderivative of a function that does not have an elementary antiderivative?***

✍ ____

____ *(See text p. 278-280.)*

We have seen in this chapter that formulas for some antiderivatives are easy to find because we recognize that the function f is the derivative of F. These are usually called elementary antiderivatives. The rules stated in 6.2 are for elementary antiderivatives. When we cannot find a formula for F, we use the definite integral to construct the antiderivative. Generally we create the new function piece by piece numerically and can use a grapher to help in this process. What allows us to do this however is the *Second Fundamental Theorem of Calculus* or

The Construction Theorem for Antiderivatives:

If f is a continuous function on an interval, and if a is any number in that interval, then the function F, defined by

$$F(x) = \int_a^x f(t)dt$$

is an antiderivative of f with $F(a) = 0$.

Example 6.4.1

Find an antiderivative of $f(x) = e^{\sin 2x}$.

a. Find the antiderivative F that satisfies $F(0) = 0$.

$F(x) = \int_0^x e^{\sin 2t}dt$. *F(x) is the antiderivative defined by the integral.*

This antiderivative has the values listed below, at $F(1)$, $F(2)$, $F(3)$, $F(4)$ and $F(5)$ over the interval $[0,5]$. The graph of $F(x)$ can be constructed from the values.

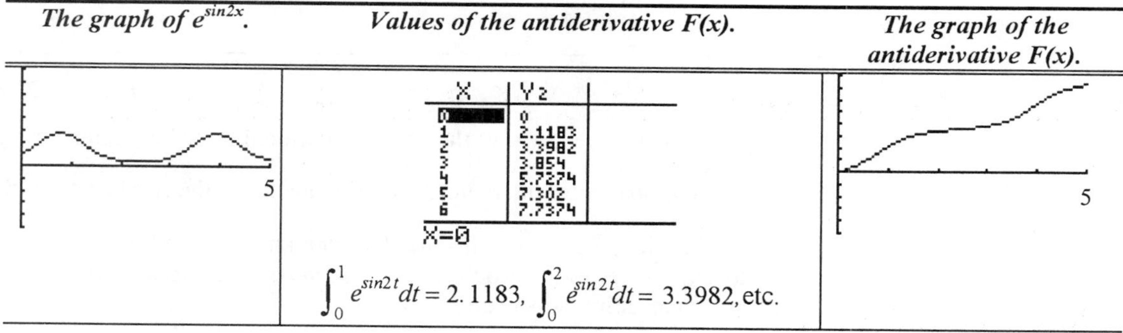

The graph of $e^{\sin 2x}$.
Values of the antiderivative F(x).
The graph of the antiderivative F(x).
5
X | Y2
0 | 0
1 | 2.1183
2 | 3.3982
3 | 3.854
4 | 5.7274
5 | 7.302
6 | 7.7374
X=0
$\int_0^1 e^{\sin 2t}dt = 2.1183$, $\int_0^2 e^{\sin 2t}dt = 3.3982$, etc.
5

The key here is that we know values for $F(x)$ and we can sketch the graph of $F(x)$, but the function $F(x)$ must be written in integral form.

Key Notation

- $F(x) = \int_a^x f(t)dt$. This is the Construction Theorem that enables us to write down the antiderivative of functions that do not have elementary antiderivatives.

Key Study Skills

- Choose at least two problems that represent this section; put each on a 3 x 5 note card with the solution on the back. Review all problems from sections 6.1-6.4.
- Use the ✓ ? ★ system on homework problems and get answers to all ★ problems.

6.5 THE EQUATIONS OF MOTION

Objective: To use the change in velocity to understand motion, rather than a change in position.

Preparations: Read and take notes on section 6.5.
Suggested practice problems: 3, 6, 9.

Key Ideas of the Section

You should be able to:

- Recognize that motion was difficult to understand because scientists were trying to develop ideas of motion based on change of position rather than change in velocity.
- Newton's understanding of motion was based on a *change in velocity* or *acceleration* .
- *Force* causes velocity to change; *gravity* is the force.
- Galileo showed that the *mass* of an object did not play a part in the equation of motion, i.e. all objects regardless of weight (ignoring friction) fall with constant acceleration.

Key Questions to Ask and Answer

1. What is Newton's law of gravity and how did he define force?

✍ ____

____ *(See text p. 282,283.)*

Newton's famous inverse square law says that the gravitational force between bodies is: $F = \dfrac{GMm}{r^2}$ where G is a gravitational constant m and M are the masses of the two bodies, and r is the distance between them. Near to the earth the distance between an object and the Earth hardly changes so the force is a constant. Newton also defined force as *mass* times *acceleration* or $F = ma$, so the gravitational force acting downward is:

$$\frac{GMm}{r^2} = ma \text{ or } \frac{GM}{r^2} = a .$$

Since *acceleration* is the second derivative of position s

$$a = \frac{d^2s}{dt^2} = \frac{GM}{r^2} = g ,$$

the gravitational constant.

Example 6.5.1

What is the acceleration due to gravity 100,000 meters from the ground, if the radius of the earth is 6.4 x 10^6 meters, and the acceleration due to gravity 2 ft above the surface of the earth is 9.8m/sec^2?

Using $a = \frac{d^2s}{dt^2} = \frac{GM}{r^2} = g$

$$\frac{GM}{(6.4x10^6 + 2)^2} = 9.8m/sec^2$$

at 100,000 meters $g_{new} = \frac{GM}{r^2} \Rightarrow g_{new} = \frac{GM}{(6.4x10^6 + 100{,}000)^2}$

$$\frac{g_{new}}{9.8} = \left(\frac{GM}{\left(6.4x10^6 + 100{,}000\right)^2}\right) \Big/ \left(\frac{GM}{\left(6.4x10^6 + 2\right)^2}\right)$$

$$g_{new} = 9.8\left(\frac{6{,}400{,}002}{6{,}500{,}000}\right)^2 = 9.5 \text{ m/sec}^2$$

Key Notation

- The position function is $s = -\frac{gt^2}{2} + v_0t + s_0$, where g is acceleration due to gravity, v_0 the initial velocity and s_0 the initial position.

Key Study Skills

- Choose at least two problems that represent this section; put each on a 3 x 5 note card with the solution on the back. Review all problems from sections 6.1-6.4.
- Use the ✓ ? ★ system on homework problems and get answers to all ★ problems.

CUMULATIVE REVIEW CHAPTER SIX

Key Study Skills

- Review all of the **Key Ideas** from 6.1 - 6.5.
- Do the Review Exercise problems from the text Chapter Six p. 284-287.
- Do Check Your Understanding from the text p. 287-288.
- Make up your own test from the 3 x 5 note card problems.
- Go back and redo all the problems that you had ★ (starred).
- Do the Chapter Six Practice Test.

Chapter Six Practice Test

1. For the function f below, sketch two functions F such that $F' = f$. In one case let $F(0) = 0$ and in the other let $F(0)=2$.

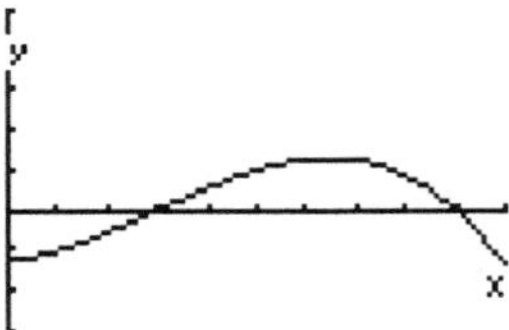

2. a. Find the total area, without regard to sign, bounded by the graph of the function $f(x) = -x^3 + 7x^2 - 10x$ and the x-axis.

   b. Evaluate the definite integral: $\int_0^5 \left(-x^3 + 7x^2 - 10x\right)dx$

3. Consider the initial value problem: $\dfrac{dy}{dt} = \cos^2 t,\ y(0) = 0$

   a. Is $y(t) = \dfrac{1}{3}\cos^3 t$ a solution to this problem? Why or why not?

   b. From the graph of $\dfrac{dy}{dt} = \cos^2 t,\ y(0) = 0$, sketch a graph of $y(t)$.

   c. Use the trig identity: $\cos^2 t = \dfrac{1}{2}(1 + \cos 2t)$ to find a formula for $y(t)$ and show that it satisfies the initial value problem.

4. A stone is thrown off a 300 ft cliff with a velocity of 50 ft/sec downward.

   a. Express the stones height, $h(t)$, in feet above the ground as a function of time, t.

   b. What is the velocity function $v(t)$ and what is the velocity of the stone at $t = 1$ second?

   c. When does it hit the ground?

5. Use numerical methods to complete the table of values for the function $F(x) = \int_1^x \frac{1}{t}dt$. Describe the behavior of F for $x \geq 1$.

x	1	1.5	2	2.5	3
$F(x) = \int_1^x \frac{1}{t}dt$					

6. Match the following functions with their antiderivatives

Function	*Antiderivative*
a. 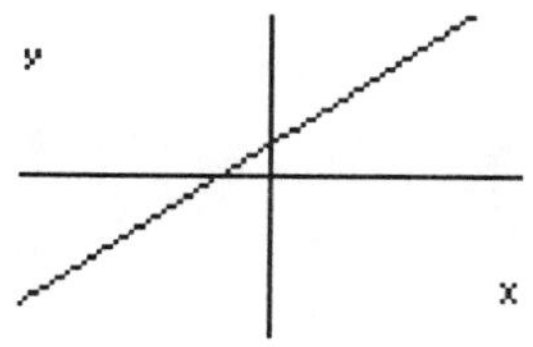	i. 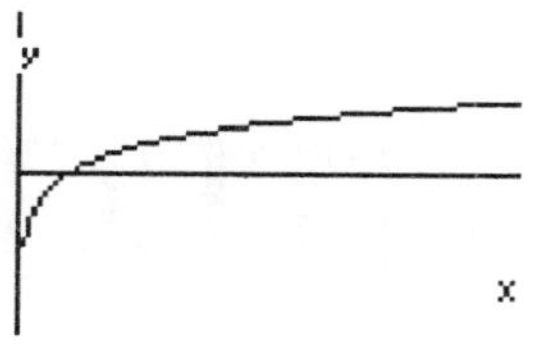
b. 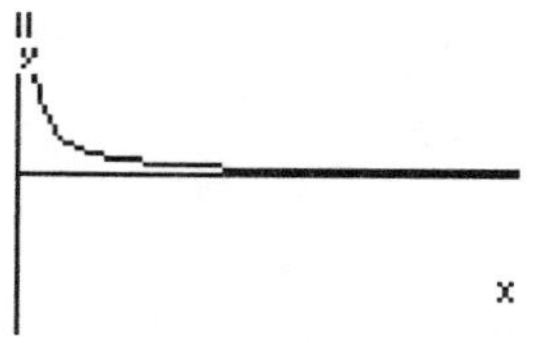	ii. 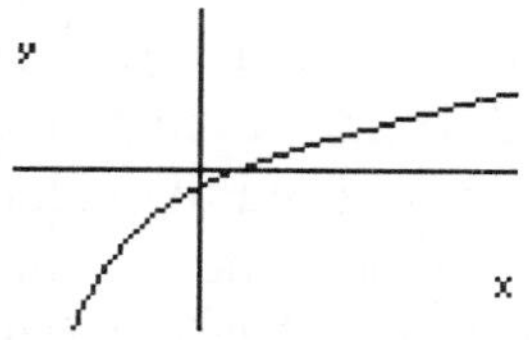
c. 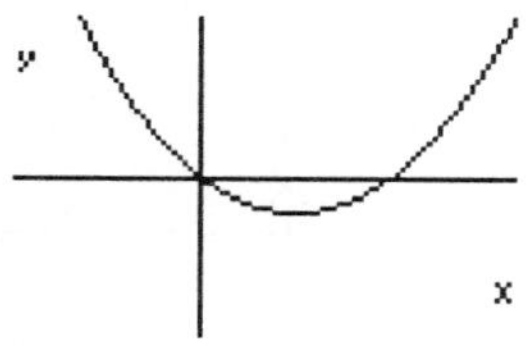	iii. 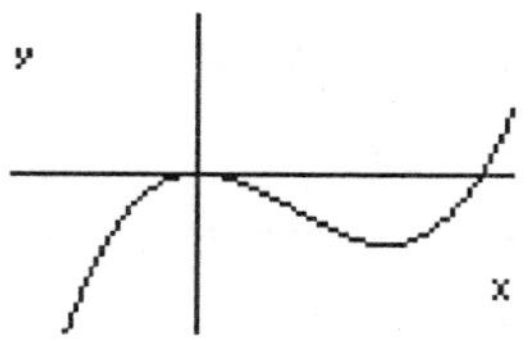
d. 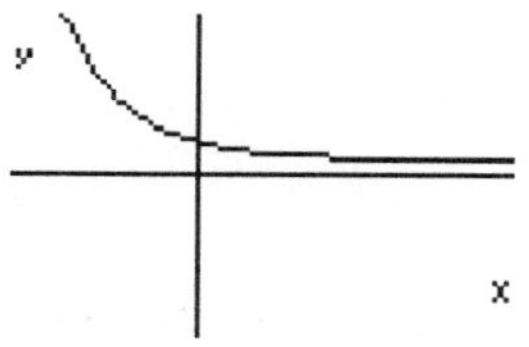	iv. 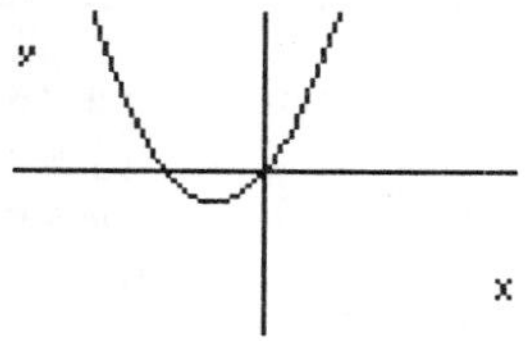

(Answers appear at the end of this study guide.)

CHAPTER SEVEN

INTEGRATION

Substitution and integration by parts are methods that can be used to find some antiderivatives. Several numerical methods are used to evaluate definite integrals, with concern about whether these give overestimates or underestimates and the size of the error. Improper integrals are such that one limit of integration is infinite or the integrand is unbounded.

7.1 INTEGRATION BY SUBSTITUTION

Objective: To find antiderivatives using substitution.

Preparations: Read and take notes on section 7.1.
Suggested practice problems: Evens 4-64, 70,75

✠ Key Ideas of the Section

You should be able to:

- Recognize patterns where antiderivatives can be found without formal substitution.
- Recognize when integrands are in the form for which the method of substitution can be used to find the antiderivative.
- Use Guess-and-Check to find some antiderivatives.
- Use formal substitution to find the antiderivatives.
- Use substitution to evaluate definite integrals by two methods.

Key Questions to Ask and Answer

1. How do I find antiderivatives of functions with a constant factor?

✍ ____

____ *(See text p. 290.)*

You are familiar with finding derivatives of functions with a constant factor. To find the antiderivative with a constant factor, you simply divide by that factor. Use the Guess-and-Check Method and look for patterns. Remember that you can always check your answer by differentiating.

Example 7.1.1

The Guess-and-Check Method is used to find the following antiderivatives. Notice that the antiderivative is divided by the factor:

a. $\int \sin(2x)dx = \frac{-1}{2}\cos(2x) + C$

b. $\int \cos(3x)dx = \frac{1}{3}\sin(3x) + C$

c. $\int e^{3x}dx = \frac{1}{3}e^{3x} + C$

d. $\int e^{1+5x}dx = \frac{1}{5}e^{1+5x} + C$

e. $\int (1-2x)^5 dx = \frac{-1}{2}\frac{(1-2x)^6}{6} = -\frac{1}{12}(1-2x)^6 + C$

f. $\int \frac{1}{\sqrt{2+5x}}dx = \int (2x+5x)^{\frac{-1}{2}}dx = \frac{1}{5}\frac{(2+5x)^{\frac{1}{2}}}{\frac{1}{2}} = \frac{2}{5}(2+5x)^{\frac{1}{2}} + C$

On Your Own 1

Use the Guess-and-Check Method to find the following antiderivatives.

1. $\int \sin(5x)dx$
2. $\int \cos(2x+3)dx$
3. $\int e^{-x}dx$
4. $\int 3e^{2x-5}dx$
5. $\int (4t+7)^{10}dt$
6. $\int \frac{1}{\sqrt[3]{6-5z}}dz$

(The answer appears at the end of this section.)

2. ***What form must the integrand be in order to use substitution to find the antiderivative?***

✍ ____

____ *(See text p. 290-293.)*

Substitution is simply reversing the Chain Rule, $\frac{d(f(g(x))}{dx} = f'(g(x))g'(x)$. Therefore, if the integrand is in the form $f'(g(x))g'(x)$, an antiderivative is $f(g(x))$. This means *g(x)* is the "inside" function, and the integrand must have the derivative of $g(x)$ as a factor

Example 7.1.2

Identify the integrals that are in the form for which the method of substitution applies.

Integral	*Inside function, g(x) and g'(x) needed*
a. $\int 2x\sin(x^2)dx$	$g(x)=x^2$ $g'(x)=2x$
b. $\int x^3\sin(x^4)dx$	$g(x)=x^4$ $g'(x)=4x^3$
c. $\int \sin z\sqrt{\cos z+1}dz$	$g(z)=\cos z+1$ $g'(z)=-\sin z$
d. $\int t^2e^{t^3}dt$	$g(t)=t^3$ $g'(t)=3t^2$
e. $\int x^3\sqrt{2+x^4}dx$	$g(x)=2+x^4$ $g'(x)=4x^3$
f. $\int x^2\sqrt{2+x^4}dx$	$g(x)=2+x^4$ $g'(x)=4x^3$ Cannot be put in the form: $f'(g(x))g'(x)$

3. *What is formal substitution and how do I use it?*

✍ ____

____ *(See text p. 290-293)*

Formal substitution is a way of systematizing the Guess-and Check Method you've been using. Substitute $w = g(x)$, the "inside" function. Then since $dw = g'(x)dx$, you get the integral exclusively in terms of w and dw, hopefully resulting in an elementary function for which you know the antiderivative.

Example 7.1.3

Use substitution to find the following general antiderivatives:

Integral	*Inside function , w*	*General antiderivative*
a. $\int 2x\sin(x^2)dx$	$w = x^2$ $dw = (2x)dx$	$\int \sin(w)dw = -\cos w + C = -\cos x^2 + C$
b. $\int x^3 \sin(x^4)dx$	$w = x^4$ $dw = (4x^3)dx$ $\frac{1}{4}dw = x^3dx$	$\int \sin(w)\frac{1}{4}dw = \frac{-1}{4}\cos w + C$ $= \frac{-1}{4}\cos x^4 + C$
c. $\int \sin z\sqrt{\cos z + 1}dz$	$w = \cos z + 1$ $dw = (-\sin z)dz$ $-dw = (\sin z)dz$	$\int \sqrt{w}(-dw) = \frac{-2}{3}w^{\frac{3}{2}} + C$ $= \frac{-2}{3}(\cos z + 1)^{\frac{3}{2}} + C$
d. $\int t^2 e^{t^3} dt$	$w = t^3$ $dw = 3t^2dt$ $\frac{1}{3}dw = t^2dt$	$\int e^w \frac{1}{3}dw = \frac{1}{3}e^w + C$ $= \frac{1}{3}e^{t^3} + C$
e. $\int x^3\sqrt{2+x^4}dx$	$w = 2 + x^4$ $dw = 4x^3dx$ $\frac{1}{4}dw = x^3dx$	$\int w^{\frac{1}{2}} \cdot \frac{1}{4}dw = \frac{1}{4}\frac{w^{\frac{3}{2}}}{3/2} + C$ $= \frac{1}{6}(2+x^4)^{\frac{3}{2}} + C$
f. $\int x^2\sqrt{2+x^4}dx$	$w = 2 + x^4$ $dw = 4x^3dx$	Cannot get the integral entirely in terms of *w* and *dw*.

✎ On Your Own 2

Use the Substitution Method, if possible, to find the following antiderivatives.

Integral
1. $\int y\sqrt{1-y^2}dy$
2. $\int \frac{1}{(4x-1)^2})dx$
3. $\int (\cos 3z)(\sqrt[3]{4-\sin 3z})dz$

4. $\int (\cos 5t)(e^{\sin t})dt$

5. $\int \frac{e^{-x}}{13+e^{-x}} dx$

(The answer appears at the end of this section)

4. *How do you evaluate a definite integral using substitution?*

✍ ____

____ *(See text p. 294.)*

To evaluate the definite integral you should either:

- compute the indefinite integral, getting the antiderivative in terms of the original variable, and then evaluate using the original limits, or
- change the original limits to limits with respect to the new variable, and evaluate in terms of this new variable.

Example 7.1.4

Evaluate the following definite integrals using each of the two methods:

Definite Integral	*Method 1*	*Method 2*
$\int_1^2 2x\sin(x^2)dx$ $w = x^2$ $dw = (2x)dx$	• Find the antiderivative in terms of the original variable. $\int 2x\sin(x^2)dx = \int \sin(w)dw$ $= -\cos w + C = -\cos x^2 + C$ • Evaluate in terms of the original limits. $\int_1^2 sin(x^2)(2xdx) = -cos(x^2)\Big\vert_1^2$ $= -cos(2^2) + cos(1^2) \approx 1.19$	• Change the limits to limits for new variable. if $x = 1$, then $w = 1$ if $x = 2$, then $w = 4$ • Evaluate with new limits for new variable. $\int_1^2 2x\sin(x^2)dx =$ $\int_1^4 \sin(w)dw = -\cos w\Big\vert_1^4$ $= -\cos 4 + \cos 1 \approx 1.19$

$\int_{-3}^{2} t^2 e^{t^3} dt$ $w = t^3$ $dw = 3t^2 dt$ $\frac{1}{3} dw = t^2 dt$	• Find the antiderivative in terms of the original variable $\int t^2 e^{t^3} dt = \int e^w \frac{1}{3} dw$ $= \frac{1}{3} e^w + C = \frac{1}{3} e^{t^3} + C$ • Evaluate in terms of the original limits. $\int_{-3}^{2} t^2 e^{t^2} dt = \frac{1}{3} e^{t^3} \Big\vert_{-3}^{2}$ $= \frac{1}{3} e^{2^3} - \frac{1}{3} e^{(-3)^3} \approx 993.65$	• Change the limits to limits for new variable if t=–3, then w=–27 if t= 2, then w = 8 • Evaluate with new limits for new variable $\int_{-3}^{2} t^2 e^{t^3} dt =$ $\int_{-27}^{8} \frac{1}{3} e^w dw = \frac{1}{3} e^w \Big\vert_{-27}^{8}$ $= \frac{1}{3} e^8 - \frac{1}{3} e^{-27} \approx 993.65$

✎ On Your Own 3

Evaluate the following definite integral using both methods.

$$\int_0^{\pi} \sin z \sqrt{\cos z + 1}\, dz$$

(The answer appears at the end of this section)

5. *What are some of the less obvious substitutions?*

✍ ____

____ *(See text p. 295-296.)*

There are some “tricky” integration situations that occur again and again. You need to practice looking for the patterns, which tell when these "trick" techniques might be used. Often, it isn't obvious that substitution might work.

Example 7.1.5

Evaluate the following definite integrals using substitution if possible. If not possible, use numerical techniques.

Definite Integral	*Solutions*
a. $\int_0^1 \frac{1}{x^2 + 6x + 9} dx$	• Substitution won't work directly, but the denominator is a perfect square, thereby setting it up for substitution. $\int_0^1 \frac{1}{x^2 + 6x + 9} dx = \int_0^1 \frac{1}{(x+3)^2} dx$ $= -(x+3)^{-1} \Big\vert_0^1 = \frac{-1}{4} + \frac{1}{3} = \frac{1}{12}$

b. $\int_1^5 x\sqrt{x-1}\,dx$	• A guess is to let $w = x - 1$. But, then you have a factor of x, instead of the derivative of $x - 1$. Let $w = x - 1,\ dw = dx,\ and\ \ x = w + 1$ $\int_1^5 x\sqrt{x-1}\,dx = \int_0^4 (w+1)\sqrt{w}\,dw$ multiply to distribute $\sqrt{w}$, $= \int_0^4 w^{\frac{3}{2}} + w^{\frac{1}{2}}\,dw = \left(\frac{2}{5}w^{\frac{5}{2}} + \frac{2}{3}w^{\frac{3}{2}}\right)\Bigg\|_0^4$ $= \frac{2}{5}4^{\frac{5}{2}} + \frac{2}{3}4^{\frac{3}{2}} - 0 \approx 18.13$
c. $\int_1^2 \frac{x^3 - x + 1}{x^2}\,dx$	• Since there is only one term in the denominator you can simplify first. $\int_1^2 \frac{x^3 - x + 1}{x^2}\,dx = \int_1^2 x - \frac{1}{x} + \frac{1}{x^2}\,dx$ $= \left(\frac{1}{2}x^2 - \ln x - \frac{1}{x}\right)\Bigg\|_1^2 = \left(2 - \ln 2 - \frac{1}{2}\right) - \left(\frac{1}{2} - \ln 1 - 1\right)$ $= 2 - \ln 2$
$\int_0^1 e^{x^2}\,dx$	• Substitution doesn't help here and numerical methods will need to be used. y x ∫f(x)dx=1.4626517 $\int_0^1 e^{x^2}\,dx$ is approximately 1.46

On Your Own 4

Evaluate the following definite integral using both methods.

$$\int_0^{\frac{\pi}{2}} \cos^3 x \sin^2 x\,dx$$

Hint: $\sin^2 x + \cos^2 x = 1$

(The answer appears at the end of this section.)

Key Algebra Skills

- Review composition of functions: $f(g(x))$.

Key Notation

- Differential notation: $dw = g'(x)dx$.

Key Study Skills

- Review the Chain Rule, Section 3.4 of the text.
- Choose at least two problems that represent this section; put each on a 3 x 5 card with the solution on the back. Review 7.1.
- Use the ✓ ? ★ system on homework problems and get answers to all ★ problems.

Answers to On Your Own 1:

1.
$$\int \sin(5x)dx = \frac{-1}{5}\cos(5x) + C$$

2.
$$\int \cos(2x+3)dx = \frac{1}{2}\sin(2x+3) + C$$

3.
$$\int e^{-x}dx = -e^{-x} + C$$

4.
$$\int 3e^{2x-5}dx = \frac{3}{2}e^{2x-5} + C$$

5.
$$\int (4t+7)^{10}dt = \frac{1}{44}(4t+7)^{11} + C$$

6.
$$\int \frac{1}{\sqrt[3]{6-5z}}dz = \int (6-5z)^{\frac{-1}{3}}dz = \frac{-3}{10}(6-5z)^{\frac{2}{3}} + C$$

Answer to On Your Own 2:

Integral	*Inside function, w*	*General antiderivative*
1. $\int y\sqrt{1-y^2}\,dy$	$w = 1-y^2$ $dw = (-2y)dy$ $\frac{-1}{2}dw = ydy$	$\int y\sqrt{1-y^2}\,dy = \int \sqrt{w}\left(\frac{-1}{2}dw\right)$ $= \frac{-1}{2}\int w^{\frac{1}{2}}dw = \frac{-1}{3}w^{\frac{3}{2}} + C = \frac{-1}{3}(1-y^2)^{\frac{3}{2}} + C$
2. $\int \frac{1}{(4x-1)^2}dx$	$w = 4x-1$ $dw = (4)dx$ $\frac{1}{4}dw = dx$	$\int \frac{1}{(4x-1)^2})dx = \int w^{-2}\left(\frac{1}{4}dw\right)$ $= \frac{-1}{4}w^{-1} + C = \frac{-1}{4}(4x-1)^{-1} + C$ $= \frac{-1}{4(4x-1)} + C$
3. $\int \cos(3z)\sqrt[3]{4-\sin(3z)}\,dz$	$w = 4-\sin 3z$ $dw = (-3\cos 3z)dz$ $\frac{-1}{3}dw = (\cos 3z)dz$	$\int \cos 3z\sqrt[3]{4-\sin 3z}\,dz = \int \sqrt[3]{w}\left(\frac{-1}{3}dw\right)$ $= \frac{-1}{4}w^{\frac{4}{3}} + C = \frac{-1}{4}(4-\sin 3z)^{\frac{4}{3}} + C$

4. $\int cos(5t)(e^{\sin t})dt$	$w = \sin t$ $dw = -\cos t dt$	Cannot get the integral entirely in terms of *w* and *dw*. We need *–cos t dt*, but instead have *cos(5t)dt.*
5. $\int \frac{e^{-x}}{13+e^{-x}}dx$	$w = 13 + e^{-x}$ $dw = -e^{-x}dx$ $-dw = e^{-x}dx$	$\int \frac{e^{-x}}{13+e^{-x}}dx = \int \frac{-dw}{w} = -ln\|w\| + C$ $= -ln(13+e^{-x}) + C$

Answer to On Your Own 3

$\int_0^{\pi} \sin z\sqrt{\cos z + 1}dz$	• Find the antiderivative in terms of the original variable. $w = \cos z + 1$ $dw = (-\sin z)dz$ $-dw = (\sin z)dz$ $\int \sin z\sqrt{\cos z + 1}dz =$ $\int \sqrt{w}(-dw) = \frac{-2}{3}w^{\frac{3}{2}} + C =$ $-\frac{2}{3}(\cos z + 1)^{\frac{3}{2}} + C$ • Evaluate in terms of the original limits. $\int_0^{\pi} \sin z\sqrt{\cos z + 1}dz = \frac{-2}{3}(\cos z + 1)^{\frac{3}{2}}\Big\|_0^{\pi}$ $\frac{-2}{3}(0)^{\frac{3}{2}} - \frac{-2}{3}(2)^{\frac{3}{2}} = \frac{2^{\frac{5}{2}}}{3} \approx 1.886$	• Change the limits to limits for new variable. if $z = 0$, then $w = 2$ if $t = \pi$, then $w = 0$ • Evaluate with new limits for new variable. $\int_0^{\pi} \sin z\sqrt{\cos z + 1}dz =$ $\int_2^0 -\sqrt{w}dw = \frac{-2}{3}w^{\frac{3}{2}}\Big\|_2^0 = \frac{2^{\frac{5}{2}}}{3}$

Answer to On Your Own 4

$\int_0^{\frac{\pi}{2}} \cos^3 x \sin^2 x dx$	• This could be done easily by substitution if there had been only one factor of either *sin x* or *cos x*; since the power of cosine is odd, using the trig identity $\sin^2 x + \cos^2 x = 1$, then $\cos^2 x = 1 - \sin^2 x$, we will be left with a factor of *cos x*, the derivative of *sin x*.

$$\int_0^{\frac{\pi}{2}} \cos^3 x \sin^2 x dx = \int_0^{\frac{\pi}{2}} \cos x\left(\cos^2 x\right)\sin^2 x dx$$

$$= \int_0^{\frac{\pi}{2}} \cos x\left(1 - \sin^2 x\right)\sin^2 x dx$$

$$= \int_0^{\frac{\pi}{2}} \cos x\left(\sin^2 x - \sin^4 x\right)dx$$

$$= \int_0^{\frac{\pi}{2}} \cos x \sin^2 x dx - \int_0^{\frac{\pi}{2}} \cos x \sin^4 x dx$$

- Each of these can be evaluated letting $w = \sin x$, using the new limits of integration because $\sin(\pi/2) = 1$ and $\sin(0) = 0$

$$\int_0^1 w^2 dw - \int_0^1 w^4 dw = \frac{w^3}{3} - \frac{w^5}{5}\Big|_0^1 \approx 0.133$$

∫f(x)dx=.13333333

7.2 INTEGRATION BY PARTS

Objective: To find antiderivatives using integration by parts.

Preparations: Read and take notes on section 7.2
Suggested practice problems: 1,2,3,6,10,12,16,19,21,24,26,30,36,44,51,54.

✠ Key Ideas of the Section

You should be able to:

- Recognize when integrands are in the form for which integration by parts can be used to find the antiderivative.
- Use integration by parts to evaluate definite integrals.
- Use integration by parts for more complex integrands.

Key Questions to Ask and Answer

1. ***What types of functions have antiderivatives which can be found using integration by parts?***

✍ ____

____ *(See text p. 298-300.)*

Finding antiderivatives is reversing differentiation and integration by parts is based on the product rule. The functions will be products of the form uv' where u and v are functions of x and the product is a function u times an unrelated derivative $v'dx$: $\int uv'dx$

Example 7.2.1

Which of the following indefinite integrals might qualify for integration by parts and which for substitution?

Indefinite Integral	*Answers*
$\int xe^{-x}dx$	• Use parts method; substitution won't apply.
$\int x^2e^{-x^3}dx$	• Substitution is appropriate here; let $w=-x^3,\ dw=-3x^2dx$.
$\int x\cos xdx$	• Use parts method; substitution won't apply.
$\int x\cos x^2dx$	• Substitution is appropriate here; let $w=x^2,\ dw=2xdx$.
$\int \frac{\ln x}{x}dx$	• Substitution is appropriate here; let $w=\ln x,\ dw=\frac{1}{x}dx$.
$\int x^2\ln xdx$	• Use parts method; substitution won't apply.

✎ On Your Own 1

For which of the following indefinite integrals might integration by parts be used and for which might substitution be used?

1. $\int x^4\ln\left(x^5\right)dx$
2. $\int sin^2 x\ dx$
3. $\int \cos x\sqrt{1+\sin x}dx$

(The answer appears at the end of this section)

2. *How do you choose u and v'?*

✍ ____

____ *(See text p. 300.)*

Integration by parts says that $\int uv'dx = uv - \int u'vdx$. This suggests that u and v' should be chosen in such a way that

- u' should be simpler than u.
- v should be simpler than v'.
- If you know v', you must be able to determine v.

- You can find the antiderivative of $u'v$, or at least, it looks like it may be simpler. The antiderivative of uv' perhaps can be determined in additional steps.

Note: Some students find it easier to use integration by parts if they first organize the parts into a table that allows them to visualize the formula:

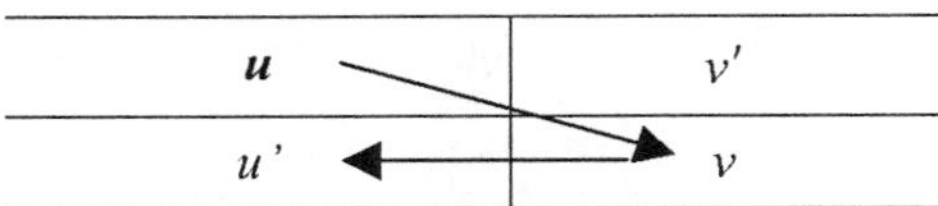

Step 1: Choose u and take its derivative u', put the results in column one.

Step 2: Choose the derivative v' then take it's antiderivative v, put the results in column two.

Step 3: Multiply u by v.

Step 4: Multiply the v by u' then take the integral of the product.

Step 5 Put it all together and integrate $\int uv'dx = uv - \int u'vdx$.

Example 7.2.2

Find the following indefinite or definite integrals using integration by parts.

Indefinite Integral	*Solutions*
a. $\int xe^{-x}dx$	• Let $u = x$, since $u' = 1dx$ is simpler. Then $v' = e^{-x}$ and $v = -e^{-x}$. $u = x$ → $v' = e^{-x}$ $u' = 1dx$ ← $v = -e^{-x}$ $\int xe^{-x}dx = (x)(-e^{-x}) - \int (1)(-e^{-x})dx$ $= -xe^{-x} - e^{-x} + C$
b. $\int x^2 \ln xdx$	• Let $u = lnx$, then $u' = \frac{1}{x}dx$ and $v' = e^{-x}$; then $v = \frac{x^3}{3}$. $u = lnx$ → $v' = x^2$ $u' = \frac{1}{x}dx$ ← $v = \frac{x^3}{3}$ • Since $u'v = \frac{1}{x}\cdot\frac{x^3}{3} = \frac{x^2}{3}$, finding the antiderivative is easy. (Notice that we cannot choose $v' = \ln x$, since we do not know its antiderivative directly.) $\int x^2 \ln xdx = (\ln x)\left(\frac{x^3}{3}\right) - \int \left(\frac{1}{x}\right)\left(\frac{x^3}{3}\right)dx$ $= \frac{x^3}{3}\ln x - \frac{1}{3}\int x^2dx = \frac{x^3}{3}\ln x - \frac{x^3}{9} + C$

c. $\int x^2 e^x dx$	• Let $u = x^2$, since $u' = 2x$ which is simpler. Then $v' = e^x$ and $v = e^x$. $\int x^2 e^x dx = (x^2)(e^x) - \int (2x)e^x dx$ $= x^2 e^x - 2\int xe^x dx$ • We can use parts again, letting $u = x,\ v' = e^x$. $= x^2 e^x - 2\left(xe^x - \int 1e^x dx\right)$ $= x^2 e^x - 2\left(xe^x - e^x\right) + C$ $= x^2 e^x - 2xe^x + 2e^x + C$
d. $\int \arcsin x dx$	• Although the integrand does not look like a product, let $u = arcsinx$ (since you know its derivative) and $v' = 1$. Then $u' = \dfrac{1}{\sqrt{1-x^2}}$ and $v = x$. $\int \arcsin x dx = (\arcsin x)(x) - \int \left(\dfrac{1}{\sqrt{1-x^2}}\right)(x)dx$ • Substitution can be used, letting $u = 1 - x^2$ and $du = -2xdx$. $= (\arcsin x)(x) - \int \dfrac{1}{\sqrt{u}}\left(\dfrac{-du}{2}\right)$ $= x\arcsin x + \dfrac{1}{2}\int u^{\frac{-1}{2}} du$ $= x\arcsin x + u^{\frac{1}{2}} = x\arcsin x + \sqrt{1-x^2} + C$
e. $\int e^x \cos x dx$	Here, the choice for u and v' isn't obvious. Try letting $u = e^x, v' = \cos x$. Then $u' = e^x, v = \sin x$. $\int e^x \cos x dx = e^x \sin x - \int e^x \sin x dx$ which isn't any better! If we use parts again and can get the original expression back, we can perhaps solve for the original indefinite integral. These types of integrals then "boomerang" back.

	Apply integration by parts to $\int e^x \sin x dx$ also. Let $u = e^x$ and $v' = \sin x dx$, then $u' = e^x$, $v = -\cos x$ $\int e^x \cos x dx = e^x \sin x - \int e^x \sin x dx$ $= e^x \sin x - \left(e^x(-\cos x) - \int e^x(-\cos x)dx\right)$, therefore the integral is now $\int e^x \cos x dx = e^x \sin x + e^x(\cos x) - \int e^x \cos x dx$, combining like integrals gives $2\int e^x \cos x dx = e^x \sin x + e^x(\cos x)$ $\int e^x \cos x dx = \frac{1}{2}\left(e^x \sin x + e^x(\cos x)\right) + C$
f. $\int_0^\pi e^x \cos x dx$	$\int_0^\pi e^x \cos x dx =$ $\frac{1}{2}\left(e^x \sin x + e^x \cos x\right)\Big\|_0^\pi$ $= \frac{1}{2}\left(e^\pi \sin \pi + e^\pi \cos \pi\right) - \frac{1}{2}\left(e^0 \sin 0 + e^0 \cos 0\right)$ $= \frac{1}{2}\left(-e^\pi - 1\right)$

✎ On Your Own 2

Find the following indefinite integrals using integration by parts.

1. $\int x \cos x dx$
2. $\int e^x \sin x dx$
3. $\int_0^2 te^t dt$

(The answer appears at the end of this section)

🗀 Key Study Skills

- Review the Product Rule, Section 3.3 of the text.
- Choose at least two problems that represent this section; put each on a 3 x 5 card with the solution on the back. Review 7.1-7.2.
- Use the ✔ ? ★ system on homework problems and get answers to all ★ problems.

Answer to On Your Own 1

	Indefinite Integral	*Answers*
1.	$\int x^4 \ln(x^5)dx$	• Substitution is appropriate here, let $w = x^5,\ dw = 5x^4dx$.
2.	$\int \sin^2 xdx$	• Parts; substitution won't work here.
3.	$\int \cos x\sqrt{1+\sin x}dx$	• Substitution is appropriate here, let $w = 1+\sin x,\ \ dw = \cos xdx$.

Answer to On Your Own 2

	Integral	*Answers*
1.	$\int x\cos xdx$	$\int x\cos xdx = (x)(\sin x) - \int (1)(\sin x)dx$ $= x\sin x + \cos x + C$
2.	$\int e^x \sin xdx$	$\int e^x \sin xdx = (e^x)(-\cos x) - \int e^x(-\cos x)dx$ $\int e^x \sin xdx = -e^x \cos x + \int e^x(\cos x)dx$ $\int e^x \sin xdx = -e^x \cos x + \left[e^x \sin x - \int e^x \sin xdx\right]$ $2\int e^x \sin xdx = -e^x \cos x + e^x \sin x$ $\int e^x \sin xdx = \frac{1}{2}\left(-e^x \cos x + e^x \sin x\right) + C$
3.	$\int_0^2 te^t dt$	$\int_0^2 te^t dt = te^t\Big\vert_0^2 - \int_0^2 1e^t dt = te^t\Big\vert_0^2 - e^t\Big\vert_0^2$ $= 2e^2 - (e^2 - 1)$

7.3 TABLES OF INTEGRALS

Objective: To recognize an integral as a form in a table of integrals.

Preparations: Read and take notes on section 7.3
Suggested practice problems: 1,3,7,8,10,14,22,25,29,31,32.

✠ Key Ideas of the Section

You should be able to:

- Recognize when it is useful to use the table of integrals.

- Recognize which form in the table an integrand resembles.
- Put the integrand in the form necessary to be able to use the table.
- Use the tables to find the antiderivative.

Key Questions to Ask and Answer

1. What are the various classes of functions for which the Tables of Integrals can be used?

✍ ____

__________________________________ *(See text p. 304-306.)*

In a calculus class you sometimes get the impression that all integrals can be found using algebraic techniques or specific integral rules, when in fact the opposite is true. In the applied sciences many integrals are found using Tables of Integrals. Thick books are published with integral tables. Computer algebra systems now have the tables built into the programs. Your text has a short table of indefinite integrals on the inside back cover. Here is how the table is organized.

Table	*Form of integrand*	*Comments and Examples*
Part I	Basic functions which have elementary antiderivatives, such as: $a^x, x^n, \ln x, \sin x, \cos x, \tan x$	• You should know the antiderivatives of these without using the table . (Antiderivatives of the combination of these various functions can be found by substitution or integration by parts). $\int 2^x dx$ $\int \sin(5x+1)dx$
Part II	Products of $e^x, \sin x, \cos x$.	• These are all derived using integration by parts. $\int e^{3x}\cos 5x dx$ $\int \sin 2x \cos 5x dx$
Part III	Products of a polynomial and $e^x, \sin x, \cos x$.	• Use formula found in the tables repeatedly to reduce the degree of the polynomial until degree 0. $\int (x^2-3x+1)e^{5x}dx$
Part IV	Integer powers of *sin x* and *cos x* .	• May need to use formula in the tables repeatedly. • May need to use substitution and/or trig identity: $\sin^2 x + \cos^2 x = 1$ $\int \cos^4 3x dx$ $\int \cos^4 x \sin^5 x\, dx$
Part V	Expressions with quadratic denominators.	• May need to compete the square or do long division first. $\int \frac{1}{x^2+2x+4}dx$

Part VI	Expressions with square roots of quadratics.	Quadratics are of the form: $x^2 \pm a^2$, $a^2 - x^2$, $(x-a)(x-b)$, $a \neq b$ $\int \frac{1}{\sqrt{4-x^2}} dx$

On Your Own 1

Identify the formula in the tables that might be useful in finding the following antiderivatives.

1. $\int x^3 \ln(x) dx$
2. $\int x^3 \ln(x^4) dx$
3. $\int \frac{1}{\cos^3 x} dx$

(The answer appears at the end of the section.)

Example 7.3.1

Use the tables of integrals (found in your text) to find the following antiderivatives.

Integrals	*Answers using Tables*
a. $\int \cos^4 x dx$	(IV-18) $n = 4$ $\int \cos^4 x dx = \frac{1}{4}\cos^3 x \sin x + \frac{3}{4}\int \cos^2 x dx$ (IV-18) $n = 2$ $= \frac{1}{4}\cos^3 x \sin x + \frac{3}{4}\left[\frac{1}{2}\cos x \sin x + \frac{1}{2}\int dx\right]$ $= \frac{1}{4}\cos^3 x \sin x + \frac{3}{8}\cos x \sin x + \frac{3}{8}x + C$
b. $\int \frac{1}{x^2 + 2x + 5} dx$	(V-24) complete the square: $x^2 + 2x + 5 = (x^2 + 2x + 1) - 1 + 5 = (x+1)^2 + 4$ let $w = x+1$, $dw = dx$ so $\int \frac{1}{x^2 + 2x + 5} dx = \int \frac{1}{(x+1)^2 + 4} dx = \int \frac{1}{w^2 + 2^2} dw$ $= \frac{1}{2}\arctan\left(\frac{w}{2}\right) + C = \frac{1}{2}\arctan\left(\frac{x+1}{2}\right) + C$

c. $\int (x^2 - 3x + 1)e^{5x} dx$	(III-14) $\int (x^2 - 3x + 1)e^{5x} dx = \frac{1}{5}(x^2 - 3x + 1)e^{5x} - \frac{1}{5}\int (2x - 3)e^{5x} dx$ $= \frac{1}{5}(x^2 - 3x + 1)e^{5x} - \frac{1}{5}\left[\frac{1}{5}(2x - 3)e^{5x} - \frac{1}{5}\int 2e^{5x} dx\right]$ $= \frac{1}{5}(x^2 - 3x + 1)e^{5x} - \frac{1}{25}(2x - 3)e^{5x} + \frac{1}{125}(2)e^{5x} + C$
d. $\int e^{3x} \cos 5x dx$	(II-9) $a = 3$, $b = 5$ $\int e^{3x} \cos 5x dx = \frac{1}{3^2 + 5^2} e^{3x}[3\cos 5x + 5\sin 5x] + C$
e. $\int \frac{1}{\sqrt{4 - x^2}} dx$	(VI-28) $\int \frac{1}{\sqrt{4 - x^2}} dx = \arcsin \frac{x}{2} + C$

Key Algebra Skills

- Review completing the square: $x^2 + bx + c = (x + \frac{b}{2})^2 - (\frac{b}{2})^2 + c$
- Review long division.

Key Study Skills

- Choose at least two problems that represent this section; put each on a 3 x 5 card with the solution on the back. Review note card problems 7.1 - 7.3.
- Make up your own quiz.
- Use the ✓ ? ★ system on homework problems and get answers to all ★ problems.

Answer to On Your Own 1

1. $\int x^3 \ln(x) dx$	III-13
2. $\int x^3 \ln(x^4) dx$	use substitution
3. $\int \frac{1}{\cos^3 x} dx$	IV-21

7.4 ALGEBRAIC IDENTITITES AND TRIGONOMETRIC SUBSTITUTIONS

Objective: To find antiderivatives using partial fractions and trigonometric substitutions.

Preparations: Read and take notes on section 7.4.
Suggested practice problems:. 1,5,9,12,16,20,24,26,30,36,38,41,46.

✠ Key Ideas of the Section

You should be able to:

- Split rational functions into partial fractions
- Use substitution and/or integral tables to calculate the resulting integrals.
- Recognize integrands for which the substitution of sin θ or tan θ can be used.
- Determine the domain of the substitution $x = a\sin\theta$ or $x = a\tan\theta$.

Key Questions to Ask and Answer

1. *How do you split a rational function* $f(x)=\dfrac{P(x)}{Q(x)}$ *into partial fractions?*

✍ ____

____ *(See text p. 312.)*

Many integrals are in fraction form and we cannot find the antiderivative unless we change the integral into a sum of fractions. This process is called finding the partial fractions.

Recall from arithmetic that division of two numbers in improper fraction form can be written as the quotient plus the remainder. Thus 25/7= 3 + 4/7. Given a rational function *f(x)*, let the degree of the polynomial numerator *P(x)* be deg$P(x)$ and the degree of the polynomial denominator be deg$Q(x)$. If deg$P(x) \geq$ deg$Q(x)$, use long division to write as $S(x)+\dfrac{R(x)}{Q(x)}$. If deg$R(x) <$ deg$Q(x)$ factor, if possible, $Q(x)$ into linear and/or quadratic factors.

- Each distinct linear factor $(x - c)$ of $Q(x)$ gives a partial fraction of the form: $\dfrac{A}{(x-c)}$
- Each repeated linear factor, $(x - c)^n$ gives partial fractions of the form:
  $$\frac{A_1}{(x-c)}+\frac{A_2}{(x-c)^2}+\ldots+\frac{A_n}{(x-c)^n}$$
- Each quadratic factor, q(x), gives a partial fraction of the form: $\dfrac{Ax+b}{q(x)}$

Example 7.4.1

Split the following rational functions into partial fractions:

Rational Expression	*Partial Fractions*
a) $\dfrac{2x^4+3x^3-x^2+x-1}{x^3-x}$	Since the *degP(x)>degQ(x),* divide: $\dfrac{2x^4+3x^3-x^2+x-1}{x^3-x}=2x+3+\dfrac{x^2+4x-1}{x^3-x}$ Factor the denominator of the remainder , x^3-x: $=2x+3+\dfrac{x^2+4x-1}{x(x-1)(x+1)}$ Write as the sum of linear factors: $=2x+3+\dfrac{A}{x}+\dfrac{B}{(x-1)}+\dfrac{C}{(x+1)}$
b) $\dfrac{x^2+x+1}{(2x+1)(x^2+1)}$	Since the denominator is factored, write as the sum of the factors: $\dfrac{x^2+x+1}{(2x+1)(x^2+1)}=\dfrac{A}{2x+1}+\dfrac{Bx+C}{x^2+1}$
c) $\dfrac{x}{(x-1)^2}$	Since the denominator is raised to a power other than one, write as the sum of the linear factor and the linear factor raised to the power. $\dfrac{x}{(x-1)^2}=\dfrac{A_1}{(x-1)}+\dfrac{A_2}{(x-1)^2}$

2. *How do I solve for the constants in the partial fraction?*

✍ ____

____ *(See text p. 309-311.)*

To find the constants of a partial fraction you must first eliminate the fractions by multiplying both sides of the equation by the least common denominator and, since the equation is true for all x, the coefficients can be equated. That is, the 1st degree terms can be equated, the 2nd degree terms equated, the 3rd degree terms equated, etc. This will give a system of equations that can be solved.

Example 7.4.2

Solve for the constants from **Example 7.4.1.**

Explanation
a) Look at only the rational term for this first example: $\dfrac{x^2+4x-1}{x(x-1)(x+1)}=\dfrac{A}{x}+\dfrac{B}{(x-1)}+\dfrac{C}{(x+1)}$ Eliminate the fraction by multiplying through by the least common denominator.

$$x^2+4x-1=A(x-1)(x+1)+Bx(x+1)+Cx(x-1)$$

$$=Ax^2-A+Bx^2+Bx+Cx^2-Cx$$

$$x^2+4x-1=(A+B+C)x^2+(B-C)x-A$$

On the left side the coefficient of x^2 is 1, the coefficient of x is +4 and the constant term is –1. Equating coefficients gives the system of equations:

1). $A+B+C=1$

2). $B-C=+4$

3). $-A=-1$

Solving the system, since A = 1,substituting in equation 1) gives B = -C, substituting in equation 2) gives – C – C= 4, so C = -2. Back substituting we get and solving the system, we get

$A=1$, $B=-2$, $C=-2$, or

The completed partial fraction is:

$$\frac{x^2-4x-1}{x(x-1)(x+1)}=\frac{1}{x}+\frac{-2}{(x-1)}+\frac{-2}{(x+1)}$$

b) $$\frac{x^2+x+1}{(2x+1)(x^2+1)}=\frac{A}{2x+1}+\frac{Bx+C}{x^2+1}$$

Eliminate the denominator by multiplying through by the least common denominator.

$$x^2+x+1=A(x^2+1)+(Bx+C)(2x+1)$$

$$=Ax^2+A+2Bx^2+Bx+2Cx+C$$

$$x^2+x+1=(A+2B)x^2+(B+2C)x+(A+C)$$

Equating coefficients gives: $A+2B=1$. $B+2C=1$, $A+C=1$, and solving the system, we get $A=3/5$, $B=1/5$, $C=2/5$, or

$$\frac{x^2+x+1}{(2x+1)(x^2+1)}=\frac{\frac{3}{5}}{(2x+1)}+\frac{\frac{1}{5}x+\frac{2}{5}}{x^2+1}=\frac{3}{5(2x+1)}+\frac{x+2}{5(x^2+1)}$$

c) Write as the sum of linear and second power factors:

$$\frac{x}{(x-1)^2}=\frac{A_1}{(x-1)}+\frac{A_2}{(x-1)^2}$$

Eliminate the denominator by multiplying through by the least common denominator.

$$x=A_1(x-1)+A_2$$
$$x=A_1x+(A_2-A_1)$$

Equating coefficients gives $A_1=1$, $A_2-A_1=0$, so $A_2=1$, or

$$\frac{x}{(x-1)^2}=\frac{1}{(x-1)}+\frac{1}{(x-1)^2}$$

✎ On Your Own 1

Write as partial fractions: $\dfrac{x^2}{(x^4-16)}$

(The answer appears at the end of this section.)

3. *Now that I have the Partial Fraction how do I find the integrals?*

✍ ____

____ (See text p. 309-311)

Recall the following rules for antiderivates that involve fractions:

- $\int \frac{A}{(x-c)}dx = A\,ln|x-c| + C$
- $\int \frac{A_n}{(x-c)^n}dx = \int A_n(x-c)^{-n}dx$ can be integrated directly.
- $\int \frac{Ax+b}{q(x)}dx = \int \frac{Ax}{q(x)}dx + \int \frac{b}{q(x)}dx$ where the first can be integrated with substitution and the second using the integral tables.

Example 7.4.4

Find the integrals of the rational functions from **Example 7.4.1**.

Explanation

a) Since *degP(x)>degQ(x)*, divide to get a quotient and a remainder:

$$\int \frac{2x^4+3x^3-x^2+x-1}{x^3-x}dx = \int (2x+3)\,dx + \int \frac{x^2+4x-1}{x(x-1)(x+1)}dx$$

Write the remainder as a partial fraction, then integrate each sum separately.

$$= \int (2x+3)dx + \int \frac{1}{x}dx + \int \frac{-2}{(x-1)}dx + \int \frac{-2}{(x+1)}dx$$

$$= x^2 + 3x + ln|x| - 2\,ln|x-1| - 2\,ln|x+1| + C$$

b) Find the partial fractions and integrate the sums separately:

$$\int \frac{x^2+x+1}{(2x+1)(x^2+1)}dx = \int \frac{3}{5(2x+1)}dx + \int \frac{x}{5(x^2+1)}dx + \int \frac{2}{5(x^2+1)}dx$$

$$= \frac{3}{10}ln|2x+1| + \frac{1}{10}ln|x^2+1| + \frac{2}{5}arctan\,x + C$$

c) Find the partial fractions and integrate the sums separately:

$$\int \frac{x}{(x-1)^2}dx = \int \frac{1}{(x-1)}dx + \int \frac{1}{(x-1)^2}dx$$

$$= \ln(|x-1|) - (x-1)^{-1} + C$$

On Your Own 2

Find the following integral: $\int \frac{1}{x^2-4}dx$

(The answer appears at the end of this section.)

4. *How can I use trigonometric substitutions for integrands with square roots of the form $\sqrt{a^2-x^2}$ or $\sqrt{a^2+x^2}$ or with a^2+x^2?*

____

____ *(See text p. 321-315.)*

Recall the Pythagorean Identity: $sin^2\,\theta + cos^2\,\theta = 1$

Integrand includes:	***Visualizing the Integrand***	***Substitution***
• $\sqrt{a^2-x^2}$	A, X, √(A^2-X^2) Sine substitution	Let θ be the angle in standard position then $sin\,\theta = \frac{x}{a}$ or $x = a\,sin\,\theta$. Use $x = a$ sin θ, *or θ = arc sin* (x/a), for $\frac{-\pi}{2} \le \theta \le \frac{\pi}{2}$
• $\sqrt{a^2+x^2}$ or a^2+x^2	√(A^2+X^2), X, A Tangent substitution	Let θ be the angle in standard position then $tan\,\theta = \frac{x}{a}$ or $x = a\,tan\,\theta$. Use $x = a$ tan θ, or θ = arc tan (x/a), for $\frac{-\pi}{2} \le \theta \le \frac{\pi}{2}$

Example 7.4.5

Find $\int \frac{1}{\sqrt{25-x^2}}dx$.	Visualize the integrand as 5, X, √(25-X^2)	Use $x = 5$ sin θ, or θ = arc sin ($x/5$), for $\frac{-\pi}{2} \le \theta \le \frac{\pi}{2}$, so dx = 5cosθ dθ.

Substitute for x:

$$\int \frac{1}{\sqrt{25-x^2}}dx = \int \frac{5\,sin\,\theta}{\sqrt{25-25\,sin^2\,\theta}}d\theta = \int \frac{cos\,\theta}{\sqrt{1-sin^2\,\theta}}d\theta = \int \frac{cos\,\theta}{\sqrt{cos^2\,\theta}}d\theta =$$

$\int d\theta = \theta + C = arcsin\frac{x}{5} + C$. Note that

we use the fact that for $\frac{-\pi}{2} \le \theta \le \frac{\pi}{2}$, $cos\,\theta \ge 0$ so $\sqrt{cos^2\,\theta} = cos\,\theta$.

On Your Own 3

Find the following integral:

$$\int \frac{x^2}{16+x^2}dx$$

(The answer appears at the end of this section.)

Key Algebra Skills

- Review trigonometric functions, inverse trigonometric functions and identities.
- Review algebra skills in long division, completing the square, manipulating rational expressions.

Key Study Skills

- Choose at least two problems that represent this section; put each on a 3 x 5 card with the solution on the back. Review problems from 7.1-7.4.
- Make up your own quiz.
- Use the ✓ ? ★ system on homework problems and get answers to all ★ problems.

Answer to On Your Own 1

$$\frac{x^2}{(x^4-16)}=\frac{x^2}{(x-2)(x+2)(x^2+4)}=\frac{A}{x-2}+\frac{B}{x+2}+\frac{Cx+D}{x^2+4}$$

$$x^2=A(x+2)(x^2+4)+B(x-2)(x^2+4)+(Cx+D)(x-2)(x+2)$$

$$x^2=x^3(A+B+C)+x^2(2A-2B+D)+x(4A+4B-4C)+(8A-8B-4D)$$

Equating coefficients: $A+B+C=0$, $2A-2B+D=1$, $4A+4B-4C=0$, $8A-8B-4D=0$

Solving the system: $A = 1/8$, $B = -1/8$, $C = 0$, $D = 1/2$

$$\frac{x^2}{(x^4-16)}=\frac{1}{8(x-2)}+\frac{-1}{8(x+2)}+\frac{1}{2(x^2+4)}$$

Answer to On Your Own 2

Use partial fractions

$$\frac{1}{(x-2)(x+2)}=\frac{A}{(x-2)}+\frac{B}{(x+2)}$$

$$1=A(x+2)+B(x-2)$$

$$1=(A+B)x+(2A-2B)$$

Equating coefficients:

$$A+B=0$$

$$2A-2B=1$$

or $A=\frac{1}{4}$, $B=\frac{-1}{4}$

$$\int\frac{1}{x^2-4}dx=\int\frac{1}{4(x-2)}-\frac{1}{4(x+2)}dx$$

$$=\frac{1}{4}\left[ln|x-2|-ln|x+2|\right]+C$$

Answer to On Your Own 3

Use $x = 4\tan\theta$, or $\theta = \arctan(x/4)$, for $\frac{-\pi}{2}<\theta<\frac{\pi}{2}$

$$dx = 4\sec^2\theta\, d\theta$$

$$\int\frac{x^2}{16+x^2}dx=\int 1dx-\int\frac{16}{16+x^2}dx=\int 4\sec^2\theta d\theta-\int\frac{16}{16+(4\tan\theta)^2}(4\sec^2\theta d\theta)=$$

$$4\int\sec^2\theta d\theta-4\int\frac{\sec^2\theta d\theta}{1+\tan^2\theta}=4\tan\theta-4\int d\theta=4\tan\theta-4\theta=$$

since $x = 4\tan\theta$, and $\theta = \arctan(x/4)$,

$$=x-4\arctan\left(\frac{x}{4}\right)+C$$

7.5 APPROXIMATING DEFINITE INTEGRALS

Objective: To compare different numerical methods in approximating definite integrals.

Preparations: Review text section 5.2 on the definite integral.
Read and take notes on section 7.5.
Suggested practice problems: 2,4,6,7,9,10,11-15,19,20.

✠ Key Ideas of the Section

You should be able to:

- Approximate definite integrals using the Left Rule, Right Rule, Midpoint Rule and Trapezoid Rule.
- Determine whether you have an underestimate or an overestimate by knowing the behavior of the function on the interval.

Key Questions to Ask and Answer

1. What are some numerical methods of approximating the value of a definite integral?

✍ ____

____ *(See text p. 317-319.)*

Review the rules for approximating the area under a curve over interval $[a,b]$, where the interval width is $\Delta x = \dfrac{b-a}{n}$ and n is the number of subdivisions:

- **Left rule:** Riemann sum of n subdivisions using left-hand endpoints of subintervals, the resulting area denoted LEFT(n). The formula is $\sum_{i=0}^{n-1} f(x_i)\Delta x$.
- **Right rule**: Riemann sum of n subdivisions using right-hand endpoints of subintervals, the resulting area denoted RIGHT(n). The formula is $\sum_{i=1}^{n} f(x_i)\Delta x$.
- **Midpoint rule**: Riemann sum of n subdivisions using midpoints of the left and right endpoints of the subintervals, the result area denoted MID(n). Let the midpoint be $\bar{x}_i = \frac{1}{2}(x_{i-1} + x_i)$. The formula is $\sum_{i=1}^{n} f(\bar{x}_i)\Delta x$.
- **Trapezoid rule**: use trapezoids to approximate the value of the definite integral, the resulting area denoted TRAP(n). The formula is based upon the area of a trapezoid

$$A_T = \frac{1}{2}(height)(base_1 + base_2)$$

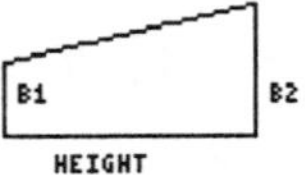

$$\sum_{i=1}^{n}\frac{1}{2}f(x_{i-1})\Delta x+\sum_{i=1}^{n}\frac{1}{2}f(x_i)\Delta x=\frac{\Delta x}{2}\left(f(x_0+2f(x_1)+2f(x_2)+\ldots+2f(x_{n-1})+f(x_n)\right).$$

It follows that the Trapezoid rule is nothing more than the average of the Left and Right Sums: TRAP(n)= (LEFT(n) + RIGHT(n))/2

Example 7.5.1

For the following integrals, compute LEFT(2), RIGHT(2), MID(2), and TRAP(2) and compare your answers.

Definite Integral	***Graph***	***Answers***
a. $\int_1^8 x^{\frac{2}{3}}dx$	Left sum: 1 4.5 8 Right Sum: Midpoint: 2.75 6.25 Trapezoid: 1 4.5 8 ∫f(x)dx=18.6	If n = 2, on [1,8], then Δx =(8-1)/2= 3.5, and $x_0=1,\ x_1=4.5,\ x_2=8.$ LEFT(2) = $f(1)(3.5)+f(4.5)(3.5)=$ LEFT(2) $=1^{\frac{2}{3}}(3.5)+4.5^{\frac{2}{3}}(3.5)=13.04$ RIGHT(2) = $f(4.5)(3.5)+f(8)(3.5)$ RIGHT(2) = $4.5^{\frac{2}{3}}(3.5)+8^{\frac{2}{3}}(3.5)=23.54$ MID(2) = $f(2.75)(3.5)+f(6.25)(3.5)$ MID(2) = $2.75^{\frac{2}{3}}(3.5)+6.25^{\frac{2}{3}}(3.5)=18.75$ The area of a trapezoid= $\frac{1}{2}(base_1+base_2)(height)$ TRAP(2) = $\frac{1}{2}(f(1)+f(4.5))(3.5)+\frac{1}{2}(f(4.5)+f(8))(3.5)$ $=\frac{1}{2}(3.5)(f(1)+2f(2)+f(8))=\left(\frac{\text{LEFT(2)}+\text{RIGHT(2)}}{2}\right)$ TRAP(2) =(LEFT(2)+ RIGHT(2))/2 = (13.04+23.54)/2 = 18.29 Exact value = 18.6

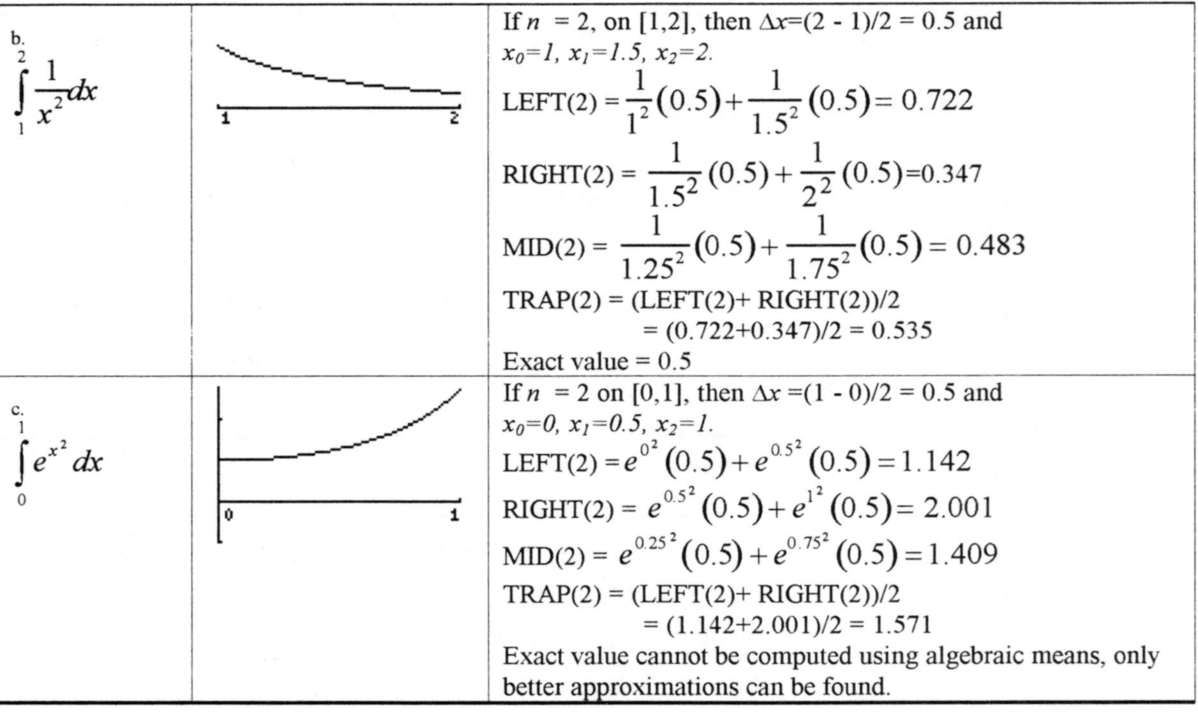

b. $\int_1^2 \frac{1}{x^2}dx$		If $n = 2$, on [1,2], then $\Delta x=(2-1)/2 = 0.5$ and $x_0=1,\ x_1=1.5,\ x_2=2.$ LEFT(2) = $\frac{1}{1^2}(0.5)+\frac{1}{1.5^2}(0.5)= 0.722$ RIGHT(2) = $\frac{1}{1.5^2}(0.5)+\frac{1}{2^2}(0.5)$=0.347 MID(2) = $\frac{1}{1.25^2}(0.5)+\frac{1}{1.75^2}(0.5) = 0.483$ TRAP(2) = (LEFT(2)+ RIGHT(2))/2 = (0.722+0.347)/2 = 0.535 Exact value = 0.5
c. $\int_0^1 e^{x^2}\,dx$		If $n = 2$ on [0,1], then $\Delta x=(1-0)/2 = 0.5$ and $x_0=0,\ x_1=0.5,\ x_2=1.$ LEFT(2) = $e^{0^2}(0.5)+e^{0.5^2}(0.5)=1.142$ RIGHT(2) = $e^{0.5^2}(0.5)+e^{1^2}(0.5)= 2.001$ MID(2) = $e^{0.25^2}(0.5)+e^{0.75^2}(0.5)=1.409$ TRAP(2) = (LEFT(2)+ RIGHT(2))/2 = (1.142+2.001)/2 = 1.571 Exact value cannot be computed using algebraic means, only better approximations can be found.

On Your Own 1

For the following integrals, compute LEFT(2), RIGHT(2), MID(2), and TRAP(2) and compare your answers.

1. $\int_4^{16} \sqrt{x}\,dx$
2. $\int_0^1 e^{-x}\,dx$

(The answer appears at the end of this section)

2. *How do you know whether your approximation will be an overestimate or an underestimate?*

✍ ____

____ *(See text p. 319-320.)*

Try to visualize the graph of the function *f*, noting the concavity and the areas of the rectangles or trapezoids used.

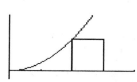

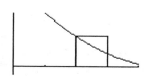

f is *increasing* on [a,b], Left. 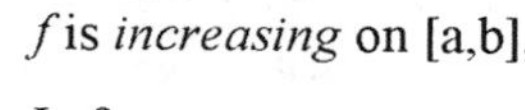	• LEFT(n) will be an *underestimate* and RIGHT(n) will be an *overestimate* : • $LEFT(n) \le \int_a^b f(x)dx \le RIGHT(n)$.
f is *decreasing* on [a,b], Left.	• LEFT(n) will be an *overestimate* and RIGHT(n) will be an *underestimate* : • $RIGHT(n) \le \int_a^b f(x)dx \le LEFT(n)$.
f is *concave up* on [a,b], Midpoint. 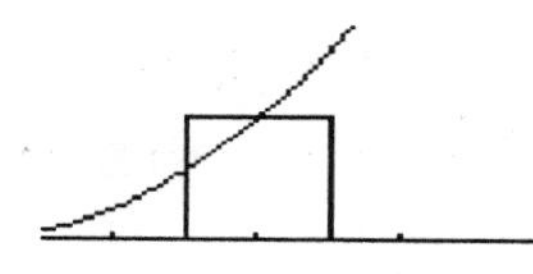	• TRAP(n) will be an *overestimate*; MID(n) will be an *underestimate* : • $MID(n) \le \int_a^b f(x)dx \le TRAP(n)$.
f is *concave down* on [a,b], Midpoint. 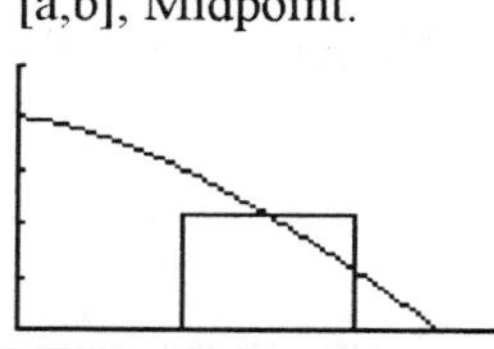	• TRAP(n) will be an *underestimate*; MID(n) will be an *overestimate* : • $TRAP(n) \le \int_a^b f(x) \le MID(n)$.

Example 7.5.2

Consider the definite integrals in **Example 7.5.1** above. Arrange LEFT(2), RIGHT(2), and TRAP(2) in ascending order and explain how you could predict without first calculating.

Definite Integral	***Graph***	***Answers***
a. $\int_1^8 x^{\frac{2}{3}} dx$	1 8	• f is increasing on [1,8]so, • LEFT(2)$\le \int_1^8 x^{\frac{2}{3}} dx \le$ RIGHT(2) • f is concave down on [1,8], so *(continued on next page)*

		• TRAP(2)$\le \int_1^8 x^{\frac{2}{3}}dx$, but TRAP(2) is a better estimate than LEFT or RIGHT, so • LEFT(2) $\le$ TRAP(2) $\le \int_1^8 x^{\frac{2}{3}}dx \le$ RIGHT(2).
b. $\int_1^2 \frac{1}{x^2}dx$	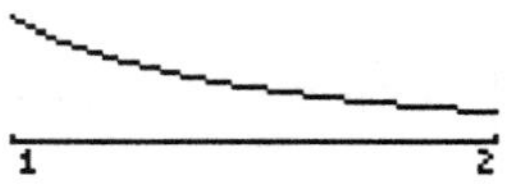	• f is decreasing on [1,2] so, • RIGHT(2)$\le \int_1^2 \frac{1}{x^2}dx \le$ LEFT(2) • f is concave up on [1,2], so • $\int_1^2 \frac{1}{x^2}dx \le$ TRAP(2), but TRAP(2) is a better estimate than LEFT(2) or RIGHT(2), so • RIGHT(2) $\le \int_1^2 \frac{1}{x^2}dx \le$ TRAP(2) $\le$ LEFT(2).
c. $\int_0^1 e^{x^2} dx$	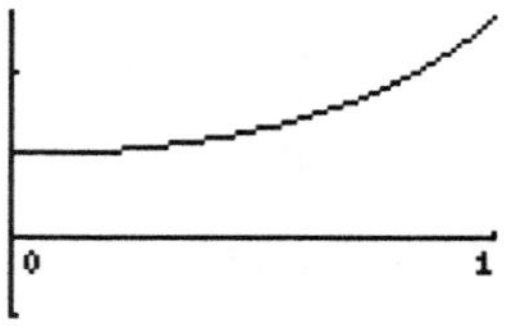	• f is increasing on [0,1]so, • LEFT(2)$\le \int_0^1 e^{x^2} dx \le$ RIGHT(2) • f is concave up on [0,1], so • $\int_0^1 e^{x^2} dx \le$ TRAP(2), but TRAP(2) is a better estimate than LEFT(2) or RIGHT(2), so • LEFT(2) $\le \int_0^1 e^{x^2} dx \le$ TRAP(2) $\le$ RIGHT(2).

On Your Own 2

Consider the definite integrals in **On Your Own 1** above. Arrange LEFT(2), RIGHT(2), and TRAP(2) in ascending order and explain how you could predict without first calculating.

1. $\int_4^{16} \sqrt{x}dx$

2. $\int_0^1 e^{-x}dx$

(The answer appears at the end of this section.)

3. *What affects the error of the approximation?*

✍ ____

____ *(See text p. 319.)*

We've seen before that the larger the number of subintervals n, the closer the approximation, and the smaller the error. However, approximations for "steeper" functions will produce a larger error than approximations for less "steep" functions.

Example 7.5.3

For the following situations, predict which approximation will produce the *smaller error?*

Situations	*Answers*
1a. RIGHT(10) for $\int_1^8 x^{\frac{2}{3}}dx$ or 1b. RIGHT(20) for $\int_1^8 x^{\frac{2}{3}}dx$	1b. produces the smallest error, since there are more subintervals and we are comparing the same function. error RIGHT(20) $\leq$ error RIGHT(10)
2a. RIGHT(10) for $\int_0^1 \sqrt{x}dx$ or 2b. RIGHT(10) for $\int_0^1 5\sqrt{x}dx$	2a. produces the smallest error, because the graph is less steep.

🗁 Key Study Skills

- Review the Riemann Sums, Section 5.2 of Text.
- Choose at least two problems that represent this section; put each on a 3 x 5 card with the solution on the back. Review problems 7.1-7.5.
- Make up your own quiz.
- Form a study group to work on practice problems.
- Use the ✓ ? ★ system on homework problems and get answers to all ★ problems.

Answer to On Your Own 1

1. $\int_4^{16} \sqrt{x}dx$	If $n = 2$, on [4,16], then $\Delta x = 6$ LEFT(2) = $\sqrt{4}(6)+\sqrt{10}(6)= 30.97$ RIGHT(2) = $\sqrt{10}(6)+\sqrt{16}(6)= 42.97$ MID(2) = $\sqrt{7}(6)+\sqrt{13}(6)= 37.51$ TRAP(2) =(LEFT(2)+ RIGHT(2))/2 = (30.97+42.97)/2 = 36.97 Exact value = 112/3 = 37.33...

2. $\int_0^1 e^{-x}dx$	If $n = 2$, on [0,1], then $\Delta x = 0.5$ LEFT(2)= $e^{-0}(0.5) + e^{-0.5}(0.5) = 0.803$ RIGHT(2)= $e^{-0.5}(0.5) + e^{-1}(0.5) + = 0.487$ MID(2) = $e^{-0.25}(0.5) + e^{-0.75}(0.5) = 0.6256$ TRAP(2) = (LEFT(2)+ RIGHT(2))/2 = (0.803+0.487)/2 = 0.645 Exact value = 0.632

Answer to On Your Own 2

Integral	*Graph*	*Answers*
1. $\int_4^{16} \sqrt{x}dx$	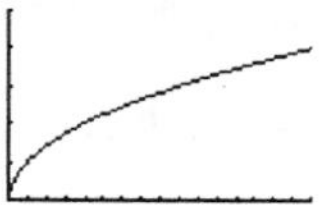	• f is increasing on [4,16] so, • LEFT(2)≤ $\int_4^{16} \sqrt{x}dx$ ≤ RIGHT(2) • f is concave down on [4,16], so • TRAP(2)≤ $\int_4^{16} \sqrt{x}dx$, but TRAP(2) is a better estimate than LEFT or RIGHT, so • LEFT(2) ≤ TRAP(2) ≤ $\int_4^{16} \sqrt{x}dx$ ≤ RIGHT(2).
2. $\int_0^1 e^{-x}dx$	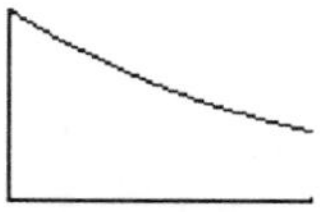	• f is decreasing on [0,1] so, • RIGHT(2) ≤ $\int_0^1 e^{-x}dx$ ≤ LEFT(2) • f is concave up on [0,1], so • $\int_0^1 e^{-x}dx$ ≤ TRAP(2), but TRAP(2) is a better estimate than LEFT or RIGHT, so • RIGHT(2) ≤ $\int_0^1 e^{-x}dx$ ≤ TRAP(2) ≤ LEFT(2)

7.6 APPROXIMATION ERRORS AND SIMPSON'S RULE

Objective: To analyze errors in using numerical methods to evaluate definite integrals.

Preparations: Read and take notes on section 7.6.
Suggested practice problems:. 8,10,11,17,19.

✠ Key Ideas of the Section

You should be able to:

- Approximate definite integrals using Simpson's Rule.
- Compare the errors in approximations using various methods.
- Understand what factors affect the size of the error of the approximation.

Key Questions to Ask and Answer

1. What happens to the size of the error as the number of subintervals increases?

✍ ____

____ *(See text p. 322-323.)*

When using numerical techniques for definite integrals, remember that:

Error = Actual Value – Approximate Value.

As you saw in Section 7.5, for most functions, the larger the number n of subintervals, the closer the approximation and the smaller the error.

Example 7.6.1

Compare the error for the left and right rule.

Definite Integral	*n*	*Error in Left Rule*	*Error in Right Rule*
$\int_1^8 x^{\frac{2}{3}} dx = 18.6$	2	18.6 – 13.04 = 5.56	18.6 – 23.54 = –4.93
	10	18.6 – 17.54 = 1.06	18.6 – 19.64 = –1.04
	50	18.6 – 18.39 = 0.21	18.6 – 18.81 = –0.21
	100	18.6 – 18.49 = 0.11	18.6 – 18.70 = –.10
	250	*18.6 –18.56 = 0.04*	*18.6 – 18.64 = -0.04*

2. *Which method usually produces the best approximation (the smallest error)?*

✍ ____

____ *(See text p. 322-323.)*

Simpson's Rule is a weighted average of the Trapezoid and Midpoint Rules.

$$SIMP(n) = \frac{2 \cdot MID(n) + TRAP(n)}{3}$$

As such, for a fixed number of subintervals n, Simpson's Rule generally gives better approximations than either the Trapezoid or Midpoint Rule. The Trapezoid and Midpoint Rules, however, generally give better approximations than the Left or Right Rules.

Example 7.6.2

For each method, calculate the errors in the approximations using $n = 50$ subintervals for the following integrals.

Integral	*Error in Left rule*	*Error in Right rule*	*Error in Midpoint rule*	*Error in Trapezoid rule*	*Error in Simpson's rule*
$\int_1^8 x^{\frac{2}{3}} dx$	0.21054	–0.20946	–0.00027	0.00054	0.00001
$\int_4^{16} \sqrt{x} dx$	0.24060	–0.23940	–0.0003	0.00060	–0.0000033
$\int_0^1 e^{-x} dx$	–0.00634	0.00630	0.00001	–0.00002	–0.00000056

3. *How does the integrand effect the error of the approximation?*

✍ ____

____ *(See text p. 324-325)*

The "steepness" of functions effects the size of errors using the Left and Right rule. The magnitude of the concavity of functions affects the size of errors using the Midpoint and Trapezoid rules. For most functions, we have the following:

- The larger the magnitude of the first derivative f', the larger the error of approximation using the Left and Right rule.
- The larger the magnitude of the second derivative f'', the larger the error of the approximation using the Midpoint or Trapezoid rule.
- It can be shown that the size of the fourth derivative $f^{(4)}$, affects the size of the error using Simpson's rule.

Example 7.6.3

Compare the expected size of the errors for the following integrals.

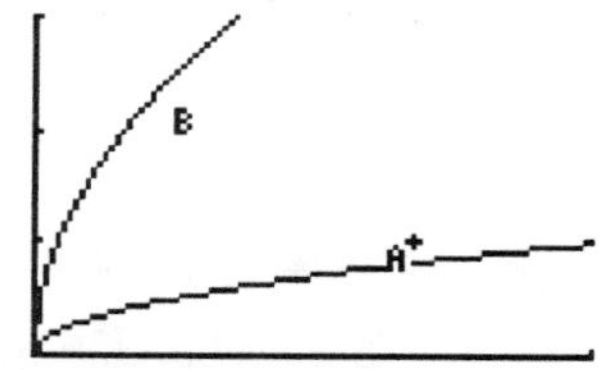

- Function B is "steeper" than function A and therefore has a "larger" f'. The Left and Right rules will produce a larger error for function B than for function A.

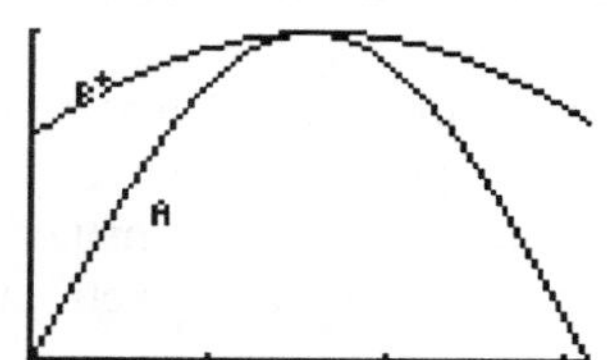

- Function A is more concave or "bent" and therefore has a "larger" f''. The Midpoint and Trapezoid rules will produce a larger error for function A than for function B.

Key Notation

- Simpson's Rule: $SIMP(n) = \dfrac{2 \cdot MID(n) + TRAP(n)}{3}$.
- Review the Midpoint and review the Trapezoid rules from section 7.5 and review for Left and Right rules from section 5.2.

Key Study Skills

- Review the Riemann Sums, Section 5.2 of Text.
- If you are using a graphing calculator, you should add the program for numerical integration to compute approximations to definite integrals using the Left and Right rules, the Midpoint, the Trapezoid and Simpson's Rule. See your instructor for these programs.
- Choose at least two problems that represent this section; put each on a 3 x 5 card with the solution on the back. Review the problems from 7.1-7.6.
- Use the ✓ ? ★ system on homework problems and get answers to all ★ problems.

7.7 IMPROPER INTEGRALS

Objective: To understand convergence and divergence of improper integrals.

Preparations: Read and take notes on section 7.7.
Suggested practice problems:. 4,9,10,14,15,19,24,32,34,39.

✠ Key Ideas of the Section

You should be able to:

- Recognize an improper integral.
- Determine the convergence or divergence of an improper integral.
- Determine the behavior of various classes of improper integrals.

Key Questions to Ask and Answer

1. *What makes an integral improper?*

✍ ____

____ *(See text p. 326-330.)*

Definite integrals are defined over an interval of finite length, and for our discussion thus far in Chapter 7, the functions have been continuous on a closed interval.

- *Improper integrals* may have an *infinite limit of integration* , or
- may have an *unbounded integrand.*

An unbounded integrand is such that the interval of integration contains a value that makes the function undefined at that value and in a way that the function has a vertical asymptote. Notice that the function $y = \frac{|x|}{x}$ is not defined at $x = 0$, but it is not unbounded. We assume here that the functions are positive.

Example 7.7.1

Which of the following are improper integrals and why?

Integral	*Answer*	*Explanation*
$\int_1^\infty \frac{1}{\sqrt{x}}dx$	improper	• Infinite limit of integration since $x \to \infty$.
$\int_1^\infty \frac{1}{\sqrt{x-1}}dx$	improper	• Infinite limit of integration since $x \to \infty$. • Unbounded integrand at $x = 1$, since $\frac{1}{\sqrt{x-1}}$ is undefined at $x = 1$.
$\int_2^{100} \frac{1}{\sqrt{x-1}}dx$	not improper	• Finite interval of integration since [2,100] is a closed interval. • Integrand bounded on interval [2,100].
$\int_{-1}^{1} \frac{1}{x^3}dx$	improper	• Integrand unbounded at x = 0; since $\frac{1}{x^2}$ is undefined at $x = 0$ is contained in the interval of integration. $\int_{-1}^{1} \frac{1}{x^3}dx = \int_{-1}^{0} \frac{1}{x^3}dx + \int_{0}^{1} \frac{1}{x^3}dx$
$\int_0^\infty \cos x dx$	improper	• Infinite limit of integration since $x \to \infty$.

2. *What is convergence and divergence of improper integrals?*

✍ ____

____ *(See text p. 327-331.)*

We can determine convergence and divergence for two types of improper integrals.

- **Infinite limit of integration** $\int_a^\infty f(x)dx$ **:**

Assume $f(x) > 0$ for $x \geq a$. If the $\lim_{b\to\infty}\int_a^b f(x)dx$ is a finite number, then the improper integral $\int_a^\infty f(x)dx$ *converges* ; otherwise it *diverges*.

Because we are assuming $f(x) > 0$, convergence means that the limit of the areas under the curve approach a finite number. For an infinite limit of integration, as $x \to \infty$, if the function gets small enough, fast enough, the limit of the areas approach a finite area and thus converges.

- **Unbounded integrand** $\int_a^b f(x)dx$:

Assume $f(x) > 0$ and continuous on $a \leq x < b$ and tends to infinity as x $x \to b$ If the $\lim_{c\to b^-}\int_a^c f(x)dx$ is a finite number, then the improper integral $\int_a^b f(x)dx$ *converges*; otherwise it *diverges*.

For an unbounded integrand, as x gets close to the vertical asymptote, if the function grows fast enough, the areas approaches a finite number and thus converges.

Example 7.7.2

Explore the convergence or divergence of the following improper integrals numerically and state whether you expect them to converge or to diverge.

Integral	*Graphs*	*Answers*
a. $\int_1^\infty \frac{1}{x^3}dx$	Graph shown on $1 \leq x \leq 5$.	$\int_1^{10} \frac{1}{x^3}dx = 0.495$ $\int_1^{100} \frac{1}{x^3}dx = 0.49995 \Rightarrow \int_1^{1000} \frac{1}{x^3}dx = 0.4999995$ The area appears to converges to ≈ 0.5
b. $\int_1^\infty \frac{1}{\sqrt{x}}dx$	Graph shown on $1 \leq x \leq 5$.	$\int_1^{10} \frac{1}{\sqrt{x}}dx = 4.3246$ $\int_1^{100} \frac{1}{\sqrt{x}}dx = 18 \Rightarrow \int_1^{1000} \frac{1}{\sqrt{x}}dx = 61.2456$ The area appears to diverge.

Example 7.7.3

Investigate the convergence of the following improper integrals analytically.

Integral	*Graphs*	*Answers*
a. $\int_1^\infty \frac{1}{x^3}dx$	Graph shown on $1 \leq x \leq 5$.	$\int_1^\infty \frac{1}{x^3}dx = \lim_{b\to\infty}\int_1^b \frac{1}{x^3}dx$ $\int_1^b \frac{1}{x^3}dx = \int_1^b x^{-3}dx = \frac{-1}{2}x^{-2}\Big\vert_1^b = \frac{-1}{2b^2} + \frac{1}{2}$ $\int_1^\infty \frac{1}{x^3}dx = \lim_{b\to\infty}\left(\frac{-1}{2b^2} + \frac{1}{2}\right) = \frac{1}{2}$

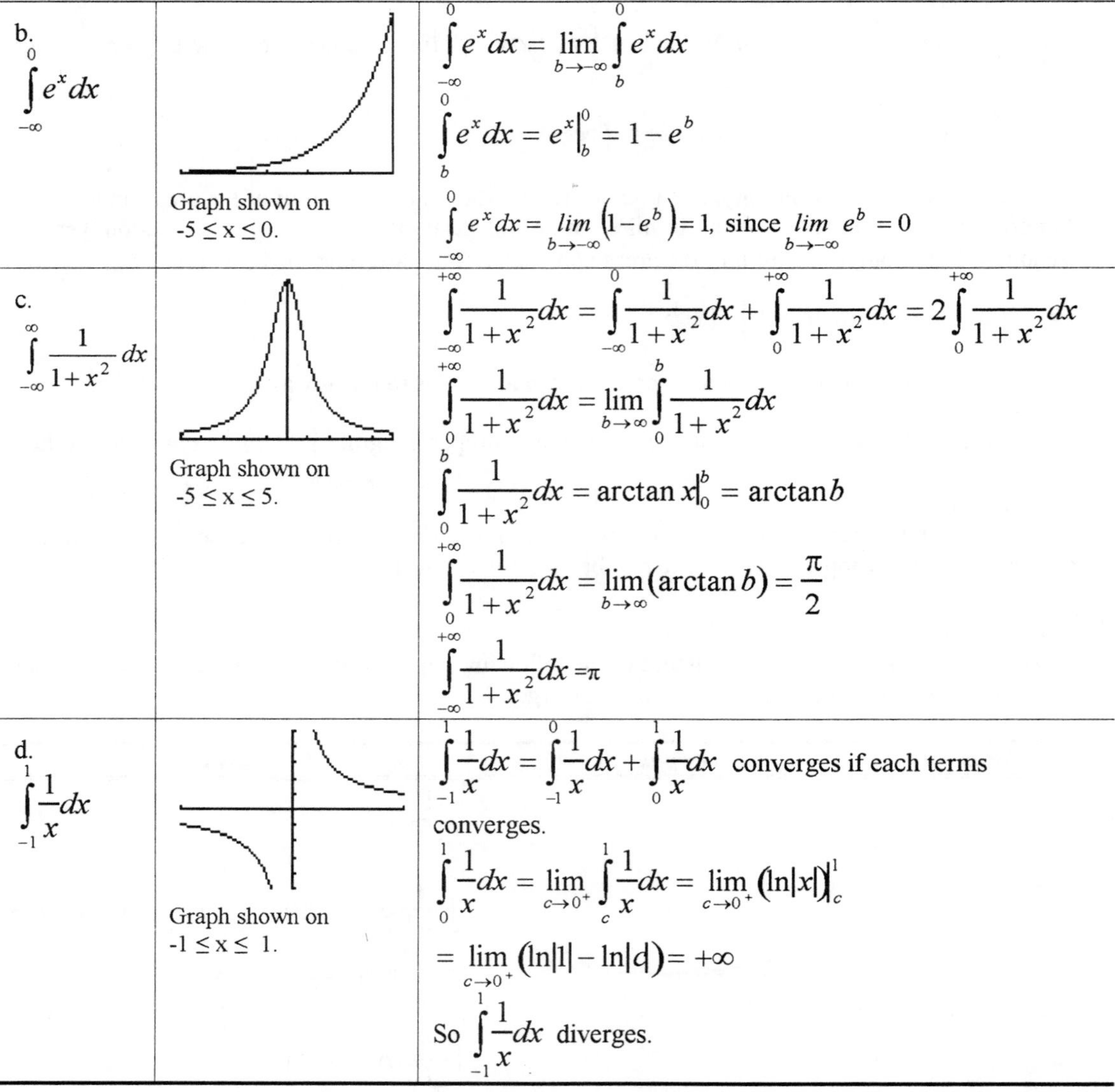

b. $\int_{-\infty}^{0} e^x dx$	Graph shown on $-5 \le x \le 0$.	$\int_{-\infty}^{0} e^x dx = \lim_{b \to -\infty} \int_{b}^{0} e^x dx$ $\int_{b}^{0} e^x dx = e^x \Big\|_b^0 = 1 - e^b$ $\int_{-\infty}^{0} e^x dx = \lim_{b \to -\infty} \left(1 - e^b\right) = 1$, since $\lim_{b \to -\infty} e^b = 0$
c. $\int_{-\infty}^{\infty} \frac{1}{1+x^2} dx$	Graph shown on $-5 \le x \le 5$.	$\int_{-\infty}^{+\infty} \frac{1}{1+x^2} dx = \int_{-\infty}^{0} \frac{1}{1+x^2} dx + \int_{0}^{+\infty} \frac{1}{1+x^2} dx = 2\int_{0}^{+\infty} \frac{1}{1+x^2} dx$ $\int_{0}^{+\infty} \frac{1}{1+x^2} dx = \lim_{b \to \infty} \int_{0}^{b} \frac{1}{1+x^2} dx$ $\int_{0}^{b} \frac{1}{1+x^2} dx = \arctan x \Big\|_0^b = \arctan b$ $\int_{0}^{+\infty} \frac{1}{1+x^2} dx = \lim_{b \to \infty} (\arctan b) = \frac{\pi}{2}$ $\int_{-\infty}^{+\infty} \frac{1}{1+x^2} dx = \pi$
d. $\int_{-1}^{1} \frac{1}{x} dx$	Graph shown on $-1 \le x \le 1$.	$\int_{-1}^{1} \frac{1}{x} dx = \int_{-1}^{0} \frac{1}{x} dx + \int_{0}^{1} \frac{1}{x} dx$ converges if each terms converges. $\int_{0}^{1} \frac{1}{x} dx = \lim_{c \to 0^+} \int_{c}^{1} \frac{1}{x} dx = \lim_{c \to 0^+} \left(\ln\|x\|\right)\Big\|_c^1$ $= \lim_{c \to 0^+} \left(\ln\|1\| - \ln\|c\|\right) = +\infty$ So $\int_{-1}^{1} \frac{1}{x} dx$ diverges.

3. *What is the behavior of $\int_{1}^{\infty} \frac{1}{x^p} dx$ for various values of p?*

✍ ____

____ *(See text p. 328-329.)*

Look at the function $\frac{1}{x^p}$ for two cases:

- **For p >1, the integral will converge.**

$$\int_{1}^{\infty} \frac{1}{x^p} dx = \frac{1}{p-1} \quad \text{since}$$

$$\int_{1}^{\infty} \frac{1}{x^p} dx = \lim_{b \to \infty} \int_{1}^{b} \frac{1}{x^p} dx = \lim_{b \to \infty} \left(\frac{1}{-p+1} x^{-p+1} \right) \Bigg|_1^b = \lim_{b \to \infty} \frac{1}{1-p} \left(\frac{1}{b^{p-1}} - 1 \right) = \frac{1}{p-1}.$$

- **For 0< p ≤ 1, the integral will diverge.**

  If $0 < p < 1$,

  $$\int_1^\infty \frac{1}{x^p}dx \text{ diverges since, } \int_1^\infty \frac{1}{x^p}dx = \lim_{b\to\infty}\frac{1}{1-p}\left(\frac{1}{b^{p-1}} - 1\right) \to \infty .$$

  If $p = 1$,

  $$\int_1^\infty \frac{1}{x^p}dx \text{ since, } \int_1^\infty \frac{1}{x}dx = \lim_{b\to\infty} \ln b - \ln 1 \to \infty$$

Example 7.7.4

Confirm the convergence or divergence of the improper integrals from Example 7.7.3 above.

Integral	*Answers*
a. $\int_1^\infty \frac{1}{x^3}dx$	p=3, hence, $\int_1^\infty \frac{1}{x^3}dx$ converges to 1/2, since $\lim_{b\to\infty}\int_1^b x^{-3}dx = \frac{x^{-2}}{-2}\Big\vert_1^b = \lim_{b\to\infty}\frac{b^{-2}}{-2} - \frac{1}{-2} = \frac{1}{2}$.
b. $\int_1^\infty \frac{1}{\sqrt{x}}dx$	p=1/2, hence, $\int_1^\infty \frac{1}{\sqrt{x}}dx$ diverges, since $\lim_{b\to\infty}\int_1^b x^{\frac{-1}{2}}dx = \frac{x^{\frac{1}{2}}}{\frac{1}{2}}\Big\vert_1^b = \lim_{b\to\infty}\frac{b^{\frac{1}{2}}}{\frac{1}{2}} - \frac{1}{\frac{1}{2}} = \infty$.

Key Notation

- Be careful not to confuse the notation for the definite integral with the notation for the improper integral. $\int_1^\infty \frac{1}{x^p}dx = \lim_{b\to\infty}\int_1^b \frac{1}{x^p}dx \neq \int_1^b \frac{1}{x^p}dx$.

🗁 Key Study Skills

- Choose at least two problems that represent this section; put each on a 3 x 5 card with the solution on the back. Review problems from 7.1-7.7.
- Make up your own quiz.
- Use the ✓ ? ★ system on homework problems and get answers to all ★ problems.

7.8 COMPARISON OF IMPROPER INTEGRALS

Objective: To apply the comparison test for convergence or divergence of improper integrals.

Preparations: Read and take notes on section 7.8.
Suggested practice problems: 2,6,12,15,18,22,28,31.

✠ Key Ideas of the Section

You should be able to:

- Know the convergence or divergence of certain improper integrals useful for comparison.
- Test whether or not you expect convergence or divergence of an improper integral numerically.
- Determine the convergence or divergence of an improper integral using the comparison test.

Key Questions to Ask and Answer

1. ***What is the comparison test for convergence or divergence of improper integrals?***

✍ ____

____ *(See text p. 335.)*

For improper integrals, compare the function to ones that we are certain converge or diverge. Let us assume $f(x)>0$.

a. If you suspect $\int_a^\infty f(x)dx$ *converges*, then look for a function $g(x)$ which is *above* $f(x)$ for $x \geq a$ and such that $\int_a^\infty g(x)dx$ converges.

b. If you suspect $\int_a^\infty f(x)dx$ diverges, then look for a function $g(x)$ that is below $f(x)$ for $x \geq$ a, and such that $\int_a^\infty g(x)dx$ diverges.

Useful functions for comparison are:

- $\int_1^\infty \frac{1}{x^p}dx$ converges for $p > 1$ and diverges for $p \leq 1$.
- $\int_0^1 \frac{1}{x^p}dx$ converges for $p < 1$ and diverges for $p \geq 1$.
- $\int_0^\infty e^{-ax}dx$ converges for $a > 0$.

Example 7.8.1

Investigate the convergence of the following improper integrals.

Integral	*Reasoning*	*Graphs*
$\int_4^{\infty} \frac{1}{x^4+e^x}dx$	• Testing numerically: $\int_4^{10} \frac{1}{x^4+e^x}dx \approx 0.0037$, $\int_4^{100} \frac{1}{x^4+e^x}dx \approx 0.003743$ • Guess: convergent. • For $x \geq 4$ $\frac{1}{x^4+e^x} < \frac{1}{x^4}$ and $\int_4^{\infty} \frac{1}{x^4}dx$ converges, so therefore $\int_4^{\infty} \frac{1}{x^4+e^x}dx$ converges.	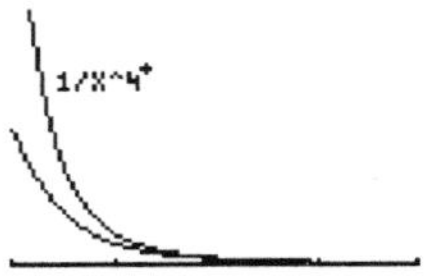 The graph of $1/x^4$ *lies above* the graph of $1/(x^4+e^x)$.
$\int_1^{\infty} \frac{\ln x}{\sqrt{x}}dx$	• Testing numerically: $\int_1^{10} \frac{\ln x}{\sqrt{x}}dx \approx 5.91$, $\int_1^{100} \frac{\ln x}{\sqrt{x}}dx \approx 56.09$ • Guess: divergent. for x>e , $\frac{1}{\sqrt{x}} < \frac{\ln x}{\sqrt{x}}$ and $\int_1^{\infty} \frac{1}{\sqrt{x}}dx$ diverges, so therefore $\int_1^{\infty} \frac{\ln x}{\sqrt{x}}dx$ diverges.	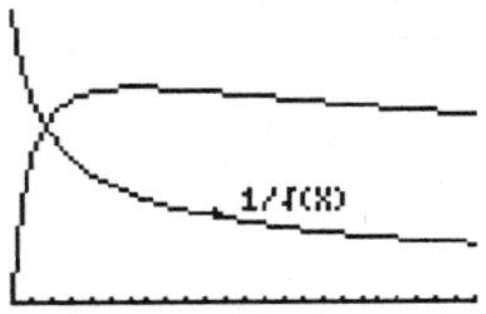 The graph of $\frac{1}{\sqrt{x}}$ *lies below* the graph of $\frac{\ln x}{\sqrt{x}}$ .for $x > e$.
$\int_0^{1} \frac{1}{\sqrt{x^3+x}}dx$	• Testing numerically: $\int_{0.1}^{1} \frac{1}{\sqrt{x^3+x}}dx \approx 1.222$, $\int_{0.01}^{1} \frac{1}{\sqrt{x^3+x}}dx \approx 1.6545$ $\int_{0.001}^{1} \frac{1}{\sqrt{x^3+x}}dx \approx 1.79899$ • Guess: convergent for 0 < x < 1, $\frac{1}{\sqrt{x^3+x}} < \frac{1}{\sqrt{x}}$ and $\int_0^{1} \frac{1}{\sqrt{x}}dx$ converges, so therefore $\int_0^{1} \frac{1}{\sqrt{x^3+x}}dx$ converges.	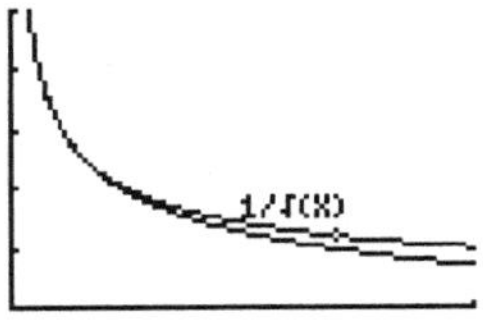 The graph of $\frac{1}{\sqrt{x}}$ on (0,1] is *above* $\frac{1}{\sqrt{x^3+x}}$.

On Your Own 1

Investigate the convergence of the following improper integrals.

1. $\int_1^\infty \frac{5+\sin\theta}{\theta^2}d\theta$

2. $\int_1^\infty \frac{1}{\sqrt{x^2+x}}dx$

(The answer appears at the end of this section.)

Key Study Skills

- Choose at least two problems that represent this section; put each on a 3 x 5 card with the solution on the back. Review the note card problems 7.1-7.8.
- Use the ✓ ? ★ system on homework problems and get answers to all ★ problems.

Answers to On Your Own 1

Integral	*Reasoning*	*Graphs*
$\int_1^\infty \frac{5+\sin\theta}{\theta^2}d\theta$	• Testing numerically: $\int_1^{10} \frac{5+\sin\theta}{\theta^2}d\theta \approx 5.0145$ $\int_1^{100} \frac{5+\sin\theta}{\theta^2}d\theta \approx 5.648$ • Guess: convergent. for $\theta \geq 1$, $\frac{5+\sin\theta}{\theta^2} \leq \frac{5+1}{\theta^2} = \frac{6}{\theta^2}$ and $\int_1^\infty \frac{6}{\theta^2}d\theta$ converges, so therefore $\int_1^\infty \frac{5+\sin\theta}{\theta^2}d\theta$ converges.	6/x²

$\int_1^{\infty} \frac{1}{\sqrt{x^2+x}}dx$	• Testing numerically: $\int_1^{10} \frac{1}{\sqrt{x^2+x}}dx \approx 1.974$ $\int_1^{100} \frac{1}{\sqrt{x^2+x}}dx \approx 4.2388$ $\int_1^{1000} \frac{1}{\sqrt{x^2+x}}dx \approx 7.51247$ • Guess: divergent. for $x \geq 1$, $\frac{1}{\sqrt{x^2+x}} \geq \frac{1}{\sqrt{x^2+x^2}} = \frac{1}{\sqrt{2}x}$ and $\int_1^{\infty} \frac{1}{\sqrt{2}x}dx$ diverges, so therefore $\int_1^{\infty} \frac{1}{\sqrt{x^2+x}}dx$ diverges.	1/√(2)x

CUMULATIVE REVIEW CHAPTER 7

🗁 Key Study Skills

- Review all of the **Key Ideas** from 7.1 - 7.8.
- Make up you own test using your 3 x 5 note cards. Review the ★ problems.
- Choose some Review Exercises and Problems for Chapter 7 in text, p. 338-341.
- Do Check Your Understanding problems in text, p. 342.
- See your instructor and discuss what sections will be most emphasized on the next exam.
- Know all the formulas of the various integral approximations. Practice, practice, practice.
- Do the practice test which follows.

Review of Approximation Rules:

Let f be a function defined on [a, b] of n subdivisions of length Δx where $a = x_0 < x_1 < \ldots < x_n = b$	
Riemann sum: Left rule	LEFT(n)= $\sum_{i=0}^{n-1} f(x_i)\Delta x$
Riemann sum: Right rule	RIGHT(n)= $\sum_{i=1}^{n} f(x_i)\Delta x$
Riemann sum: Midpoint rule	MID(n)= $\sum_{i=1}^{n} f(x_{i-mid})\Delta x$, where x_{i-mid} is midpoint of ith interval

Trapezoid rule	TRAP(n)= $\sum_{i=1}^{n}\left[\frac{f(x_i)+f(x_{i-1})}{2}\right]\Delta x$ = (LEFT(n)+ RIGHT(n))/2
Simpson's rule	SIMP(n)=(2MID(n)+TRAP(n))/3

Chapter Seven Practice Test

Integrate:

1. $\int \frac{dx}{x^2+6x+12}$

2. $\int e^{3x}\sin 3x dx$

3. $\int_4^{12} \frac{x}{\sqrt{x-3}}dx$

4. Use numerical methods to approximate: $\int_0^1 \sin(x^2)dx$.

5. Test for convergence or divergence: $\int_0^{\frac{\pi}{2}} \tan^2\theta d\theta$

5. Find the following integral: $\int \frac{x+1}{x^2-x}dx$.

7. Find the following integral: $\int \frac{1}{\sqrt{4-x^2}}dx$

(The answers appear at the end of this Study Guide)

CHAPTER EIGHT

USING THE DEFINITE INTEGRAL

The definite integral can be used to solve
problems in geometry, physics, economics and probability.

8.1 AREAS AND VOLUMES

Objective: To calculate areas and volumes using Riemann sums.

Preparations: Read and take notes on section 8.1.
Suggested practice problems: 2,5,8,10,12,14,18,27,28.

✠ Key Ideas of the Section

You should be able to:

- Use use horizontal or vertical slices to set up a Riemann sum to approximate an area or volume of a geometric figure or solid.
- Represent this area or volume as a definite integral.
- Use the definite integral to calculate the area or volume.

Key Questions to Ask and Answer

1. ***How can a geometric region be sliced to construct a Riemann sum to approximate the area?***

✍ ____

____ *(See text p.346-350.)*

Slice the figure horizontally (or vertically) so that you can estimate the area of each slice. Find the approximate areas of each strip in terms of one variable and then sum the areas of the strips to approximate the area of the region. Taking the limit as you have more and more strips expresses the area of the region as a definite integral.

Example 8.1.1

Use both horizontal and vertical slices to set up the definite integral to calculate the area of a right triangle of height 5 inches and base 10 inches.

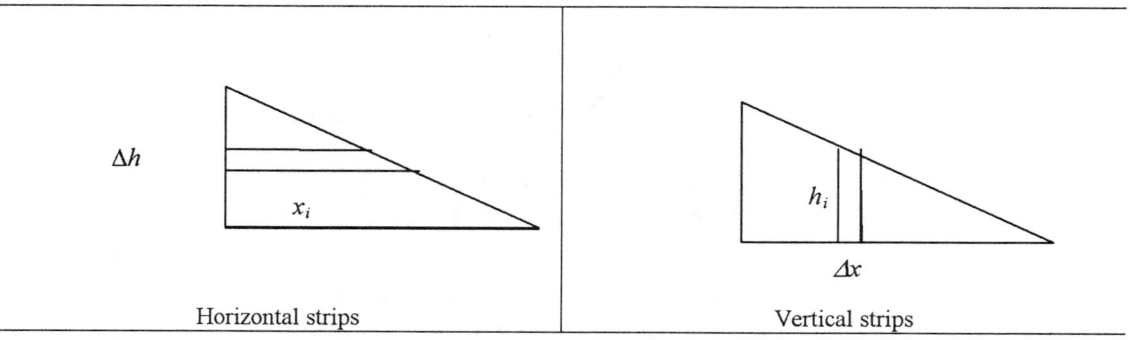

Horizontal slices have width, x_i and thickness, Δh. The area of each strip $\approx x_i \Delta h$ in^2. Because we want to sum the areas of the strips for $0 \le h \le 5$ inches, get x_i in terms of h_i using similar triangles: $\frac{5}{10} = \frac{5 - h_i}{x_i}$ or $x_i = 2(5 - h_i)$. This gives the area of the strip $\approx 2(5 - h_i)\Delta h$ in^2. The Area of the triangle $\approx \sum_{i=1}^{n} 2(5 - h_i)\Delta h$ in^2 or

$$\text{Area} = \approx \lim_{n\to\infty} \sum_{i=1}^{n} 2(5 - h_i)\Delta h \text{ in}^2 = \int_0^5 2(5-h)\,dh = 2(5h - \frac{h^2}{2})\Big|_0^5 = 25in^2$$

Vertical slices have width Δx and height, h_i. Here, because we want to add up the areas of the strips for $0 \le x \le 10$, use similar triangles to get h_i in terms of x_i: $\frac{5}{10} = \frac{h_i}{x_i}$ or $h_i = 0.5\,x_i$. The area of each strip $\approx h_i \Delta x$ in^2 giving the Area of the triangle $\approx \sum_{i=1}^{n} 0.5 x_i \Delta x$ in^2 or

$$\text{Area} = \approx \lim_{n\to\infty} \sum_{i=1}^{n} 0.5 x_i \Delta x \text{ in}^2 = \int_0^{10} 0.5x dx = 0.25x^2\Big|_0^{10} = 25in^2.$$

2. ***How can a solid be sliced to construct a Riemann sum to approximate the volume?***

✍ ____

____ (See text p.346-349.)

Approximate the volume of a solid by dividing the solid into small slices for which it will be easy to compute an approximate volume. You will choose either vertical slices or horizontal slices.

Example 8.1.2

For each of the solids given below, decide how to slice it and calculate the volume of the typical *i-th* slice:

a. A solid cylinder of height h cm. and radius r cm,
b. A right circular cone of height 5 cm and radius 2 cm and
c. A solid triangular pyramid whose base is a right triangle with sides of length 3, 4, and 5 in. and the vertical altitude above the vertex of the right angle has a height of 10 in.

Here are some possible approaches:

Solid	*Division and Volume of each Slice*
a. A solid cylinder of height h cm and radius r cm.	• Divide the cylinder into n parallel horizontal slices. Each slice is "coin-shaped" of radius r and of thickness Δh. • The Volume of each slice $\approx \pi r^2 \Delta h$ cm^3.
b. A right circular cone.	• Divide the cone into n parallel horizontal slices, which also give "coin-shaped" slices of radius r_i and of thickness Δh. Notice that here, the radius of the piece depends on the height of that slice above the base. • The Volume of an *i-th* slice$\approx \pi r_i^2 \Delta h$ cm^3.
c. A solid triangular pyramid whose base is a right triangle and a vertical altitude above the vertex of the right angle.	• Divide the pyramid into n horizontal slices parallel to the base. Each piece is the shape of a right triangle of thickness Δx. with Base b_i and height h_i. • The Volume of each slice $\approx \frac{1}{2} b_i \Delta x$ in^3.

3. *How can Riemann sums be used to approximate the total volume?*

✍ ____

____ *(See text p.346-349.)*

The Riemann sum adds up the volumes of all these small pieces, giving an approximate total volume.

Example 8.1.3

Write the expression for the approximate Total Volume of each of the solids used in **Example 8.1.2**.

Solid	*Total Volume*
a. A solid cylinder of height h cm and radius r cm.	*The Total Volume* $\approx \sum_{i=1}^{n} \pi r^2 \Delta h$ cm^3.
b. A right circular cone	*The Total Volume* $\approx \sum_{i=1}^{n} \pi r_i^2 \Delta h$ cm^3.
c. A solid triangular pyramid whose base is a right triangle and a vertical altitude above the vertex of the right angle.	*The Total Volume* $\approx \sum_{i=1}^{n} \frac{1}{2} b_i h_i \Delta x$ in^3.

4. *How can the integral be used to compute the volume?*

✍ ____

____ *(See text p.346-349.)*

First, write this volume in terms of the variable over which you're summing. The definite integral for the total volume is the limit of the Riemann sum as the number of terms tends towards infinity.

Example 8.1.4

Write the expression for the *Total Volume* for each of the solids below.

Solid	*Volume*
a. A solid cylinder of height h cm. and radius r cm.	Here the volumes of the slices are all the same, so the *Total Volume* $V=\int_0^h \pi r^2 dh = \pi r^2 h$ cm^3.
b. A right circular cone of height 5 cm and radius 2 cm.	The radius r of a typical *i-th* slice depends on the height from the base of the cone. By similar triangles, $\frac{r}{2} = \frac{5-h}{5}$ or $r = \frac{2}{5}(5-h)$. The *Total Volume* : $V = \int_0^5 \pi r^2 dh = \int_0^5 \pi\left(\frac{2}{5}(5-h)\right)^2 dh = \frac{20}{3}\pi$ cm^3.
c. A solid triangular pyramid whose base is a right triangle with sides of length 3, 4, and 5 inches and the vertical altitude above the vertex of the right angle is height of 10 in.	Take a typical triangular slice x–units from the base of the pyramid with the base b and height h and of thickness Δx. Get both the variables b and h in terms of x. By similar triangles: $\frac{b}{3} = \frac{10-x}{10}$ *and* $\frac{h}{4} = \frac{10-x}{10}$ or $b = \frac{3}{10}(10-x)$ *and* $h = \frac{4}{10}(10-x)$ *Total Volume*: $V = \int_0^{10} \frac{1}{2} bh dx = \int_0^{10} \frac{1}{2}\left(\frac{3}{10}(10-x)\right)\left(\frac{4}{10}(10-x)\right)dx$ $= \left(\frac{1}{2}\right)\left(\frac{3}{10}\right)\left(\frac{4}{10}\right)\int_0^{10}(10-x)^2 dx$ $= 20$ in^3.

On Your Own 1

Find the volume of a sphere of radius 10 cm.

(The answer appears at the end of this section.)

Key Notation

- Riemann sum notation: $\sum_{i=1}^{n} f(x_i)\Delta x$ is the sum of the products of $f(x_i)$ and Δx from 0 to n.
- To determine volume use the integral as the limit of Riemann sums: $V = \lim_{n\to\infty} \sum_{i=1}^{n} f(x_i)\Delta x = \int_a^b f(x)dx$

Key Algebra

- Review summation notation.
- Review various area formulas such as the area of circle of radius r, $A = \pi r^2$.
- Review similar triangles.

Key Study Skills

- Choose two problems that represent this section and put each on a 3 x 5 note card with the solution on the back.
- Use the ✓ ? ★ system on homework problems and get answers to all ★ problems.
- Learn to use the technology preferred by your instructor. The text assumes you have a graphing calculator available or a computer algebra system.

Answer to On Your Own 1

This can be done by taking horizontal slices as **Example 2** on page 347 of the text. See the graph below.

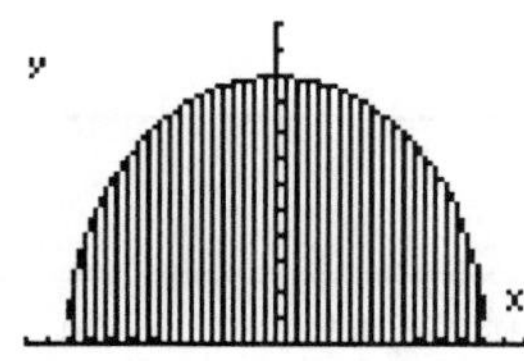

Volume of each slice $\approx \pi\left(\sqrt{10^2 - x_i^2}\right)^2 \Delta x$.

Total Volume $\approx \sum_{i=1}^{n} \pi\left(\sqrt{10^2 - x_i^2}\right)^2 \Delta x$.

$$V = \lim_{n\to\infty} \sum_{i=1}^{n} \pi(10^2 - x_i^2)\Delta x =$$

$$\int_{-10}^{10} \pi(10^2 - x^2)dx = 10^2\pi x - \frac{\pi x^3}{3}\bigg|_{-10}^{10} = \frac{4}{3}(10^3)\pi \text{ cm}^3$$

8.2 APPLICATIONS TO GEOMETRY

Objective: To compute volumes of more complicated solids or lengths of curves using the definite integral.

Preparations: Read and take notes on section 8.2.
Suggested practice problems: 2,6,7,10,17,25,26.

✠ Key Ideas of the Section

You should be able to:

- Slice the solid (or curve) into pieces for which you can approximate the volume (or length).
- Use Riemann sums to approximate the total volume (or length) of a solid (or curve).
- Represent this volume (or length) as a definite integral.
- Find volumes of revolution.
- Find volumes of regions of a known cross-section.
- Find arc length of curves.

Key Questions to Ask and Answer

1. *How can you find the volume of a solid of revolution?*

✍ ____

____________________________________ *(See text p. 353-354.)*

Revolving a region in the plane about a line creates a solid of revolution. To find the volume, slice the region perpendicular to the axis of revolution. The slices are therefore circular and hence have volume of the form $V \approx \pi r^2 \Delta x$ (if revolving about the x-axis or Δy if revolving about the y-axis) where r is the radius and Δx (or Δy) is the thickness. The radius will be expressed as a function of the variable over which you're rotating.

Example 8.2.1

Find the volumes of the solids of revolution below.

Solid of revolution	*Graph of region*	*Volume*
A region bounded by the curve $y = x^3$, $x = 0$, $x = 1$ and the x-axis is revolved around the x-axis. *This will create a "trumpet" like figure.*		Slice this solid into pieces perpendicular to the x-axis. Each slice will be coin-shaped of thickness Δx and with radius the distance from the x-axis to the curve $y = x^3$. The volume of a slice is $\approx \pi y^2 \Delta x$. The total volume of the solid $V \approx \sum \pi y^2 \Delta x = \sum \pi \left(x^3\right)^2 \Delta x$. As the thickness tends toward zero $V = \int_0^1 \pi x^6 dx = \frac{\pi}{7}$ cubic units.

A region bounded by the curve $y = x^3$, $y = 0$, $y = 1$ and the y-axis, and is revolved around the y-axis. *This will create a solid "bowl" type figure.*	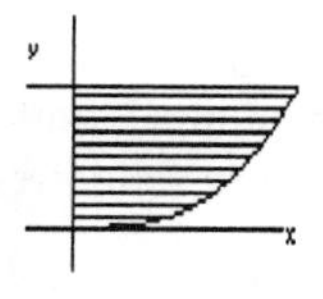	Slice this solid into pieces perpendicular to the y-axis. Each slice will be coin-shaped of thickness Δy and with radius the distance from the y-axis to the curve which is x, such that $y = x^3$. The volume of a slice is $\approx \pi x^2 \Delta y = \pi\left(y^{\frac{1}{3}}\right)^2 \Delta y$. The total volume of the solid $V \approx \sum \pi x^2 \Delta y = \sum \pi\left(y^{\frac{1}{3}}\right)^2 \Delta y$. As the thickness tends toward zero $V = \int_0^1 \pi y^{\frac{2}{3}} dy = \frac{3\pi}{5}$ cubic units.
A region bounded by the curve $y = x^3$, $x = 0$, $x = 1$ and the x-axis is revolved around the line $y = -2$.	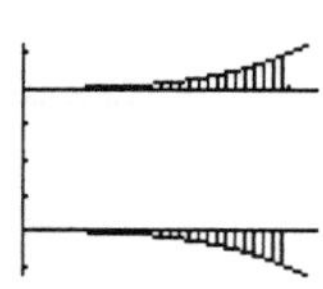	Revolve this region about the line $y = -2$ and slice the solid into pieces perpendicular to the x-axis of thickness Δx. Each piece is "washer" shaped. To find the volume of the "washer" we need to find the volume of the disk and subtract the volume of the "hole". If r_{out} is the radius of the outside and r_{in} is the radius of the "hole", the volume of the slice is $\approx \pi r_{out}^2 \Delta x - \pi r_{in}^2 \Delta x$ where r_{out} is the distance from the line $y = -2$ to the curve $y = x^3$ or $r_{out} = 2 + x^3$ and r_{in} is the distance from the line $y = -2$ to the x-axis or $r_{in} = 2$. So the volume of the slice is $\approx \pi(2 + x^3)^2 \Delta x - \pi(2)^2 \Delta x$. The total volume of the solid is $V \approx \sum \pi[(2 + x^3)^2 - 2^2]\Delta x$. As the thickness tends toward zero $V = \int_0^1 \pi\left[(2 + x^3)^2 - 2^2\right]dx = \frac{8\pi}{7}$ cubic units.

On Your Own 1

Find the volume of the solid created by revolving the region bounded by $y = x^2$ and $y = x$ about the x-axis.

(The answer appears at the end of this sections)

2. *How can you find the volume of regions of a known cross-section?*

✍ ____

____ *(See text p.356.)*

These solids are of standing squares, triangles or semicircles with the base being a region in the xy-plane. Cross-sections taken perpendicular to the xy-plane will be squares, semicircles or triangles standing vertically on edge. Find a typical slice and compute the approximate volume. Sum these up and as you take the limit the volume can be expressed as a definite integral.

Example 8.2.2

Find the volume of the solid whose base is the region in the *xy*-plane bounded by the curves $y = x^2$ and $y = x$ and whose cross-sections perpendicular to the *x*-axis are semicircles. The radius of each slice is $r = \frac{1}{2}\left(x - x^2\right)$ and the thickness Δx. The volume of each slice is

$\approx \frac{\pi}{2} r^2 \Delta x = \frac{\pi}{2}\left(\frac{1}{2}\left(x - x^2\right)\right)^2 \Delta x$. The Total volume V$\approx \sum \frac{\pi}{2}\left(\frac{1}{2}\left(x - x^2\right)\right)^2 \Delta x$ and as $\Delta x \to 0$,

$$V = \int_0^1 \frac{\pi}{2}\left(\frac{1}{2}\left(x - x^2\right)\right)^2 dx = \frac{\pi}{8}\int_0^1 \left(x - x^2\right)^2 dx = \frac{\pi}{240}$$

On Your Own 2

Find the volume of the solid whose base is the region in the xy-plane bounded by the curves $y = x^2$ and $y = x$ and whose cross-sections perpendicular to the x-axis are squares.

(The answer appears at the end of this section.)

3. *How can a curve be divided into small pieces in order to find an arc length?*

____

____ *(See text p.356-357.)*

Just as the definite integral is used to compute volumes, it can be used to compute the lengths of curves. Cut the curve into small pieces, approximate the length of each piece using a straight line approximation. The Riemann sum adds up the lengths of all of these small pieces and then as the lengths tend toward zero, we get the arc length in terms of the definite integral.

Example 8.2.3

If you walk along a path described by the curve $y = sin(x)$ for $x = 0$ to $x = \pi$, how far have you walked?

The arclength of a curve $f(x) = sin(x)$ from $x=0$ is given by:

$$\int_0^{\pi} \sqrt{1 + \left(f'(x)\right)^2}\, dx = \int_0^{\pi} \sqrt{1 + \left(\cos(x)\right)^2}\, dx = 3.8 \text{ units, where } f'(x) = \cos(x).$$

Most often these integrands do not have elementary antiderivatives and you will need to use numerical methods. Computer algebra systems and calculators are most helpful here.

On Your Own 3

Compare the length of the curves $L_1 : y = x$ and $L_2 : y = x^4$, for $x = 0$ to $x = 1$.

(The answer appears at the end of this section.)

Key Notation

- Riemann sum notation: $\sum_{i=0}^{n-1} f(x_i)\Delta x$ is the sum of the product of *f(x)* and Δx from 0 to *n-1*.
- To determine volume use the integral as the limit of Riemann sums:
  $V = \lim_{n\to\infty} \sum_{i=0}^{n-1} f(x_i)\Delta x = \int_a^b f(x)dx$

Key Algebra

- Review summation notation. *(See text p. 229.)*
- Review various area formulas such as the area of circle of radius r, $A = \pi r^2$.
- Review similar triangles.

Key Study Skills

- Choose two problems that represent this section and put each on a 3 x 5 note card with the solution on the back.
- Use the ✓ ? ★ system on homework problems and get answers to all ★ problems.
- Learn to use the technology preferred by your instructor. The text assumes you have a graphing calculator available or a computer algebra system.

Answer to On Your Own 1

Graph the two functions and notice that they intersect at x = 0 and x = 1. This region to be revolved is bounded above by $y = x$ and below by $y = x^2$. When revolved, the slices perpendicular to the x-axis are shaped like washers with outside radius $r_{out} = x$ and inside radius $r_{in} = x^2$. The approximate volume of each slice is $\pi\left(r_{out}^2 - r_{in}^2\right)\Delta x$. The Volume of the solid of revolution $\approx \sum \pi\left(r_{out}^2 - r_{in}^2\right)\Delta x$ and

Volume$= \int_0^1 \pi\left(x^2 - \left(x^2\right)^2\right)dx = \pi\left(\frac{x^3}{3} - \frac{x^5}{5}\right)\Big|_0^1 = \frac{2\pi}{15}$

Answer to On Your Own 2

The length of each side of the square slice is $s = x - x^2$ and the thickness Δx. The volume of each slice is $\approx s^2\Delta x = \left(x - x^2\right)^2 \Delta x$. The Total volume V$\approx \sum \left(x - x^2\right)^2 \Delta x$ and as Δx $\to$ 0,

$$V = \int_0^1 \left(x - x^2\right)^2 dx = \int_0^1 (x^2 - 2x^3 + x^4)dx = \frac{1}{30}$$

Answer to On Your Own 3

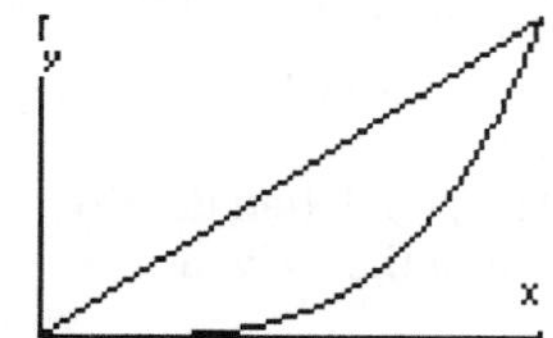

L1: arclength = $\sqrt{2} \approx$ 1.414 units by Pythagorean Theorem

L2: arclength $= \int_0^1 \sqrt{1 + \left(f'(x)\right)^2}\,dx = \int_0^1 \sqrt{1 + \left(4x^3\right)^2}\,dx \approx 1.60$ units

8.3 DENSITY AND MASS

Objective: To compute total population, total mass, or center of mass of a substance from density using the definite integral.

Preparations: Read and take notes on section 8.3.
Suggested practice problems:. 5,6,12,18,23,27

✠ Key Ideas of the Section

You should be able to:

- Use Riemann sums to approximate the total population or total mass of a substance over a line, a region or a solid, given the density.
- Represent this total population or total mass as a definite integral.
- Find the center of mass of a system with given density over a line, a region or a solid.

Key Questions to Ask and Answer

1. How do you calculate the total mass or total population when the density is not constant?

✍ ____

____ *(See text p.360-362.)*

To calculate the total mass we use Riemann sums similar to sections 8.1 and 8.2, where volumes or arclength were found by dividing the region into small pieces and summing.

- Divide the region into small pieces so that the density is approximately constant on each piece. These pieces may be small segments along a line, thin rectangular strips, thin concentric circles, or circular slices.
- Calculate the approximate population or mass of each piece and then find the sum of all the estimates.
- Find the definite integral as the limit of this sum as the "size" of the pieces goes to zero.

As with all definite integral applications, it is helpful to use the units of measurement in setting up a problem. Density of population is measured in people per unit area or bacteria per cubic centimeter, for example. The density of a substance is the mass of a unit volume that might be measured in grams per cubic centimeter.

Example 8.3.1

Suppose you have City A, City B, and City C as follows:

- City A is a 2 mile square region bounded on the east by a river. The population density, P, is a function of the distance r miles from the river, or $P = f(r)$ people/square mile.
- City B is a circular city of radius 2 miles where the population density $P = g(x)$ people/square mile is a function of the distance x miles from the center.
- City C is in a very narrow canyon along a highway of length 6 km. The population density, $P = h(q)$ people/km, is a function of the distance q km from the mouth of the canyon

Express the total population as a definite integral.

City		Total Population
A		• Divide the region into vertical rectangular strips of width Δr, and height 2. So the area of a strip is approximately $2\Delta r$ sq miles. • The population per rectangular strip is approximately: (area of slice in sq miles)(density in people/sq. mile) $=2\Delta r\, f(r)$. • The total population of City A $= \lim\limits_{n\to\infty} \sum\limits_{i=1}^{n} 2f(r)\Delta r = \int_0^2 2f(r)dr$ people.
B		• Divide the region into circular rings of width Δx, and length of circumference is $2\pi x$ which can be stretched out into rectangles. • The area of each ring is approximately the area of a rectangle $2\pi x\Delta x$ sq miles. • The population per ring is approximately: (area of slice in sq miles)(density in people/sq mile) = $2\pi x\Delta x\, g(x)$ • Total Population of City B $= \lim\limits_{n\to\infty} \sum 2\pi x(\Delta x)(g(x)) = \int_0^2 2\pi x(g(x))dx$ people.
C		• Divide line into subintervals of length Δq km. • The population per segment is approximately: (length in km)(density in people/km) = $(\Delta q)(h(q))$ • Total Population of City C $= \lim\limits_{n\to\infty} \sum h(q)\Delta q = \int_0^6 h(q)dq$ people.

On Your Own 1

Find the total mass of a straight rod of length a cm whose density, in gm/cm, varies as the square of the distance from one end

(The answer appears at the end of this section.)

2. *What is the center of mass of a mechanical system?*

____

____ *(See text p.362-366.)*

Finding the center of mass of a system of n discrete masses m_i along the x-axis each located at coordinate x_i can be thought similar to finding where to put the fulcrum for a see-saw having n people of weights m_i located at position x_i.

$$\textit{The center of mass } \bar{x} = \frac{\text{the sum of the moments of the mass}}{\text{the total mass of the system}}$$

Let $\delta(x)$ be the mass density (mass per unit length) where an object is lying on the x-axis at point x. Just as in the applications in previous sections, to extend this to the continuous case for mass density we use Riemann sums.

- Divide the object into n small pieces of length (width, or thickness) Δx.
- Approximate the mass of each piece, $\bar{x}$
- Form the Riemann sums for each and as $n \to \infty$ we get the formula for the center of mass:

$$\text{The center of mass } = \bar{x} = \frac{\text{Moment}}{\text{Total Mass}} = \frac{\int_a^b x\delta(x)dx}{\int_a^b \delta(x)dx}$$

Remember that ***mass*** = (*density*) (***length***) or (*density*) (***area***) or (*density*) (***volume***).

Example 8.3.2

Find the center of mass of a uniform circular half-disk of radius 5 meters and mass 10 kg with center at the origin and diameter along the y-axis.

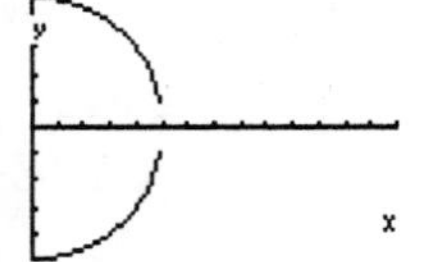

Because the mass of the semi-circle is symmetrically distributed with respect to the x-axis, we expect the center of mass to be closer to the origin than to $x = 5$.

The function for a semi circle is $y = \sqrt{5^2 - x^2}$

- The area of half-disk= $\frac{1}{2}\pi r^2 = \frac{1}{2}\pi \cdot 5^2 = \frac{25\pi}{2}$ m^2.

  Density of half-disk = mass/area = $\frac{10}{\frac{25\pi}{2}} = \frac{4}{5\pi}$ kg/m^2

- Slice into vertical strips of with Δx, area of strip$\approx 2\sqrt{5^2 - x_i^2}\,\Delta x$ m^2.
- Mass of strip = (density)(area)$\approx \frac{4}{5\pi} 2\sqrt{5^2 - x_i^2}\,\Delta x$.
- Center of Mass $\bar{x} \approx \dfrac{\sum x_i\left(\frac{8}{5\pi}\sqrt{25 - x_i^2}\,\Delta x\right)}{10}$.
- Center of Mass $\bar{x} = \dfrac{\frac{8}{5\pi}\int_0^5 x\left(\sqrt{25 - x^2}\right) dx}{10} \approx 2.1$ m.

On Your Own 2

Find the center of mass of a rod of length 10 cm., lying along the x-axis with density directly proportional to the square of the distance from one end.

(The answer appears at the end of this section.)

Key Notation

- Riemann sum notation: $\sum_{i=1}^{n} f(x_i)\Delta x$ is the sum of the product of *f(x)* and Δx from 0 to *n*.

Key Algebra

- Review summation notation.
- Review various area formulas such as the area of circle of radius r is $A = \pi r^2$

Key Study Skills

- Choose two problems that represent this section and put each on a 3 x 5 note card with the solution on the back.
- Use the ✓ ? ★ system on homework problems and get answers to all ★ problems.
- Use a computer algebra to help find the definite integral.

Answer to On Your Own 1

Suppose you visualize the rod lying on the x-axis, with the one end at $x = 0$ and the other end of the rod at $x = a$ cm. Slice into segments of width Δx cm.

Slice	*Mass of each slice*	*Total Mass of Rod*
x cm from left-hand end	The approximate uniform density of this slice is $f(x) = kx^2 gm/cm$ The total mass of this slice is: $\approx (kx^2 \text{ gm / cm })(\Delta x \text{ cm})$.	Add up the mass of each slice and as Δx tends toward zero the total mass is: $\int_0^a kx^2 dx = (k)\frac{1}{3}a^3$ gm

Answer to On Your Own 2

- Place the left end of the rod at the origin. The density function is $\delta(x) = kx^2$. We can expect the center of mass to be more than 5 cm since the rod is more dense the further away from the left end.
- Divide the rod into n pieces of thickness Δx.
- The mass of each slice is approximately $\delta(x)\,\Delta$x.
- The Riemann sum is $\sum_{i=1}^{n} \delta(x_i)\Delta x = \sum_{i=1}^{n} kx_i^2 \Delta x$.
- As $n \to \infty$ this sum approximates the definite integral:

$$Total\ mass = \int_0^{10} kx^2 dx = \frac{10^3 k}{3}\ \text{gm}$$

$$Moment = \int_0^{10} x(kx^2) dx = \frac{10^4 k}{4}\ \text{gm - cm}$$

$$Center\ of\ Mass = \frac{moment}{total\ \ mass} = \frac{\dfrac{k10^4}{4} gm - cm}{\dfrac{10^3 k}{3} gm} = 7.5cm$$

8.4 APPLICATIONS TO PHYSICS

Objective: To calculate the work done on an object using the definite integral.
To calculate the force exerted by a liquid on a surface.

Preparations: Read and take notes on section 8.4
Suggested practice problems: 2,9,12,15,16,19.

✠ Key Ideas of the Section

You should be able to:

- Use Riemann sums to approximate the work done on an object and to represent this total work as a definite integral.
- Use Riemann sums to approximate force exerted by a liquid on a surface and to represent this total force as a definite integral.

Key Questions to Ask and Answer

1. What is meant by the work done on an object?

____

____ *(See text p.368.)*

*Work done = Force * Distance = F*d* where a constant force *F* is applied to some object to move it a distance *d*, in the same or opposite direction of the force.

	Units of force or weight	Distance	Units of work
British	*pound*	*foot*	*foot-pound*: work done when a 1-pound object moves up 1 foot
International (SI)	1 *newton* $=1kg\ m/\sec^2$	*meter*	1 *newton-meter* = 1 *joule*: work done when force of 1 newton moves an object 1 meter.

*See the conversion table between the two sets in text on page 369.

2. What is the difference between weight and mass?

✍ ____

____ *(See text p.370-371.)*

Mass and weight are often confused. Mass always refers to the quantity of matter an object comprises, while weight is the force exerted on an object due to gravity. The unit of mass is kilogram in the International system.

Example 8.4.1

How much work is done to lift a 10 pound box from the floor to a table 3 feet high?

The force due to gravity is 10 pounds, so *Work = F*d* = (10 pounds)(3 feet) = 30 foot-pounds

Example 8.4.2

How much work is done to lift a 10 kilogram box from the floor to a table 1 meter high?

Here you are given the mass instead of the force (or weight) and need to convert it to weight or the force that gravity exerts on an object. The gravitational force acting on a mass, m, due to gravity is $m\,g$, where $g = 9.8$ m/sec^2, the acceleration due to gravity. The force due to gravity is (10kg)(9.8 m/sec^2) and therefore

Work =[(10kg)(9.8 m/sec^2)](1 m)= [(10) (9.8) newtons](1m)
= 98 newton-meters = 98 joules.

3 *How do you calculate the work done when the force and distance moved vary?*

✍ ____

____ (See text p.369.)

As before, slice the object in such a way to find the work done on each piece and then add the contributions for each piece. Again, as the length or thickness of the slices tends towards zero, we arrive at the definite integral giving the total work.

Example 8.4.3

How much work is done to lift a 50 pound bucket from a 100-foot well with a rope which weighs 0.25 pounds/foot?

The work done by lifting the bucket is (50)(100) foot-pounds; however, the work done by lifting the rope varies as the length of the rope varies. Divide the rope into pieces each of length Δy feet, each of which weighs $0.25\Delta y$ lb. If y_i is the distance of this typical piece of rope from the top of the well, the work done on each small piece is approximately
($0.25\Delta y$ lb) (y_i feet).

The Total Work is:

$$W \approx \sum_{i=1}^{n} 0.25 y_i \Delta y + (50)(100)\,foot-pounds$$, which as $\Delta y \to 0$ gives:

$$W = \int_0^{100} (0.25y)dy + 50(100) = \frac{0.25y^2}{2}\Big|_0^{100} + 5000 = 1250 + 5000 = 6250 \text{ foot - pounds}$$

✎ On Your Own 1

How much work is done to lift a 100-pound bucket from a 50-foot well with a rope which weighs 0.25 pounds/foot?

(The answer appears at the end of this section.)

Example 8.4.4

How much work is done to pump water from a square tank of dimensions 5 feet by 5 feet and depth of 2 feet if the water is pumped to the top of the tank?

Solution:

Take a thin slice of water at height y_i from the top of the tank of thickness Δy feet. The work done to pump this slice is approximately the weight of the layer times the distance pumped. Since the density of water is 62.4 lb/ft^3, the weight of the slice is approximately
((5)(5)Δy ft^3)(62.4 lb/ft^3) and hence the work for that slice is:

$$W \approx ((5)(5)(\Delta y \text{ ft}^3)(62.4 \text{ lb/ft}^3)(y_i \text{ ft})$$

Again, add the work for each slice and take the limit as $\Delta y \to 0$, giving the definite integral:

$$W = \int_0^2 5 \cdot 5 \cdot 62.4 y dy = \frac{1562.5y^2}{2}\Big|_0^2 = 3120 \text{ foot - pounds} .$$

On Your Own 2

How much work is done to pump water from a tank that is an inverted cone with radius 5 feet and height 2 feet filled with water? Set up the definite integral that gives the work done to pump water over the top.

(The answer appears at the end of this section.)

4. *How do you calculate the force exerted by a liquid on a surface, provided the pressure is constant?*

____

____ *(See text p372-375.)*

Pressure in a liquid is the force per unit area exerted by the liquid. Pressure P, exerted by a volume of liquid at a particular depth is the total weight, or p = (mass density)(g)(depth), measured in newtons/m^2.

To calculate the force exerted by a liquid you need to know

- δ, density , which is mass per unit volume, for example in kg/m^3,
- g , the acceleration due to gravity, usually in m/sec^2 and
- h , the depth of the liquid, usually in meters.

Notice the units are (kg/m^3)(m/sec^2)(m)=(kg m/sec ^{2})/m ^{2} = nt/ m ^{2}.

Note: If the density is given as weight per unit volume, for example measured in lb/ft^3, then you do not need to multiply by g as it has already been done.

If the pressure is constant over a given area, then *Force = (Pressure)(Area)*. When pressure isn't constant, then we divide the area into strips (most likely horizontal since pressure increases with depth) where the pressure is nearly constant on each one. Riemann sums add up the contributions of the force on all strips, and as the width of the strips tends towards zero, you get the related definite integral. The limits of integration are the range of the depths.

Example 8.4.5

Find the force on one end of a water-filled trough 10 feet long with ends that are semi-circles of radius 2 feet.

Take a rectangular strip on the end with width Δy ft at depth y_i ft. The dimensions of this rectangle are Δy ft by $2\sqrt{4-y_i^2}$ ft with area $2\sqrt{4-y_i^2}\,\Delta y$ ft^2. Water weighs 62.4 lb/ft^3. The force against this strip is:

(the pressure at that depth)(area of strip) or $F = \left(62.4\,\text{lb}/ft^3\right)(y\text{ft})\left(2\sqrt{4-y_i^2}\,\Delta y\text{ft}^2\right)$.

Therefore the force on one end is $F = \int_0^2 124.8y\sqrt{4-y^2}\,dy = 332.8$. (This calculation can be done using numerical methods, calculator or computer algebra).

Key Study Skills

- Choose two problems that represent this section and put each on a 3 X 5 note card with the solution on the back. review problems 8.1-8.4.
- Use the ✓ ? ★ system on homework problems and get answers to all ★ problems.
- Use technology such as graphing calculator or a computer algebra system to verify your answers.
- When setting up problems, pay attention to the units being used and the units of the answer.

Answer to On Your Own 1

$$W = \int_0^{50} 0.25y\,dy + (100)(50) = 5312.5\,foot - pounds$$

Answer to On Your Own 2

Take horizontal slices of thickness Δy a distance of y feet from the top of the cone. The radius r is found by using similar triangles and the proportion $\frac{r}{2-y} = \frac{5}{2} \Rightarrow r = \frac{10-5y}{2}$. The volume of a slice is $\approx \pi r^2 \Delta y = \pi\left(\frac{10-5y}{2}\right)^2 \Delta y$. The density of water is $62.5\frac{lb}{ft^3}$ so the weight of that slice is $\left(\pi\left(\frac{10-5y}{2}\right)^2 \Delta y\ ft^3\right)\left(62.5\ lb/ft^3\right)$ and the work done to lift the slice a distance of y feet is $W \approx \left(\pi\left(\frac{10-5y}{2}\right)^2 \Delta y\ ft^3\right)\left(62.5\ lb/ft^3\right)(y_i\ ft)$. Sum all the slices and take the limit as $n \to \infty$ giving the work done: $W = \int_0^2 62.5\pi r^2 y\,dy = \int_0^2 62.5\pi\left(\frac{10-5y}{2}\right)^2 y\,dy \approx 1636$ ft - lb.

8.5 APPLICATIONS TO ECONOMICS

Objective:	To make financial comparisons using the present value and future value of a continuous income stream. To express consumer and producer surplus as definite integrals.
Preparations:	Read and take notes on section 8.5. Suggested practice problems: 1-4,10,11,13,18.

✠ Key Ideas of the Section

You should be able to:

- Understand the future value and the present value of discrete or continuous payments and know when to use them.
- Set up Riemann sums and the related definite integral to calculate the future value and present value of continuous income streams.
- Use the present values or future values to compare alternatives and to make economic decisions.
- Interpret consumer and producer surplus in terms of definite integrals.

Key Questions to Ask and Answer

1. ***What are the future value and the present value of a discrete payment?***

✍ ____

____ *(See text p.378.)*

It makes sense that P dollars (\$$P$) today is worth more than P dollars in the future, since P dollars today can be invested earning interest and growing into, say, \$$B$, in the future. That is, \$$B$ is the Future Value of \$$P$; on the other hand, \$$P$ is the Present Value of \$$B$. Note that:

Present Value of a payment < Future Value of a payment.

The formula we are most likely to use is for interest compounded continuously, if r is the continuous rate and t is the number of years, then

$$(FV) = (PV)e^{rt} \text{ or } (PV) = (FV)e^{-rt}$$

Example 8.5.1

\$1000 is put into an account paying 8% compounded continuously. What is the future value of this investment in 10 years?

$$FV = \$1000e^{(0.08)(10)} = \$2225.54$$

Example 8.5.2

Suppose you are offered \$70,000 today, January 1, 1999, for the right to have your \$100,000 inheritance you will receive exactly 5 years from now, on January 1, 2004. Should you take the offer? (Assume interest rates remain at 8% compounded continuously).

You can compare the future values of each option or the present values of each option.

Option 1: receive \$70,000 today	Present Value (Option 1) = \$70,000	Future Value (Option 1) = $\$70,000e^{(0.08)(5)}$ =\$104,428
Option 2: receive \$100,000 in 5 years	Present Value (Option 2) = $\$100,000e^{-(0.08)(5)} =$ \$67,032	Future Value (Option 2)= \$100,000

Analysis with either present or future values leads to the same decision: take the offer today.

On Your Own 1

Suppose you are offered the choice: take \$70,000 today, January 1, 1999, or have the \$100,000 inheritance paid in five \$20,000 payments, the first on January 1, 2000 and the last on January 1, 2004. Should you take the offer? (Assume interest rates remain at 8% compounded continuously).

(The answer appears at the end of this section.)

2. *What are the future value and the present value of a continuous income stream, P(t)?*

____

____ *(See text p. 379-380.)*

Example 8.5.3

You project that your company will generate profit at a rate of $P(t) = 10{,}000 + 3000t$ dollars per year t years after the base year 1998. For how much would you sell the right to the profit for the next 5 years? (Assume 5% interest rate compounded continuously).

Solution:

Your asking price would be the Present Value of this profit stream for the next 5 years. Divide the 5 year period into subintervals of length Δt years, where you assume that the profit is approximately constant on that subinterval. For this typical small interval of time, profit is approximately:

(rate of profit on that interval)(length of time of interval) ≈ ($P(t_i)$ dollars/years)(Δt years)

$$PValue(profit) \approx [P(t_i)\Delta t]e^{-(0.05)t_i}$$

$$TotalPV \approx \sum_{i=1}^{n} P(t_i)e^{-(0.05)t_i} \text{ and as } n \to \infty$$

$$PV = \int_0^5 P(t)e^{-.05t}dt = \int_0^5 (10,000 + 3000t)e^{-.05t}dt =$$

$$= 10,000(4.424) + 3000(10.5996) = \$76038.70$$

Example 8.5.4

This same company generates profit at a rate of $P(t)=10,000 + 3,000t$ dollars per year t years after the base year 1998. What is the Future Value of the total profit in 5 years? (Assume 5% interest rate compounded continuously).

$$FV = \int_0^5 (10,000 + 3,000t)e^{0.05(5-t)}dt = \$97,637.30$$

Since these small subintervals of profit are earning 5% compounded continuously for $(5-t)$ years. Alternately we can use the result of **Example 8.4.3**. The Future Value after 5 years of $76,040 is 76,040($e^{.05(5)}$)=$97,637.29

On Your Own 2

You are working on a novel. Publisher X offers you an advance of $50,000 and 5% royalties on sales, which are expected to increase exponentially according to $S(t) = 100,000e^{0.10t}$ dollars/year for 5 years and then abruptly end. Publisher Z offers you no cash advance, but 12% royalties with the same expectation of sales. Which deal is better, assuming a constant 9% continuous interest rate for the entire 5 years?

(The answer appears at the end of this section.)

3. *How do you express consumer surplus and producer surplus as definite integrals?*

✍ ____

____ *(See text p. 381-382.)*

Example 8.5.5

Given the supply equation p= S(q)= $0.04q$ and demand equation p= $D(q)$ =$50-0.05q$ for a product, explain the meaning of the equilibrium price $p*$ and quantity $q*$	At the equilibrium price of $22, the quantity demanded equals the quantity supplied equals 555 units.	

Find the Consumer Surplus and represent it on the graph.	The shaded region represents the Consumer Surplus. The Consumer Surplus is the total amount gained by consumers by buying at the current price rather than at the price they would have been willing to pay. It is the Consumer expenditure (area under the Demand curve) up until equilibrium minus the expenditure at equilibrium (area p^*q^*). Consumer Surplus $= \left(\int_0^{555} D(q)dq \right) - p^* q^*$ $= \int_0^{555} (50 - 0.05q)dq - (555)(22) = \$7,840$	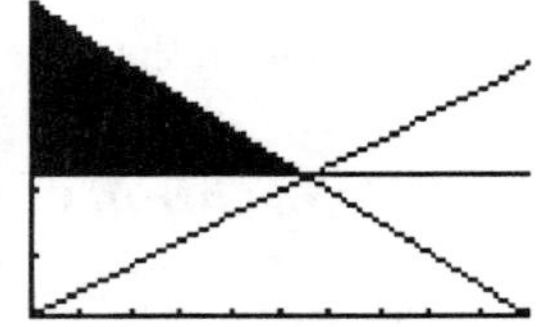
Find the Producer Surplus and represent it on the graph.	The shaded region represents the Producer Surplus. The Producer Surplus is the total amount gained by producers by selling at the current price, rather than at the price they would have been willing to accept. It is the expenditure at equilibrium (area p^*q^*) minus the area under the Supply curve up until equilibrium. Producer Surplus $= p^* q^* - \left(\int_0^{555} S(q)dq \right)$ $= (555)(22) - \int_0^{555} 0.04q\,dq = \$6050$	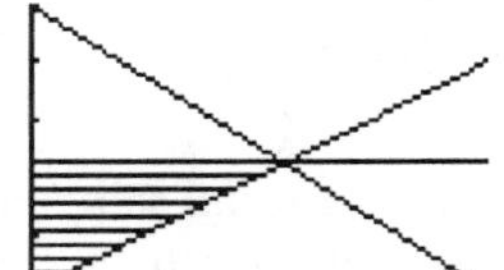

Your Turn

Find the Present Value of each of the given continuous income streams, with $t=0$ at the start of the income stream. Assume an annual interest rate of 5%. Complete the table.

Income stream (dollars/year)	***Number of years***	***Present Value***	***Answers***
$P(t)$=\$10,000	Beginning in 5 years and lasting for 10 years	?	$PV = \int_0^{10} 10000e^{-0.05(t+5)}dt$=\$61,290
$P(t)$=\$10,000+ 1,000t	Beginning now and lasting 15 years	?	$PV = \int_0^{15} (10,000 + 1,000t)e^{-0.05t}dt =$ \$174,870
$P(t)$= $\$5,000(1+\sqrt{t})$	Beginning in year 10 and ending in year 15	?	$PV = \int_0^{5} 5,000(1+\sqrt{t})e^{-0.05(t+10)}dt =$ \$32,914

$P(t) = \$10{,}000e^{0.03t}$	Beginning in 3 years and lasting for 10 years	?	$PV = \int_0^{10} (10{,}000e^{0.03t})e^{-0.05(t+3)}dt =$ \$78,000

Key Study Skills

- Choose two problems that represent this section and put each on a 3 x 5 note card with the solution on the back. Review the problems for 8.1-8.5.
- Use the ✓ ? ★ system on homework problems and get answers to all ★ problems.
- You may need to review exponential functions.
- Learn to use the technology preferred by your instructor. The text assumes you have a graphing calculator available or a computer algebra system.

Answer to On Your Own 1

One approach is to compare present values:

Present Value (Option 1) = \$70,000

Present Value (Option 2) = $\sum PV(each\ \ payment)$ =

$\$20{,}000e^{-(0.08)(1)} + \$20{,}000e^{-(0.08)(2)} + \ldots + \$20{,}000e^{-(0.08)(5)} = \$79{,}167$

Choose the five \$20,000 payments.

Answer to On Your Own 2

$$PV(X) = \$50{,}000 + \int_0^5 (0.05)(100{,}000e^{0.1t})e^{-.09t}dt = \$75{,}636$$

$$PV(Z) = \int_0^5 (0.12)(100{,}000e^{0.1t})e^{-.09t}dt = \$61{,}525.$$ *Publisher X offers the better deal.*

8.6 DISTRIBUTION FUNCTIONS

Objective: To represent distributions of various quantities by density functions.

Preparations: Read and take notes on section 8.6.
Suggested practice problems: 8,10,11,17,19.

✠ Key Ideas of the Section

You should be able to:

- Understand the density function as describing a "smoothed out" histogram.
- Use the density function to find the proportion of a population with a characteristic falling between two values.
- Create a cumulative distribution function from a density function.

Key Questions to Ask and Answer

1. ***What is a density function p(x) for a population where x is the characteristic being measured?***

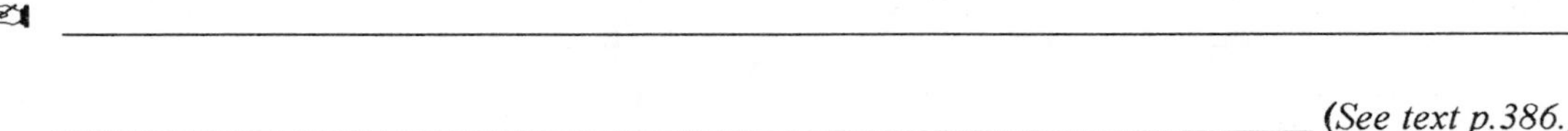

(See text p.386.)

Recall that histograms of discrete random variables x, allow us to calculate the fraction of a population with this characteristic assuming a value or values by calculating certain areas of rectangles of the histogram. Also remember that the total area is one. This idea is extended to density functions. A density function, $p(x)$, can be thought of as a function which "smoothes out" the histogram. The fraction of the population with the characteristic falling between values a and b will be the area under the curve or the definite integral. Similarly, the total area under the curve is one.

Formally, the function $p(x)$, is a density function if:

Fraction of population for which x is between a and b = Area under the graph of p between a and b $= \int_a^b p(x)dx$

and, $\int_{-\infty}^{\infty} p(x)dx = 1$ and $p(x) \geq 0$.

Be careful to interpret the density function evaluated at a point correctly: $p(x_0)$ means that for a small interval of length Δx about x, the fraction of the population with values within this interval is approximately $p(x_0)\,\Delta x$; it does not mean that this is the fraction of the population with exactly the characteristic x_0.

Example 8.6.1

Suppose a city observes x the number of trips per day each ambulance makes for a large number of days and gets the following.

Number of trips per day`	Fraction of days
0	0.05
1	0.10
2	0.15
3	0.25
4	0.21
5	0.14
6	0.05
7	0.03
8	0.02

Table of values of data

Histogram of data

The fraction of days with 3 trips or more per day is $0.25 + 0.21 + 0.14 + 0.05 + 0.03 + 0.02 = 0.70$. The density function, $p(x)$, is a function which would "smooth out" the histogram, and would be such that the fraction of days where the ambulance had 3 calls or more would be the area under the graph of this function p to the left of $x = 2.5$ (because x is discrete assume the dividing point is between $x = 2$ and $x = 3$ is $x = 2.5$) or $\int_{2.5}^{8.5} p(x)dx$, this assumes the maximum number of calls per day is 8.

On Your Own 1

For the data in **Example 8.6.1**, find the fraction of days where $2 \le x \le 5$ trips per day and express this as a definite integral of the density function, $p(x)$.

(The answer appears at the end of this section.)

Example 8.6.2

The data in the table below is of x, the commute time in minutes for a large number of students at a University. Find formulas to approximate the density function for the commute times.

x=commute time (min)`	Fraction of days
0 - 10	0.22
10 - 20	0.23
20 - 30	0.25
30 - 40	0.10
40- 50	0.15
50 - 60	0.05

Histogram of commute time

We want the total area to be one. For the first 30 minutes, the fraction is nearly constant and has a total of 70%. To find the fraction of students per minute of commute time, $0.70/30 = 0.023$, so we can have $p(x) = 0.023$ for $0 \le x \le 30$. For $30 \le x \le 60$, the fractions are decreasing at a constant rate. This "smoothed out" histogram can be approximated by a decreasing linear function, the area of which must be 0.30. This gives a triangle, with base $60 - 30 = 30$ minutes and height b, whose area is 0.30. Solving for b, gives $b = 0.02$, so $p(30) = 0.02$ and $p(60) = 0$. The equation of the line through these two points is approximately: $p(x) = -0.007x + 0.04$. The density function is as follows:

$$p(x) = \begin{cases} 0.023 \text{ if } 0 \le x < 30 \\ -0.0007x + 0.04 \text{ if } 30 \le x \le 60 \end{cases}$$

window: $0 \le x \le 60$, $0 \le y \le 0.03$

1. ***What is a cumulative distribution function, P(t) for a density function, p?***

✍ ____

____ (See text p.387.)

Cumulative distribution functions give the percentage of time the characteristic of a population falls below a certain value. From **Example 8.6.1**, 30% of the time the number of calls per day by an ambulance was less than 3 (that is 2 or less).

$P(t)$	=	Fraction of population with number of trips fewer than t	$= \int_0^t p(x)\,dx$

Formally, the cumulative distribution function is $P(t) = \int_{-\infty}^{t} p(x)\,dx$. Notice that the cumulative distribution function is the antiderivative of p. The properties are:

- P is nondecreasing
- $\lim_{t\to\infty} P(t) = 1$ and $\lim_{t\to-\infty} P(t) = 0$
- Fraction of population having values of x between a and b $= \int_a^b p(x)dx = P(b) - P(a)$

Example 8.6.3

Find the values of the cumulative distribution function for commute times in **Example 8.6.2** where $p(x) = \begin{cases} 0.023 \text{ if } 0 \le x < 30 \\ -0.0007x + 0.04 \text{ if } 30 \le x \le 60 \end{cases}$.

t	$P(t) = \int_0^t p(x)\,dx$.	
0	$\int_0^0 p(x)dx$	$= 0$
10	$\int_0^{10} p(x)dx$	$= 0.023x\vert_0^{10} = 0.23$
20	$\int_0^{20} p(x)dx$	$= 0.023x\vert_0^{20} = 0.46$
30	$\int_0^{30} p(x)dx$	$= 0.23x\Big\vert_0^{30} = .69$
40	$\int_0^{30} p(x)dx + \int_{30}^{40} p(x)$	$= 0.69 + \int_{30}^{40} (-0.0007x + 0.04)dx$ $= 0.69 + \left(-0.007\frac{x^2}{2} + 0.04x\right)\Big\vert_{30}^{40}$ $= 0.854$

✎ On Your Own 2

Find $P(50)$ and $P(60)$ for the density function in **Example 8.6.2** and interpret it.

(The answer appears at the end of this section.)

Key Notation

- Distinguish between $p(x)$ the density function, and its antiderivative $P(t)$ the cumulative distribution function.

Key Study Skills

- Choose two problems that represent this section and put each on a 3 x 5 note card with the solution on the back.
- Use the ✓ ? ★ system on homework problems and get answers to all ★ problems.
- Learn to use the technology preferred by your instructor. The text assumes you have a graphing calculator available or a computer algebra system.

Answer to On Your Own 1

The fraction of days is 0.15 + 0.25 + 0.21 + 0.14 = 0.75 or 75%. As a definite integral: $\int_{1.5}^{5.5} p(x)dx$.

Answer to On Your Own 2

$P(50)=\int_0^{50} p(x)dx$	$P(60)=\int_0^{60} p(x)dx$
$=\int_0^{40} p(x)dx+\int_{40}^{50} p(x)dx$	$=\int_0^{50} p(x)dx+\int_{50}^{60} p(x)dx$
$=0.854+\int_{40}^{50}(-0.0007x+0.04)dx$	$=0.939+\int_{50}^{60}(-0.0007x+0.04)dx$
$=0.854+\left(-0.007\frac{x^2}{2}+0.04x\right)\Big\vert_{40}^{50}$	$=0.939+\left(-0.007\frac{x^2}{2}+0.04x\right)\Big\vert_{50}^{60}$
$=0.939$	$=0.954$

8.7 PROBABILITY, MEAN, AND MEDIAN

Objective: To use definite integrals to describe probability distributions.

Preparations: Read and take notes on section 8.7.
Suggested practice problems: 3,7,9,14,15.

✠ Key Ideas of the Section

You should be able to:

- Find probabilities using both the cumulative distribution and the density functions.
- Find the median and mean of distributions.
- Use the normal probability distribution.

Key Questions to Ask and Answer

1. ***How do you use the density function and a cumulative distribution function to calculate probabilities?***

✍ ____

____ *(See text p.390.)*

The probability that a variable assumes values between a and b is the same as the fraction of the population that lies between a and b. Consequently, we have:

- Probability that a variable assumes values between a and b = Fraction of population between a and b $= \int_a^b p(x)dx$
- Probability that a variable assumes values less than t = Fraction of population with characteristic values less than t $P(t) = \int_{-\infty}^{t} p(x)\,dx$

Example 8.7.1

Consider the commute time data from **Example 8.6.3** where the function for commute times is given in **Example 8.6.2** :

$$p(x) = \begin{cases} 0.023 \text{ if } 0 \le x < 30 \\ -0.0007x + 0.04 \text{ if } 30 \le x \le 60 \end{cases}, \ x \text{ is minutes.}$$

Use both the density function and the cumulative distribution function to find the probability that the commute time of a randomly selected student is 40 or more minutes.

a) Probability that the student's commute time is 40 or more minutes = $\int_{40}^{60} p(x)dx =$

$$\int_{40}^{60} (-0.0007x + 0.04)dx = -0.007\frac{x^2}{2} + 0.04x\Bigg|_{40}^{60} = 0.14 \text{ (this is approximate).}$$

b) Using the cumulative distribution function, the probability that the student's commute time is 40 or more minutes = $\int_{40}^{60} p(x)dx = P(60) - P(40) = \int_{0}^{60} p(x)dx - \int_{0}^{40} p(x)dx = 1 - 0.854 = 0.146.$

On Your Own 1

For the data in **Example 8.6.1**, find the fraction of days where $2 \le x \le 5$ trips per day and express this as a definite integral of the density function, $p(x)$.

(The answer appears at the end of this section.)

2. *What are the measures of central tendency for a probability density function, p?*

____

____ *(See text p.392-393.)*

The *median* of a quantity x distributed through a population is a value such that half the population lies above and half below. The *mean* value of a quantity can be thought of as a "center of gravity." Formally

1. The *median* is a value T such that $\int_{-\infty}^{T} p(x)dx = 0.5$

2. The *mean* value of a quantity is $\int_{-\infty}^{\infty} xp(x)dx$

Example 8.7.2

Find the median and the mean commute times in **Example 8.6.2** where

$$p(x)=\begin{cases}0.023 \text{ if } 0\le x<30\\ -0.0007x+0.04 \text{ if } 30\le x\le 60\end{cases}.$$

a) We want T such that $\int_{-\infty}^{T} p(x)dx = 0.5$. Since $p(x) = 0.23$ for $0\le x \le 30$, we want the commute time T that gives the area of this rectangle to be 0.5, or $T = \frac{0.5}{0.23} = 21.7$ minutes. Half of the population has commute time less than 21.7 minutes.

b) The mean commute time is

$$\int_0^{60} tp(t)dt = \int_0^{30} t(0.023)dt + \int_{30}^{60} t(-0.0007t+0.04)dt =$$

$$\left.\frac{0.023t^2}{2}\right|_0^{30} + \left.\frac{-0.0007t^3}{3}+\frac{0.04t^2}{2}\right|_{30}^{60} \approx 20 \text{ minutes}$$

3. *What is the density function for the normal distribution?*

✍ ____

____ *(See text p.394.)*

The normal distribution is perhaps the most important distribution in that it describes the distribution of many real phenomena such as heights of people, diameters of ball bearings, and many biological and physical measurements. It is "bell shaped" and is described entirely by two parameters: the mean μ and standard deviation (measure of "spreadout-ness") σ. The density function is:

$$p(x) = \frac{1}{\sigma\sqrt{2\pi}} e^{\frac{-(x-\mu)^2}{2\sigma^2}}.$$

Example 8.7.3

Suppose that weekly demand for gasoline at a particular filling station will be approximately normally distributed, with a mean of 1000 and a standard deviation of 50 gallons. The density function is :

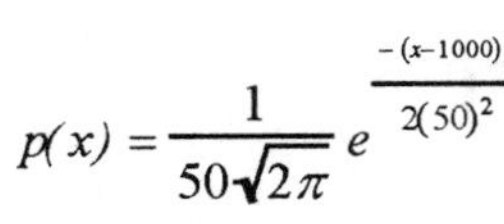

$$p(x) = \frac{1}{50\sqrt{2\pi}} e^{\frac{-(x-1000)^2}{2(50)^2}}$$

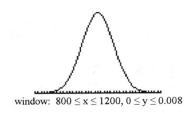

window: $800 \le x \le 1200$, $0 \le y \le 0.008$

a) Find the fraction of weeks where the demand exceeds 1100 gallons. This is

$$\int_{1100}^{+\infty} \frac{1}{50\sqrt{2\pi}} e^{\frac{-(x-1000)^2}{2(50)^2}} dx \approx 0.023$$

b) If the station is supplied with gasoline once a week, what must be the capacity of its tank if the fraction of time its supply is completely exhausted in a given week is to be no more than 0.01?

Here we want to find a demand, a, such that $\int_{a}^{+\infty} \frac{1}{50\sqrt{2\pi}} e^{\frac{-(x-1000)^2}{2(50)^2}} dx \approx 0.01$

Using a calculator or guess and check, we see that $a \approx 117$ gallons per week

On Your Own 2

Answer the same questions in **Example 8.7.3**, except assume the mean is 1000 gallons and the standard deviation is 200 gallons.

(The answer appears at the end of this section.)

Key Study Skills

- Choose two problems that represent this section and put each on a 3 x 5 note card with the solution on the back.
- Use the ✓ ? ★ system on homework problems and get answers to all ★ problems.
- Learn to use the technology preferred by your instructor. The text assumes you have a graphing calculator available or a computer algebra system.

Answer to On Your Own 1

The fraction of days is $0.15 + 0.25 + 0.21 + 0.14 = 0.75$ or 75%. As a definite integral: $\int_{1.5}^{5.5} p(x)dx$.

Answer to On Your Own 2

a) $\int \frac{1}{200\sqrt{2\pi}} e^{\frac{-(x-100)^2}{2(200)^2}} dx \approx 0.31$

b) We want $a. \approx 1465$ gallons per week.

CUMULATIVE REVIEW CHAPTER EIGHT

Key Study Skills

- Choose from some of the Review Exercises and Problems in the text for Chapter Eight p. 401.
- Do Check Your Understanding problems in text p. 401.
- Make up your own test from the 3 x 5 note card problems.
- Review the Key Ideas from all of the sections in this Study guide.
- Study with a partner and ask each other the meaning of the **Key Ideas** of Chapter Eight.
- Go back and redo all the homework practice problems that you had ★ (starred).
- Do the Chapter Eight practice test.

Chapter Eight Practice Test

1. A pyramid 100 feet tall has a square base and the perimeter of the pyramid at any height y is $(100 - y)$ feet. (For example, the perimeter of the square at the bottom is $(100 - 0) = 100$ feet. The perimeter of the square 20 feet from the bottom is 80 ft.) Find the volume of the pyramid.

2. Suppose that the density of automobiles (automobiles per mile) on Interstate 80 beginning at the San Francisco Bay Bridge is given by:
$$\rho(x) = 100 + 50\sin(\pi x) \;,$$
where x is the distance in miles from the Bay Bridge and $0 \le x \le 50$. Write the Riemann sum which estimates the total number of automobiles on this 50 mile stretch. Convert to an integral and find its value.

3. A 300 pound bucket of roofing material is to be lifted by a cable from the ground to the roof of a building 30 feet above the ground. The cable weighs 3 pounds per foot. Find the work done in lifting the bucket.

4. You are writing software for a game. Company A offers you a cash bonus of $50,000 to be paid now plus 5% royalties on sales, which, starting two years from now are expected to increase exponentially for 5 years according to the function $S(t) = 1{,}000{,}000e^{(0.1t)}$ dollars/year and then abruptly die away. Company B offers you no bonus, but 6% royalties with the same expectations for sales. Which deal is better, assuming an interest rate of 10% for the next 7 years?

5. The average grade on a calculus examination taken by a large number of students is 80. The standard deviation of the grades is 6. Suppose you assume the grades are approximately normally distributed. The instructor wants to give A's to 10 percent of the class. Above what numerical grade would he/she give an A?

(The answers appear at the end of this Study Guide.)

CHAPTER NINE

SERIES

Convergence or divergence of infinite sums, or infinite series, may be determined by various tests. Infinite series may be of constants, such as the geometric series, or in terms of x, such as the power series.

9.1 GEOMETRIC SERIES

Objective: To identify, sum and apply geometric series.

Preparations: Read and take notes on section 9.1.
Suggested practice problems: 2,4,7,12,15,17,21,25.

✠ Key Ideas of the Section

You should be able to:

- Identify finite and infinite geometric series.
- Find the sum of a geometric series.
- Apply geometric series to various situations.

Key Questions to Add and Answer

1. How can you identify a geometric series?

✍ ____

____ *(See text p. 407.)*

A geometric series has two forms:

- $a + ax + ax^2 + \ldots + ax^{n-1}$, a finite geometric series, or
- $a + ax + ax^2 + \ldots + ax^{n-1} + \ldots$, an infinite geometric series.

For geometric series, the ratio between consecutive terms is constant: $\dfrac{ax^n}{ax^{n-1}} = x$

Example 9.1.1

Which of the following are geometric series?

Series	*Answers*
$2-\frac{2}{3}+\frac{2}{9}-\frac{2}{27}+\ldots+(-1)^{n+1}\frac{2}{3^{n-1}}$	Yes (finite); ratio $=\frac{\frac{2}{9}}{\frac{-2}{3}}=\frac{\frac{-2}{27}}{\frac{2}{9}}=-\frac{1}{3}$
$1+\frac{1}{2}+\frac{1}{3}+\frac{1}{4}+\ldots$	No; the ratio between consecutive terms is not constant. $\frac{\frac{1}{3}}{\frac{1}{2}}=0.66\ldots$, $\frac{\frac{1}{4}}{\frac{1}{3}}=0.75$
$1-x^2+x^4-x^6+\ldots$	Yes (infinite); ratio $=\frac{-x^6}{x^4}=\frac{x^4}{-x^2}=-x^2$
$4+4x+8x^2+12x^3+16x^4\ldots$	No; the ratio between consecutive terms is not constant: $\frac{4x}{4}=x$, $\frac{8x^2}{4x}=2x$

Your Turn 1

Create four series: a finite geometric series, an infinite geometric series, and two series which are not geometric. Show the reasons why each is or is not a geometric series.

(There are many answers to this problem. Use the ratio test to confirm on your own..)

2. *How do you find the sum of a geometric series?*

✍ ____

____ *(See text p. 407-408.)*

The sum of a **finite** geometric series of n terms is given by

$$S_n=a+ax+ax^2+\ldots+ax^{n-1}=\frac{a(1-x^n)}{1-x},\text{ for } x\neq 1.$$

For $|x|<1$, the sum of a **infinite** geometric series is given by

$$S=a+ax+ax^2+\ldots+ax^n+\ldots=\frac{a}{1-x},$$

since the partial sums, S_n converge to $S=\frac{a}{1-x}$ as $n\to\infty$.

Example 9.1.2

For the following geometric series, find the sum (if it exists).

Series	*Answers*
$2-\frac{2}{3}+\frac{2}{9}-\frac{2}{27}+\ldots+2\left(-\frac{1}{3}\right)^{n-1}$	$a=2$, $x=-1/3$; this finite geometric series has a partial sum of: $S_n=\frac{2\left(1-\left(\frac{-1}{3}\right)^n\right)}{1-\left(\frac{-1}{3}\right)}=\frac{3}{2}\left(1-\left(\frac{-1}{3}\right)^n\right)$

1+3+9+27+81+....	$a = 1, x = 3$; This infinite geometric series does not converge, the partial sums are: $S_1 = 1, S_2 = 1 + 3 = 4$; $S_3 = 1 + 3 + 9 = 13$; $S_4 = 1 + 3 + 9 + 27 = 40$. which grow larger and larger; so the series does not converge.
$100 + 10 + 1 + 0.1 + 0.01 + \ldots$	$a = 100, x = 0.1 < 1$; this infinite geometric series converges, the partial sums are: $S_1 = 100, S_2 = 100 + 10 = 110$; $S_3 = 100 + 10 + 1 = 111$; $S_4 = 100 + 10 + 1 + 0.1 = 111.1$; or $S = \frac{100}{1-0.1} = \frac{100}{0.9} = 111.111\ldots$

✎ Your Turn 2

Find the sums (if they exist) of the series you created in **Your Turn 1.**

(Answers vary according to series created in Your Turn 1.)

3. *What are some applications of geometric series?*

✍ ____

____ *(See text p. 406,409-410.)*

Geometric series are sums of terms resulting from multiplying by a constant factor. For example, money invested at a fixed rate at a regular interval generates a geometric series; a drug injected regularly that decays at a constant rate generates a geometric series.

Example 9.1.3

Suppose you deposit \$100 each month into an account earning 6% interest compounded monthly. How much do you have after 10 years? How much do you have if you live forever?

Let n = month. Let the interest rate <u>per month</u> = (6/12)%=0.50%=0.005

Balance at month n
$B_1 = 100$
$B_2 = 100(1.005) + 100$
$B_3 = 100(1.005)^2 + 100(1.005) + 100$
$B_n = 100(1.005)^{n-1} + 100(1.005)^{n-2} + \ldots 100(1.005) + 100$
$B_n = \dfrac{100\left(1-(1.005)^n\right)}{1-1.005}$

Since B_n is a finite geometric series, in 10 years, n=120,

$$B_{120} = \frac{100\left(1-(1.005)^{120}\right)}{1-1.005} = \$16{,}387.93$$

The series does not converge (x = 1.005 > 1), so if you live forever, you will become infinitely rich!

Your Turn 3

A ball is dropped from a height of 60 feet. Each bounce is 2/3 of the height of the bounce before. Complete the following table. Find the total vertical distance the ball has traveled after it hits the floor the n-th time.

***n* = # bounce**	***Height after bounce***	***Total vertical distance D***	***Answers***
0	60	?	$D_0 = 0$
1	60(2/3) = 40 ft	?	$D_1 = 60$
2	40(2/3) = 26.666... or $(60)(2/3)^2$	?	After bounce 1, the ball goes up and then back down so the total distance when it hits the floor the 2nd time is: $D_2 = 60 + 2(60(2/3)) = 140$ ft.
3	26.666..(2/3) = 17.7776 $= (60)(2/3)^3$	?	$D_3 = 60 + 2(60(2/3)) + 2(60(2/3)^2)$ $= 193.33...$ ft.
n	$(60)(2/3)^n$	?	$D_n = 60 + 2(60)\left(\frac{2}{3}\right) + 2(60)\left(\frac{2}{3}\right)^2 + \ldots 2\left(60\left(\frac{2}{3}\right)^{n-1}\right)$ $= 60 + 2(60)\left(\frac{2}{3}\right)\left(1 + \frac{2}{3} + \left(\frac{2}{3}\right)^2 + \left(\frac{2}{3}\right)^3 + \ldots + \left(\frac{2}{3}\right)^{n-2}\right)$ $= 60 + 80\left(\frac{1-\left(\frac{2}{3}\right)^{n-1}}{1-\frac{2}{3}}\right)$ $= 60 + 240\left(1-\left(\frac{2}{3}\right)^{n-1}\right)$

Key Algebra Skills

- Practice finding the formulas for geometric series. Finding the formula for the last term is often challenging.

Key Notation

- S_n is the sum of the finite geometric series of n terms called the partial sum of the infinite geometric series. You calculate the finite geometric series using:

$$S_n = a + ax + ax^2 + \ldots + ax^{n-1} = \frac{a(1-x^n)}{1-x}, x \neq 1$$

Note: The *n-th term* has an $n-1$ power, however the sum S_n has the *n-th* power.

- S is the sum of the infinite geometric series. If $|x| < 1$, the infinite series converges to

$$S = a + ax + ax^2 + \ldots + ax^n + \ldots = \frac{a}{1-x}.$$

🗁 Key Study Skills

- Choose at least two problems that represent this section; put each on a 3 x 5 note card with the solution on the back. Review the note card problems 9.1.
- Use the ✓ ? ★ system on homework problems and get answers to all ★ problems.
- In your notebook, start a list of series and their behavior, beginning with the geometric series.

9.2 CONVERGENCE OF SEQUENCES AND SERIES

Objective: To determine the convergence of sequences and series.

Preparations: Read and take notes on section 9.2.
Suggested practice problems: 4,5,10,12,14,16,17,23,24.

✠ Key Ideas of the Section

You should be able to:

- Distinguish between a sequence and series.
- Determine the convergence or divergence of a sequence.
- Determine the convergence or divergence of a series.
- Compare some series with improper integrals to determine convergence or divergence.

Key Questions to Ask and Answer

1. ***When does a sequence converge?***

✍ ____

____ *(See text p. 412.)*

A sequence $S_1, S_2, \ldots S_n, \ldots$ is just a string of numbers. Notice each number in the sequence is separated by a comma. A sequence converges to L if $\lim_{n\to\infty} S_n = L$; otherwise the sequence diverges.

To calculate the limit of a sequence use the fact that $\lim_{n\to\infty} x^n = 0$ if $|x| < 1$ and $\lim_{n\to\infty} \frac{1}{n} = 0$. Another useful result is that if your sequence is increasing and bounded above, then the sequence converges.

✎ On Your Own 2

Show that the series $\sum_{n=1}^{\infty} \frac{1}{n^3}$ converges.

(The answer appears at the end of this section.)

Key Algebra Skills

- Practice finding the formulas for the partial sums of the series.

Key Study Skills

- Review the convergence or divergence of improper integrals in Section 7.7
  Choose at least two problems that represent this section; put each on a 3 x 5 note card with the solution on the back. Review the note card problems 9.2.
- Use the ✓ ? ★ system on homework problems and get answers to all ★ problems.
- In your notebook, start a list of series and their behavior, beginning with the geometric series.

Answer to On Your Own 1

$\lim_{n\to\infty} e^{-2n} = 0$ so, $\lim_{n\to\infty} S_n = 3$.

Answer to On Your Own 2

Graph the function $y = \frac{1}{x^3}$ and form right hand Riemann sums beginning with $x = 2$. (We're trying to avoid the improper integral with a lower limit of x=0). The areas of the first n rectangles is $\frac{1}{2^3} + \frac{1}{3^3} + \ldots + \frac{1}{n^3}$. Because $y = \frac{1}{x^3}$ is decreasing, the right hand sums are underestimates of the integral.

Therefore, $1 + [\frac{1}{2^3} + \frac{1}{3^3} + \ldots + \frac{1}{n^3}] < 1 + \int_1^n \frac{1}{x^3}dx = 1 + \frac{-1}{2}\left(\frac{1}{(n)^2} - 1\right)$. As n gets large, this has limit 1.5.

Therefore the series $1 + \frac{1}{2^3} + \frac{1}{3^3} + \ldots$ converges.

9.3 TESTS FOR CONVERGENCE

Objective: To use various tests to determine the convergence or divergence of series.

Preparations: Read and take notes on section 9.3.
Suggested practice problems: 3,6,7,11,21,22,25,26,28,29.

✠ Key Ideas of the Section

You should be able to:

- Compare two series to determine convergence or divergence.
- Use the Ratio Test to determine convergence or divergence.
- Determine the convergence or divergence of alternating series.
- Determine the upper bound of the error in approximating alternating series by a partial sum.

✎ On Your Own 2

Show that the series $\sum_{n=1}^{\infty}\frac{1}{n^3}$ converges.

(The answer appears at the end of this section.)

Key Algebra Skills

- Practice finding the formulas for the partial sums of the series.

Key Study Skills

- Review the convergence or divergence of improper integrals in Section 7.7
  Choose at least two problems that represent this section; put each on a 3 x 5 note card with the solution on the back. Review the note card problems 9.2.
- Use the ✓ ? ✯ system on homework problems and get answers to all ✯ problems.
- In your notebook, start a list of series and their behavior, beginning with the geometric series.

Answer to On Your Own 1

$\lim_{n\to\infty} e^{-2n} = 0$ so, $\lim_{n\to\infty} S_n = 3$.

Answer to On Your Own 2

Graph the function $y=\frac{1}{x^3}$ and form right hand Riemann sums beginning with $x=2$. (We're trying to avoid the improper integral with a lower limit of x=0). The areas of the first n rectangles is $\frac{1}{2^3}+\frac{1}{3^3}+\ldots+\frac{1}{n^3}$. Because $y=\frac{1}{x^3}$ is decreasing, the right hand sums are underestimates of the integral.

Therefore, $1+[\frac{1}{2^3}+\frac{1}{3^3}+\ldots+\frac{1}{n^3}]<1+\int_1^n\frac{1}{x^3}dx=1+\frac{-1}{2}\left(\frac{1}{(n)^2}-1\right)$. As n gets large, this has limit 1.5.

Therefore the series $1+\frac{1}{2^3}+\frac{1}{3^3}+\ldots$ converges.

9.3 TESTS FOR CONVERGENCE

Objective: To use various tests to determine the convergence or divergence of series.

Preparations: Read and take notes on section 9.3.
Suggested practice problems: 3,6,7,11,21,22,25,26,28,29.

✠ Key Ideas of the Section

You should be able to:

- Compare two series to determine convergence or divergence.
- Use the Ratio Test to determine convergence or divergence.
- Determine the convergence or divergence of alternating series.
- Determine the upper bound of the error in approximating alternating series by a partial sum.

Key Questions to Ask and Answer

1. *When is the comparison test used to determine the convergence or divergence of series?*

✍ ____

____ *(See text p. 417-418.)*

The comparison test for series is similar to the technique of comparing two improper integrals to determine convergence or divergence. If you suspect that the series $\sum a_n$ converges, then look for a series $\sum b_n$ which converges such that $0 \le a_n \le b_n$ for all n. In other words it is sandwiched between the known series and zero so $\sum a_n$ has to converge too.

On the other hand, if you suspect that the series $\sum a_n$ diverges, then you look for a series $\sum b_n$ which diverges such that $0 \le b_n \le a_n$ for all n. It is good to have a number of series in mind with which to compare. To name a few, remember that

- geometric series $\sum ax^n$ converge for $|x| < 1$; otherwise it diverges.
- harmonic series $\sum \frac{1}{n}$ diverges.
- $\sum \frac{1}{n^p}$ converges for $p > 1$.
- $\sum \frac{1}{n!}$ converges.

For series with both positive and negative terms, if the series of absolute values converges, then so does the series. (Note that the converse is not necessarily true).

Example 9.3.1

Use the comparison test to determine whether the following series converge or diverge.

Series	*Answers*
$\sum_{n=1}^{\infty} \frac{n^2-2}{n^5+5}$	$\frac{n^2-2}{n^5+5}$ behaves like $\frac{n^2}{n^5}=\frac{1}{n^3}$ so we expect the series to converge and want to determine a converging series "above" it. $0 \le \frac{n^2-2}{n^5+5} \le \frac{n^2}{n^5} = \frac{1}{n^3}$, $n \ge 1$. (a fraction is larger if the numerator is larger and the denominator smaller, i.e. $\frac{5}{7} > \frac{3}{7}$ and $\frac{5}{7} > \frac{5}{8}$ and $\frac{5}{7} > \frac{4}{8}$). Since $\sum \frac{1}{n^3}$ converges, then $\sum_{n=1}^{\infty} \frac{n^2-2}{n^5+5}$ converges.
$\sum_{n=1}^{\infty} \frac{n^2+2}{n^3-3n}$	$\frac{n^2+2}{n^3-3n}$ behaves like $\frac{n^2}{n^3}=\frac{1}{n}$ so we expect the series to diverge and want to determine a diverging series "below" it. $0 \le \frac{1}{n} = \frac{n^2}{n^3} \le \frac{n^2+2}{n^3-3n}$, $n > 1$. (a fraction is smaller if the numerator is smaller and the denominator larger). Since $\sum \frac{1}{n}$ diverges, then $\sum_{n=1}^{\infty} \frac{n^2+2}{n^3-3n}$ diverges.

✎ On Your Own 1

Use the comparison test to determine whether the series $\sum_{n=1}^{\infty} \left| \frac{n \cos n}{n^4+1} \right|$ converges.

(The answer appears at the end of this section.)

2. *What is the Ratio Test and how do I use it to determine the convergence or divergence of a series?*

✍ ____

____ *(See text p. 419-420.)*

Some series behave like geometric series in that the ratios of consecutive terms tend toward a limit. That is $\lim_{n\to\infty} \left| \frac{a_{n+1}}{a_n} \right| = L$. The Ratio Test tells us that if $L < 1$, then $\sum a_n$ converges; if $L > 1$ or if L is infinite, then $\sum a_n$ diverges. However if L = 1, the test doesn't give us any information about the convergence (the series may converge or it may diverge).

Example 9.3.2

Use the Ratio Test to determine the convergence of the following series, if possible.

Series	Answers
$\sum_{n=1}^{\infty} \frac{2^n}{n!}$	$\left\|\frac{a_{n+1}}{a_n}\right\| = \left\|a_{n+1} * \frac{1}{a_n}\right\| = \left\|\frac{2^{n+1}}{(n+1)!} * \frac{n!}{2^n}\right\| = \left\|\frac{2^n \cdot 2}{(n+1)(n!)} * \frac{n!}{2^n}\right\| = \left\|\frac{2}{n+1}\right\|$ and $\lim_{n\to\infty}\left\|\frac{2}{n+1}\right\| = 0 < 1$ and so the series converges.
$\sum_{n=1}^{\infty} \frac{2^n}{n^2}$	$\left\|\frac{a_{n+1}}{a_n}\right\| = \left\|\frac{2^{n+1}}{(n+1)^2} * \frac{n^2}{2^n}\right\| = \left\|2 * \frac{n^2}{n^2+2n+1}\right\|$ and $\lim_{n\to\infty}\left\|2\frac{n^2}{n^2+2n+1}\right\| = 2 > 1$ and so the series diverges.

On Your Own 2

Determine the convergence or divergence of the series $\sum_{n=1}^{\infty} \frac{3^n}{n^3+1}$.

(The answer appears at the end of this section.)

3. *What is the Alternating Series Test and how do I use it?*

____

____ *(See text p. 420.-421)*

A series is called alternating if the terms alternate in sign, the formula may include a factor of $(-1)^{n-1}$, for example. Alternating series converge if the consecutive terms are decreasing and if $\lim_{n\to\infty} a_n = 0$. We also can estimate the bound of error of using the nth partial sum, S_n, to estimate the sum S of the series. This upper bound is $|S - S_n| < a_{n+1}$.

Example 9.3.3

Determine the convergence of the series $\sum_{n=1}^{\infty} \frac{(-1)^{n-1}}{(2n-1)!} = 1 - \frac{1}{3!} + \frac{1}{5!} - \frac{1}{7!} + \ldots$	$a_{n+1} < a_n$ and $\lim_{n\to\infty} a_n = 0$ and so the series converges.
Estimate error in approximating the sum of the series $\sum_{n=1}^{\infty} \frac{(-1)^{n-1}}{(2n-1)!}$ by the sum of the first four terms.	$S_4 = 1 - \frac{1}{3!} + \frac{1}{5!} - \frac{1}{7!} = 0.84146$. The first term omitted is $a_5 = \frac{1}{9!} = 2.755732 * 10^{-6}$

On Your Own 3

Determine the convergence of the series $\sum_{n=1}^{\infty} \frac{(-1)^{n+1}}{3^n}$.

(The answer appears at the end of this section.)

Key Notation

- $a^{n+1} = a^n \cdot a$
- $n! = n(n-1)(n-2)\ldots 1$, for example $6! = (6)(5)(4)(3)(2)(1)$
- $(n+1)! = (n+1)(n!)$

🗁 Key Study Skills

- Keep a list of series whose behavior you know in your notebook.
- Choose at least two problems that represent this section; put each on a 3 x 5 note card with the solution on the back. Review the note card problems 9.3.
- Use the ✓ ? ★ system on homework problems and get answers to all ★ problems.
- In your notebook, start a list of series and their behavior, beginning with the geometric series.

Answer to On Your Own 1

Since $-1 \le \cos n \le 1$, $\left|\dfrac{n\cos n}{n^4+1}\right| \le \dfrac{n}{n^4} = \dfrac{1}{n^3}$ and so the series $\sum\limits_{n=1}^{\infty}\left|\dfrac{n\cos n}{n^4+1}\right|$ converges.

Answer to On Your Own 2

$\left|\dfrac{a_{n+1}}{a_n}\right| = \left|\dfrac{3^{n+1}}{(n+1)^3+1} * \dfrac{n^3}{3^n}\right| = \left|3 * \dfrac{n^3}{n^3+3n^2+3n+2}\right|$ and $\lim\limits_{n\to\infty}\left|3\dfrac{n^3}{n^3+3n^2+3n+2}\right| = 3 > 1$ and so the series diverges.

Answer to On Your Own 3

The series of the absolute values $\sum\limits_{n=1}^{\infty}\left|\dfrac{(-1)^{n+1}}{3^n}\right| = \sum\limits_{n=1}^{\infty}\dfrac{1}{3^n}$ which is a converging geometric series with ratio $-\dfrac{1}{3}$.

9.4 CONVERGENCE OF SERIES

Objective: To determine intervals of convergence of power series graphically, numerically and analytically.

Preparations: Read and take notes on section 9.4.
Suggested practice problems: 2,4,7,9,14,16,19,23,26.

✢ Key Ideas of the Section

You should be able to:

- Recognize a power series.
- Find the interval of convergence for a series graphically and analytically.
- Be able to use the Ratio test to compute the radius of convergence for some series.
- Determine the convergence at the endpoints of the interval of convergence.

Key Questions to Ask and Answer

1. *What is a power series?*

✍ ____

____ (See text p. 423)

A *power series* about $x = a$ is a sum of constants times powers of $(x - a)$:

$$C_0 + C_1(x-a) + C_2(x-a)^2 + \ldots + C_n(x-a)^n + \ldots = \sum_{n=0}^{\infty} C_n(x-a)^n.$$

If x is fixed, the series is a series of numbers similar to those seen in Section 9.2.

Example 9.4.1

Examine each of the following power series.

Let $x = 0.8$, $a = 0$	$C_0 + C_1(0.8) + C_2(0.8)^2 + \ldots + C_n(0.8)^n + \ldots$.
Let $x = 0.8$, $a = 1.3$	$C_0 - C_1(0.5) + C_2(0.5)^2 + \ldots + (-1)^n C_n(0.5)^n + \ldots$.
Let x be variable, $a = 0$	$C_0 + C_1(x) + C_2(x)^2 + \ldots + C_n(x)^n + \ldots$.

2. *What does it mean for a power series to converge and how does it look graphically and numerically?*

✍ ____

____ (See text p. 423.)

Power series may converge for some values of x and not for others. We want to find the values of x for which the series converges, called the interval of convergence.

Example 9.4.2

Investigate the convergence of the series $1 + x + \frac{x^2}{2!} + \frac{x^3}{3!} + \ldots + \frac{x^n}{n!} + \ldots$ for $x = 0.1$ and $x = 2.0$.

Partial Sums	$P_n(0.1)$	$P_n(2.0)$
$P_1(x) = 1 + x$	$P_1(0.1) = 1.1$	$P_1(2) = 3$
$P_2(x) = 1 + x + \frac{x^2}{2!}$	$P_3(0.1) = 1.105$	$P_2(2) = 3+2=5$
$P_3(x) = 1 + x + \frac{x^2}{2!} + \frac{x^3}{3!}$	$P_3(0.1) = 1.10516667$	$P_3(2) = 6.333$
$P_4(x) = 1 + x + \frac{x^2}{2!} + \frac{x^3}{3!} + \frac{x^4}{4!}$	$P_4(0.1) = 1.1051708$	$P_4(2) = 7$
$P_6(x) = 1 + x + \frac{x^2}{2!} + \ldots + \frac{x^6}{6!}$	$P_6(0.1) = 1.105170885$	$P_6(2) = 7.355555556$

We will see in Chapter 10 that the sequence of numbers, $P_1(0.1), P_2(0.1), \ldots, P_n(0.1), \ldots$ converges to $e^{0.1} = 1.105170917\ldots$ and that $P_1(2), P_2(2), \ldots, P_n(2), \ldots$ converges to $e^2 = 7.38905\ldots$ You can see that the latter sequence is converging more slowly the further you are from $x = 0$.

3. *What does an interval of convergence look like graphically?*

✍ ____

____ *(See text p. 424.)*

In section 9.1 we saw that the geometric series has sum $1 + x + x^2 + x^3 + x^4 + \ldots = \frac{1}{1-x}$, for $|x| < 1$. The interval $-1 < x < 1$ is called the *interval of convergence* and the series is said to have *radius of convergence* of 1.

Example 9.4.3

Examine the graphs of the partial sums P_4 and P_6 for the geometric series $1 + x + x^2 + x^3 + x^4 + \ldots = \frac{1}{1-x}$ and compare to the graph of the function $f(x) = \frac{1}{1-x}$. Compare the partial sums numerically for $x = 0.2$ (which is within the interval of convergence) and $x = 2$ (which is not within the interval of convergence). Note that $f(0.2) = \frac{1}{1-0.2} = 1.25$ and $f(2) = \frac{1}{1-2} = -1$.

Partial Sums	$f(x) = \frac{1}{1-x}$	***Graphs***
		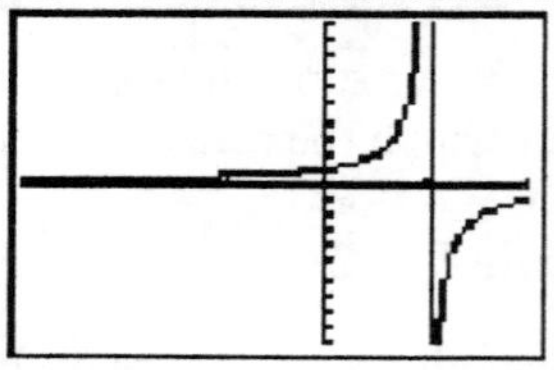 The graph of $f(x) = \frac{1}{1-x}$
$P_4(x) = 1 + x + x^2 + x^3 + x^4$	$P_4(0.2) = 1.2496$ $P_4(2) = 31$	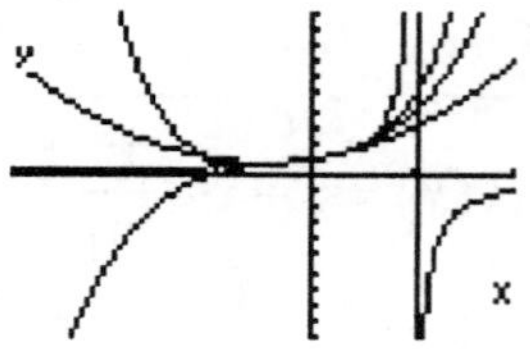 y x Between $-1 < x < 1$ the graphs are nearly the same.
$P_6(x) = 1 + x + x^2 + \ldots + x^6$	$P_6(0.2) = 1.249984$ $P_6(2) = 1 + 2 + 2^2 + \ldots + 2^6$ $= 127$	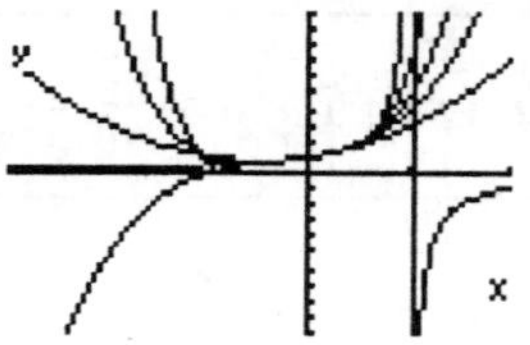 y x Adding more terms gives better convergence between $-1 < x < 1$.

Note: The sequence $P_1(0.2), P_2(0.2), \ldots, P_n(0.2), \ldots$ converges and the sequence $P_1(2), P_2(2), \ldots, P_n(2), \ldots$ diverges.

On Your Own 1

Compare the rates of convergence at $x = 0.2$ towards $f(0.2) = \dfrac{1}{1-0.2} = 1.25$ with that of $x = 0.9$ towards $f(0.9) = \dfrac{1}{1-0.9} = 10$.

(The answer appears at the end of this section.)

4. *How do you find the interval of convergence for a power series analytically?*

____

____ (See text p. 425.)

For the power series $\sum_{n=0}^{\infty} C_n(x-a)^n$, the ratio test tells us, with $a_n = C_n(x-a)^n$,

- If $\lim_{n\to\infty}\left|\dfrac{a_{n+1}}{a_n}\right|$ if infinite, then $R = 0$. The series converges only for $x = a$.
- If $\lim_{n\to\infty}\left|\dfrac{a_{n+1}}{a_n}\right| = 0$, then $R = \infty$ and the series converges for all x
- $\lim_{n\to\infty}\left|\dfrac{a_{n+1}}{a_n}\right| = K|x-a|$, where K is finite, $\neq 0$, then $R = \dfrac{1}{K}$ and the series converges for $|x-a| < 1/K$.
- If $\lim_{n\to\infty}\left|\dfrac{a_{n+1}}{a_n}\right|$ fails to exist, then the ratio test doesn't tell us anything.

Example 9.4.4

Use the Ratio test to confirm that the power series $1 - x + \dfrac{x^2}{2!} - \dfrac{x^3}{3!} + \ldots$ converges for all x.

$\lim_{n\to\infty}\left|\dfrac{a_{n+1}}{a_n}\right| = \lim_{n\to\infty}\left|\dfrac{x^{n+1}}{(n+1)!}\cdot\dfrac{n!}{x^n}\right| = \lim_{n\to\infty}\left|\dfrac{x^n x}{x^n}\cdot\dfrac{n!}{(n+1)n!}\right| = \lim_{n\to\infty}\left|\dfrac{x}{n+1}\right| = 0$ and therefore, by the Ratio Test, this series converges for all x..

On Your Own 2

Use the Ratio test to confirm that the interval of convergence for the power series $1 + x + x^2 + x^3 + x^4 + \ldots\ldots = \dfrac{1}{1-x}$ is $|x| < 1$.

(The answer appears at the end of this section.)

Example 9.4.5

Determine the interval of convergence for the series:

$$0 + 3x - \frac{9}{2!}x^2 + \frac{54}{3!}x^3 - \ldots + \frac{(-1)^{n-1}3^n(n-1)!\,x^n}{n!} + \ldots$$

To find the interval of convergence use the ratio test:

$$\lim_{n\to\infty}\left|\frac{a_{n+1}}{a_n}\right| = \lim_{n\to\infty}\left|\frac{\dfrac{(-1)^n 3^{n+1}(n+1-1)!}{(n+1)!}x^{n+1}}{\dfrac{(-1)^{n-1}3^n(n-1)!}{(n)!}x^n}\right| = \left|\frac{3^{n+1}}{3^n}\frac{n}{n+1}x\right| = 3x .$$

So R = 1/3 and the interval of convergence is $\frac{-1}{3} < x < \frac{1}{3}$ which can be seen graphically by graphing several of the partial sums.

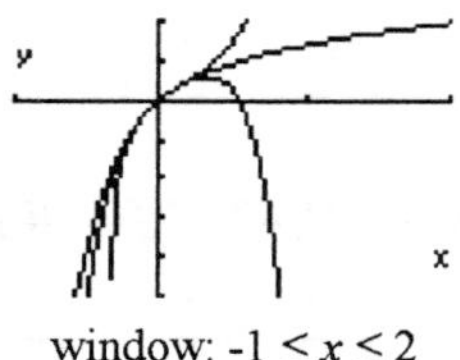

window: $-1 \le x \le 2$

5. *What about the endpoints?*

✍ ____

____ *(See text p. 427.)*

The ratio test doesn't tell whether a power series converges at the endpoints. To determine the convergence at the endpoints, substitute each endpoint into the series, giving a series of numbers and use methods from Sections 9.2 and 9.3 to determine convergence.

Example 9.4.6

We now want to determine the behavior of the series given in **Example 9.4.5** at the endpoints of the interval of convergence, that is for $x = \pm\frac{1}{3}$. At $x = +\frac{1}{3}$, the series becomes $1 - \frac{1}{2} + \frac{1}{3} - \frac{1}{4} + \ldots.$ which is an alternating series with $a_n = \frac{1}{n}$. Since $\frac{1}{n+1} < \frac{1}{n}$ and $\lim_{n\to\infty} a_n = \lim_{n\to\infty}\frac{1}{n} = 0$. This series converges, by the alternating series test.

Now let $x = -\frac{1}{3}$. The series becomes $-1 - \frac{1}{2} - \frac{1}{3} - \frac{1}{4} - \ldots.$, which diverges (this is the negative of a harmonic series). So the interval of convergence for this series $-1/3 < x \le 1/3$. The graph of the three partial sums is shown as:

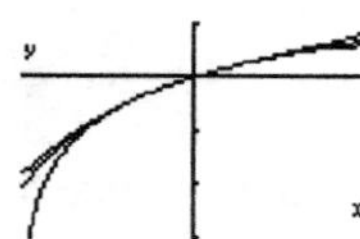

✎ On Your Own 3

Determine the behavior of the series

$$1 + x + x^2 + x^3 + x^4 + \ldots.. = \frac{1}{1-x}$$

from **Example 9.4.3** at the endpoints of the interval of convergence.

(The answer appears at the end of this section.)

Key Notation

- a_n represents the n-th term of the power series.
- The Ratio Test $\lim_{n\to\infty}\left|\frac{a_{n+1}}{a_n}\right|$, means compare the *n-th* term with the one that comes after it.

Key Study Skills

- Choose at least two problems that represent this section; put each on a 3 x 5 note card with the solution on the back. Review problems from 9.1-9.4.
- Use the ✓ ? ★ system on homework problems and get answers to all ★ problems.
- Add examples of series from this section to your list of series and their behavior in your notebook.

Answer to On Your Own 1:

$f(0.2) = 1.25$ *and* $f(0.9) = 10$

$P_n(0.2)$	$P_n(0.9)$
$P_2(0.2) = 1.24$	$P_2(0.9) = 2.71$
$P_4(0.2) = 1.2496$	$P_4(0.9) = 4.0951$
$P_6(0.2) = 1.249984$	$P_6(0.9) = 5.217031$
$P_8(0.2) = 1.24999936$	$P_8(0.9) = 6.12579511$

Notice that the convergence for $x = 0.2$ is much faster than the rate for $x = 0.9$.

Answer to On Your Own 2:

$\lim_{n\to\infty}\frac{|a_n|}{|a_{n+1}|} = \lim_{n\to\infty}\frac{1}{1} = 1$. Therefore the radius of convergence is 1.

Answer to On Your Own 3:

$\frac{1}{1-x} = 1 + x + x^2 + x^3 + x^4 + \ldots..$ for $|x| < 1$. $R = 1$. For $x = 1$, the series becomes 1 + 1 +...which diverges. Let $x = -1$. The series becomes: $1 - 1 + 1 - 1 + 1 - \ldots..$ $S_1 = 1$, $S_2 = 0$, $S_3 = 1$, $S_4 = 0$ and you can see that these partial sums do not converge. The interval of convergence is $-1 < x < 1$.

CUMULATIVE REVIEW CHAPTER NINE

Key Study Skills

- Review all of the **Key Ideas** from 9.1 - 9.4.
- Choose from Review Exercises and Problems from the text, p. 429-430.
- Do Check Your Understanding from text, p. 431.
- Review the note card problems from 9.1 -9.4.
- Make up your own test.
- Review homework problems and get answers to all ★ (starred) problems.
- Take the Chapter Nine Practice Test.

Chapter 9 Practice Test

1. Determine the convergence or divergence of the series $\sum_{n=1}^{\infty} \frac{1}{2+5^n}$.

2. Determine the convergence or divergence of the series:

   $\cos 1 + \frac{\cos 2}{2^2} + \frac{\cos 3}{3^2} + .. + \frac{\cos n}{n^2} + ..$ (Hint: notice that this is not an alternating series).

3. Find the interval of convergence of the power series $1x + \frac{1}{\sqrt{2}}x^2 + \frac{1}{\sqrt{3}}x^3 + \ldots$

(The answers appear at the end of this Study Guide)

CHAPTER TEN

APPROXIMATING FUNCTIONS

Functions can be approximated by using polynomial functions for local approximations whereas Fourier Series use sines and cosines to give generally good global approximations.

10.1 TAYLOR POLYNOMIALS

Objective: To create Taylor polynomials which approximate functions near a point.

Preparations: Read and take notes on section 10.1.
Suggested practice problems: 2,3,11,12,16,18,24,27,31.

✠ Key Ideas of the Section

You should be able to:

- Construct Taylor polynomials of degree n approximating $f(x)$ for x near 0.
- Construct Taylor polynomials of degree n approximating $f(x)$ for x near a.

Key Questions to Ask and Answer

1. How does a Taylor polynomial approximate a function f(x) for x near a?

✍ ____

____ *(See text p. 439)*

Polynomials are relatively nice functions. They are continuous and are differentiable everywhere. Because of that they can be used to approximate more complicated functions such as e^x, $sin(x)$, or ln$x$, for example. Recall the tangent line approximation for a function at a given point:

Linear approximation for x near a : $f(\mathrm{x}) \approx f(a) + f'(a)\,(x - a)$.

And, specifically for x near 0, $f(x) \approx f(0) + f'(0)\,x$.

The tangent line passes through the same point and has the same slope as the graph of f at the point $x = a$. Consequently, we saw that the tangent line is a good approximation for the function for values very near this point.

Suppose now, we want to have the same rate of bending at that point. This involves using the value of the second derivative at this point, giving a second degree polynomial approximation. In general, the Taylor polynomial of *n-th* degree adds an additional term to the polynomial of degree n–1, and takes the same value of the *n-th* derivative at the given point on the function.

The Taylor Polynomial of Degree n Approximating $f(x)$ near 0 is:

$$f(x) \approx P_n(x) = f(0) + f'(0)x + \frac{f''(0)}{2!}x^2 + \ldots + \frac{f^{(n)}(0)}{n!}x^n.$$

More generally, a Taylor Polynomial of Degree n Approximation of $f(x)$ for x near a:

$$f(x) \approx P_n(x) = f(a) + f'(a)(x-a) + \frac{f''(a)}{2!}(x-a)^2 + \ldots + \frac{f^{(n)}(a)}{n!}(x-a)^n$$

This is a Taylor polynomial centered at $x = a$.

Example 10.1.1

Find Taylor polynomials of degrees 1 and 3 as approximations to $f(x) = \sin x$ for values of x near 0.

Function	***Taylor Polynomial***	***Graphs***
$f(x) = \sin(x)$ x centered at $x = 0$		
$f(0) = \sin(0) = 0$ $f'(x) = \cos x$ so $f'(0) = \cos 0 = 1$ $f''(x) = -\sin x$ so $f''(0) = -\sin 0 = 0$ $f'''(x) = -\cos x$ so $f'''(0) = -\cos 0 = -1$	degree 1: $f(x) \approx P_1(x) = 0 + 1x = x$	
	degree 3: $f(x) \approx P_3(x) = 0 + 1(x) + \frac{0}{2!}x^2 + \frac{-1}{3!}x^3$ $= x - \frac{x^3}{3!}$	

Example 10.1.2

Find Taylor polynomials of degrees 1 and 3 as approximations to $f(x) = \sin x$ for values of x near $\pi/2$.

Function	***Taylor Polynomial***	***Graphs***
$f(x) = \sin(x)$ x centered at $x = \pi/2$		
$f(\frac{\pi}{2}) = \sin(\frac{\pi}{2}) = 1$ $f'(x) = \cos x$ so $f'(\frac{\pi}{2}) = \cos\frac{\pi}{2} = o$	degree 1: $f(x) \approx P_1(x) = 1 + 0(x - \frac{\pi}{2}) = 1$	
$f''(x) = -\sin x$ so $f''(\frac{\pi}{2}) = -\sin\frac{\pi}{2} = -1$ $f'''(x) = -\cos x$ so $f'''(\frac{\pi}{2}) = -\cos\frac{\pi}{2} = 0$	degree 3: $f(x) \approx P_3(x) = 1 + 0(x - \frac{\pi}{2}) - \frac{1}{2!}\left(x - \frac{\pi}{2}\right)^2 + 0\left(x - \frac{\pi}{2}\right)^3$ $= 1 - \frac{1}{2}\left(x - \frac{\pi}{2}\right)^2$	

On Your Own 1

Generate the Taylor polynomials of degree 1, 2 and 3 for the function $f(x) = \frac{1}{1-x}$. Use each to approximate the function when $x = 0.1$ and $x = 0.9$.

(The answer appears at the end of this section.)

2. *What is the significance of the degree of a Taylor Polynomial?*

___

___ (See text p. 440)

The higher the degree, the better the approximation is for values near the point and the larger the interval on which the function and the polynomial remain close. Approximations are better for values of x close to the centered value of the Taylor approximation than for values further away.

Example 10.1.3

Compare Taylor approximations of degrees 1, 2, 3, and 4 to the function $f(x) = e^x$ for x near 0.

Taylor polynomial	*Compare P_n (0.1) with* $f(0.1) = e^{0.1} = 1.105170918$
$P_1(x) = 1+x$	$P_1(0.1) = 1.1$
$P_2(x) = 1 + x + \frac{x^2}{2!}$	$P_2(0.1) = 1.1+0.005 = 1.105$
$P_3(x) = 1 + x + \frac{x^2}{2!} + \frac{x^3}{3!}$	$P_3(0.1) = 1.105 + 0.000167 = 1.10516667$
$P_4(x) = 1 + x + \frac{x^2}{2!} + \frac{x^3}{3!} + \frac{x^4}{4!}$	$P_4(0.1) = 1.10516667 + 0.0000042 = 1.1051708$

Your Turn

Fill in the chart below then compare Taylor approximations for $f(x) = \sin(x)$ for x near 0 and for x near $\pi/2$ by completing the following table. Notice that Taylor polynomials centered at $x = a$ give good approximations to $f(x)$ for x near a and perhaps, not very good approximations for x farther away from $x = a$.

Function	*Comparison of Values for x near a and for x not near a*					
$f(x) = \sin x$				*Actual Value*	$P_1(x)=x$	$P_3(x)= x-x^3/3!$
using Taylor Polynomials centered at $x=0$	$x=0.05$	$P_1(x)=$?	$P_3(x)$ =?	$\sin(0.05) =$ 0.049979	$P_1(0.05)=0.05$	$P_3(0.05)=$ 0.0499
	$x=\pi/2$			$\sin(\pi/2) = 1$	$P_1(\pi/2) =$ 1.5708	$P_3(\pi/2) =$ 0.924838

$f(x) = \sin x$ using Taylor Polynomials centered at $x=\pi/2$				Actual Value		
	$x=\pi/2 + 0.1$	$P_1(x)$ =?	$P_3(x)$ =?	$\sin(\pi/2+0.1) = 0.995004$	$P_1(x)=$ $\sin(\pi/2)=1$ $P_1(\pi/2+0.1)=1$	$P_3(x)=$ $1-0.5(x-\pi/2)^2$ $P_3(\pi/2+0.1)=$ 0.995
	$x = \pi/2 + 1$			$\sin(\pi/2+1) = 0.5403$	$P_1(\pi/2+1)=1$	$P_3(\pi/2+1) =$ 0.5

Key Notation

- $P_n(x)$ is a Taylor Polynomial of degree n.
- $n! = (n)(n-1)(n-2)\ldots(1)$. *"n factorial"*, means multiply n by all the numbers that come before n. Example: $6!=(6)(5)(4)(3)(2)(1)$.
- $\dfrac{f^{(n)}(a)}{n!}$ are numerical coefficients for the Taylor polynomial centered at $x = a$ and where $f^{(n)}(a)$ is the *n-th* derivative evaluated at a.

Key Study Skills

- Choose at least two problems that represent this section; put each on a 3 x 5 note card with the solution on the back.
- Review tangent line approximations on page 150 of the text.
- Use the ✓ ? ★ system on homework problems and get answers to all ★ problems.

Answer to On Your Own 1

$f(x) = \dfrac{1}{1-x}$, so $f(0) = 1$; $f'(x) = \dfrac{1}{(1-x)^2}$, so $f'(0) = 1$; $f''(x) = \dfrac{2}{(1-x)^3}$,
so $f''(0) = 2$; $f'''(x) = \dfrac{6}{(1-x)^4}$, so $f'''(0) = 6$

Actual $f(x)$	$f(x) \approx P_1(x) = 1 + 1x$	$f(x) \approx P_2(x) = 1 + 1x + \frac{2}{2!}x^2$	$f(x) \approx P_3(x) = 1 + 1x + \frac{2}{2!}x^2 + \frac{6}{3!}x^3$
$f(0.1)$=1. 1111....	$P_1(0.1)=1.1$	$P_2(0.1)=1.11$	$P_3(0.1)=1.111$
$f(0.9)=10$	$P_1(0.9)=1.9$	$P_2(0.9)=2.71$	$P_3(0.9)=3.439$

10.2 TAYLOR SERIES

Objective: To find a Taylor series for a function $f(x)$ and to investigate the convergence.

Preparations: Read and take notes on section 10.2.
Suggested practice problems: 3,4,7,10,14,18,22,31,36.

✠ Key Ideas of the Section

You should be able to:

- Construct Taylor series for a function $f(x)$ for x near 0.
- Construct Taylor series for a function $f(x)$ for x near a.
- Calculate the interval of convergence for a Taylor series.
- Define Taylor series for $\sin x$, $\cos x$ and e^x
- Define the Binomial Series for the function $(1+x)^p$

Key Questions to Ask and Answer

1. What's the difference between a Taylor polynomial and a Taylor series?

✍ ____

____ *(See text p. 442*

A Taylor series is a power series whose partial sums are Taylor polynomials. A Taylor series can be thought of as a Taylor polynomial that goes on forever. Any function, all of whose derivatives exist at $x = a$, can have a Taylor series centered at $x = a$, however, the series may not converge to $f(x)$.

Example 10.2.1

Generate the Taylor polynomials for the function $f(x) = e^x$ centered at $x = 0$ and give the Taylor series that represents this function.

Since $f(0) = 1 \;\; and \;\; f^n(0) = 1$,

$$f(x) \approx P_1(x) = 1 + 1x$$

$$f(x) \approx P_2(x) = 1 + 1x + \frac{1}{2!}x^2$$

$$f(x) \approx P_3(x) = 1 + 1x + \frac{1}{2!}x^2 + \frac{1}{3!}x^3$$

Each element in this sequence of polynomials gives better approximations to the function $f(x) = e^x$ for values of x near 0. The Taylor series for this function is this whole sequence of Taylor polynomials, written $e^x = 1 + x + \frac{x^2}{2!} + \frac{x^3}{3!} + \frac{x^4}{4!} + \ldots$.

2. Do all Taylor series converge to a function f(x) for all values of x?

✍ ____

____ *(See text p..443.)*

Any function f, all of whose derivatives exist at $x = a$, has a Taylor series centered at $x = a$. However for some functions, the series does not converge to $f(x)$ for all values of x. Functions $f(x)$ for which the Taylor series converges to $f(x)$ are called *analytic functions*.

Example 10.2.2

Generate the Taylor polynomials for the function $f(x)=\frac{1}{1-x}$ centered at $x = 0$ and give the Taylor series. On what interval does the series converge?

Since $f(0)=1;\ f'(0)=1;\ f''(0)=2!;\ f'''(0)=3!\ldots,$

$$f(x)\approx P_1(x)=1+1x$$

$$f(x)\approx P_2(x)=1+1x+\frac{2!}{2!}x^2$$

$$f(x)\approx P_3(x)=1+1x+\frac{2!}{2!}x^2+\frac{3!}{3!}x^3$$

Each element in this sequence of polynomials gives better approximations to the function $f(x)=\frac{1}{1-x}$ for values of x near 0. The Taylor series for this function is this whole sequence of Taylor polynomials written, $1+x+x^2+x^3+x^4+\ldots$. Using the Ratio Text (see Section 9.3), we can see $\lim_{n\to\infty}\left|\frac{a_{n+1}}{a_n}\right| = \lim_{n\to\infty}\left|\frac{x^{n+1}}{x^n}\right| = |x|$ and so R = 1. The function is not defined for $x = 1$ and the series does not converge for x = –1, so the interval of convergence is -1< x < 1.

✎ On Your Own 1

Find the Taylor series for the function $ln(1+x)$ centered at $x = 0$ and find the interval of convergence.

(The answer appears at the end of this section.)

3. *What are the Taylor series expansions for sin(x), cos(x), and e^x?*

✍ ____

____ *(See text p.442)*

It is probably a good idea to memorize these next formulas, so that they will be at your finger tips when you need them. These are analytic functions, so for all x:

$$\sin(x)=x-\frac{x^3}{3!}+\frac{x^5}{5!}-\frac{x^7}{7!}+\ldots$$

$$\cos(x)=1-\frac{x^2}{2!}+\frac{x^4}{4!}-\frac{x^6}{6!}+\ldots$$

$$e^x=1+x+\frac{x^2}{2!}+\frac{x^3}{3!}+\frac{x^4}{4!}+\ldots$$

4. *What is the Binomial series for $(1+x)^p$, where p is any real number?*

✍ ____

____ *(See text p.444.)*

Using the Taylor series about $x = 0$ one gets the expansion for the function:

$$(1+x)^p = 1+px+\frac{p(p-1)}{2!}x^2+\frac{p(p-1)(p-2)}{3!}x^3+\ldots$$

where p is any real number and for $-1<x<1$.

Example 10.2.3

Find the Binomial series expansion for $(1+x)^{-2}$.

Here $p = -2$, so $(1+x)^{-2} = 1+(-2)x+\frac{(-2)(-3)}{2!}x^2+\frac{(-2)(-3)(-4)}{3!}x^3+\ldots$ or

$(1+x)^{-2} = 1-2x+3x^2-4x^3+\ldots$ for $-1 < x < 1$.

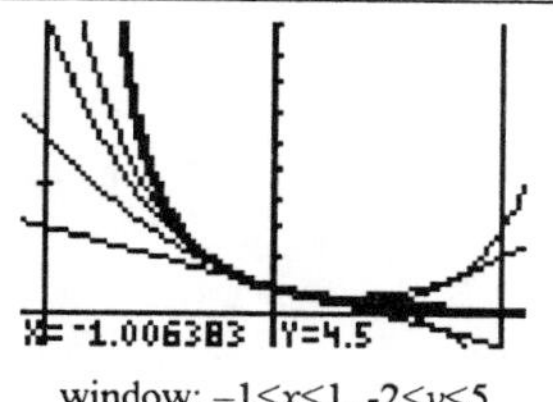 window: $-1<x<1$, $-2<y<5$	The graphs of the first 5 Taylor polynomials is given to the left. Note that the interval of convergence is $-1 < x < 1$.

On Your Own 2

Find the Binomial series for the function $\sqrt[3]{1+x}$ for $-1 < x < 1$.

(The answer appears at the end of this section.)

Key Notation

+ … means the terms in the series continue on indefinitely.

Key Study Skills

- Choose at least two problems that represent this section; put each on a 3 x 5 note card with the solution on the back.
- Review tangent line approximations on page 138 of the text.
- Use the ✓ ? ★system on homework problems and get answers to all ★ problems.

Answer to On Your Own 1

$f(x) = ln(1+x)$, $f'(x)=\frac{1}{1+x}$, $f''(x)=\frac{-1}{(1+x)^2}$, $f'''(x)=\frac{2}{(1+x)^3}$ so

$f(x) = ln(1+x) = 0+1x-\frac{1}{2}x^2+\frac{1}{3}x^3-\frac{1}{4}x^4+\ldots$. Using the Ratio Test, $\lim_{n\to\infty}\left|\frac{a_{n+1}}{a_n}\right| = \lim_{n\to\infty}\left|\frac{n-1}{n}x\right| = |x|$ and so R = 1. Examining the endpoints, the series diverges at $x = -1$, and for $x = 1$, this is the alternating harmonic series which converges. Therefore the interval of convergence is $-1< x \leq 1$.

Answer to On Your Own 2

Here, $p = 1/3$, so for $-1 < x < 1$,

$$\sqrt[3]{1+x} = 1+\frac{1}{3}x+\frac{\left(\frac{1}{3}\right)\left(\frac{1}{3}-1\right)}{2!}x^2+\frac{\left(\frac{1}{3}\right)\left(\frac{1}{3}-1\right)\left(\frac{1}{3}-2\right)}{3!}x^3+\ldots \text{ or}$$

$$\sqrt[3]{1+x} = 1+\frac{1}{3}x+\frac{-2}{3^2 2!}x^2+\frac{1\cdot 2\cdot 5}{3^3 3!}x^3+\ldots \text{ for } -1 < x < 1.$$

10.3 FINDING AND USING TAYLOR SERIES

Objective: To find new Taylor series by using substitution, differentiation and integration

Preparations: Read and take notes on section 10.3.
Suggested practice problems: 2,3,11,14,21,33.

✠ Key Ideas of the Section

You should be able to:

- Find new Taylor series by substitution.
- Find new Taylor series by differentiation or integration.
- Apply Taylor series to various situations.

Key Questions to Ask and Answer

1. *How do you find a new Taylor series from a known series by substitution?*

✍ ____

____ *(See text p.447)*

Sometimes there are easier ways to find the coefficients for a Taylor series than by differentiation. Substitution can sometimes be used to find a Taylor series for a function that is close to a series you know. For example, e^{-3x} looks close to $e^y = 1 + y + \frac{y^2}{2!} + \ldots$ and you can substitute y=-3x to find the new series for e^{-3x}.

Example 10.3.1

From Example 1, page 443 in the text, the Taylor Series for $y = ln(1+x)$ about $x = 0$ was found to be $\ln(1+x) = x - \frac{x^2}{2} + \frac{x^3}{3} - \frac{x^4}{4} + \ldots$, with interval of convergence $-1 < x \le 1$. Use this series to find the Taylor Series and interval of convergence for $y = ln(1+3x)$ about $x = 0$.

To get the series for $ln(1+3x)$ from the series for $ln(1+x)$ substitute $3x$ for x in the series to get:

$$\ln(1+3x) = (3x) - \frac{(3x)^2}{2} + \frac{(3x)^3}{3} - \frac{(3x)^4}{4} + \ldots = 3x - \frac{9}{2}x^2 + 9x^3 - \frac{81}{4}x^4 + \ldots .$$

The interval of convergence for $ln(1+x)$is $-1 < x \le 1$ and therefore, the interval of convergence for $ln(1+3x)$ is $-1 < 3x \le 1$ or $-1/3 < x \le 1/3$.

Note: The function is not defined for $x \le -1/3$. The graph below demonstrates convergence around $-1/3 < x \le 1/3$.

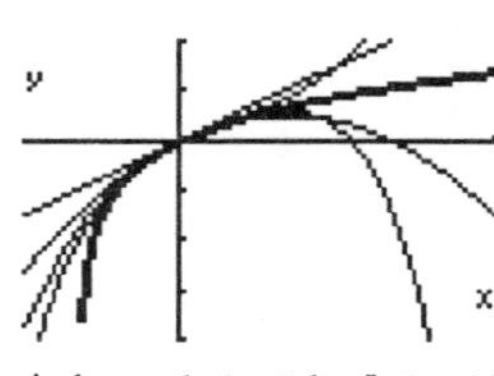

window: $-1 \le x \le 1$, $-5 < y < 2$

On Your Own 1

Use the Taylor Series you generated in **Example 10.2.2** of the Study Guide for $f(x)=\dfrac{1}{1-x}$ to find a power series representation and interval of convergence for $f(x)=\dfrac{1}{1-x^2}$. Plot the function and a polynomial representation.

(The answer appears at the end of this section.)

2. *How do you find a new Taylor series from a known series by differentiation or by integration?*

____

____ *(See text p.448)*

Differentiation or integration of a series gives a new series which will have the same open interval of convergence as the original series.

Example 10.3.2

The *sine-integral* function is defined $Si(x)=\int_0^x \frac{\sin t}{t}dt$, $Si(0)=0$. We can use the series for *sin t* to find the Taylor series about $x = 0$ for *Si(x)*. Determine accuracy by comparing values when $x = 1$.

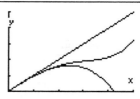

The Taylor Series for *sin t*:	$\sin(t)=t-\frac{t^3}{3!}+\frac{t^5}{5!}-\frac{t^7}{5!}+\frac{t^9}{9!}-\ldots$
Divide by *t*:	$\frac{\sin(t)}{t}=1-\frac{t^2}{3!}+\frac{t^4}{5!}-\frac{t^6}{5!}+\frac{t^8}{9!}-\ldots$
Integrate term by term:	$\int_0^x \frac{\sin(t)}{t}dt=\int_0^x 1dt-\int_0^x \frac{t^2}{3!}dt+\int_0^x \frac{t^4}{5!}dt-\ldots$ $=x-\frac{x^3}{3(3!)}+\frac{x^5}{5(5!)}-\frac{x^7}{7(7!)}+\ldots$
Si(1) = 0.94608	$P_1(1)=1$ $P_3(1)=0.94444$ $P_5(1)=0.94611$
Graphs of $Si(x)=\int_0^x \frac{\sin t}{t}dt$, $P1$ and $P3$.	window: $0 < x < 5, 0 < y < 5$

Example 10.3.3

Use differentiation to find the Taylor series for $\frac{1}{(1-x)^2}$ from the series for $\frac{1}{1-x}$, $|x| < 1$.

Recall that $\frac{1}{1-x} = 1 + x + x^2 + x^3 + \ldots$ and notice that $\frac{d}{dx}\left(\frac{1}{1-x}\right) = \frac{1}{(1-x)^2}$. So, differentiating each side gives $\frac{d}{dx}\left(\frac{1}{1-x}\right) = \frac{1}{(1-x)^2} = 1 + 2x + 3x^2 + \ldots$, $|x|<1$.

✎ On Your Own 2

Use integration to find the Taylor series about $x = 0$ for $f(x) = ln(1 + x)$.

(The answer appears at the end of this section.)

3. *What are some applications of Taylor series?*

✍ ____

____ *(See text p.449-450)*

Example 10.3.4

Approximate $\int_0^{\frac{1}{2}} \cos\sqrt{x}\,dx$ to 4 significant digits.

$\cos(x) = 1 - \frac{x^2}{2!} + \frac{x^4}{4!} - \frac{x^6}{6!} + \ldots$, for all real numbers x, and hence,

$\cos(\sqrt{x}) = 1 - \frac{\sqrt{x}^2}{2!} + \frac{\sqrt{x}^4}{4!} - \frac{\sqrt{x}^6}{6!} + \ldots$, for non-negative numbers x .

$= 1 - \frac{x}{2!} + \frac{x^2}{4!} - \frac{x^3}{6!} + \ldots$, for non-negative numbers x.

$$\int_0^{\frac{1}{2}} \cos\sqrt{x}\,dx = \frac{1}{2} - \frac{1}{2\cdot 2!}\left(\frac{1}{2}\right)^2 + \frac{1}{3\cdot 4!}\left(\frac{1}{2}\right)^3 - \frac{1}{4\cdot 6!}\left(\frac{1}{2}\right)^4 + \ldots = 0.4392.$$

This achieves 4-digit accuracy because the series is alternating and the last term considered has a value of 0.0000217 that only effects the fifth significant digit.

✎ On Your Own 3

Approximate $\sqrt[3]{e}$ to 5 decimals.

(The answer appears at the end of this section.)

Key Notation

- $Si(x) = \int_0^x \frac{\sin t}{t}\,dt$, the *sine-integral* function.

🗀 Key Study Skills

- Choose two problems that represent this section and put each on a 3 x 5 note card with the solution on the back.
- Use the ✓ ? ★ system on homework problems and get answers to all ★ problems.

- Learn to use the technology preferred by your instructor. The text assumes you have a graphing calculator available or a computer algebra system.
- If you are having difficulty with Taylor Series approximations, see your instructor immediately, as this is one of the key ideas in the course.
- Continue to make a list in your notebook of the series that you know.

Answer to On Your Own 1

$f(x) = \dfrac{1}{1-x} = 1 + x + x^2 + x^3 + \ldots$ for $-1 < x < 1$. Replace x by x^2 and get

$\dfrac{1}{1-x^2} = 1 + x^2 + x^4 + x^6 + \ldots$ for $-1 < x^2 < 1$, or $-1 < x < 1$.

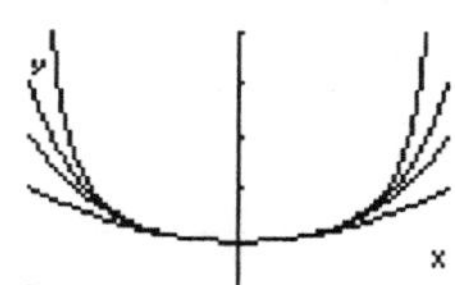

window: $-1 \le x \le 1$, $0 < y < 5$

Answer to On Your Own 2

$$\frac{1}{1+x} = 1 - x + x^2 - x^3 + \ \ldots \quad \text{for } -1 < x < 1 \ .$$

Integrate to get:

$$\int \frac{1}{1+x}\,dx = \int 1dx - \int xdx + \int x^2dx - \int x^3dx + \ldots$$

$$\ln(1+x) = x - \frac{x^2}{2} + \frac{x^3}{3} - \frac{x^4}{4} + \ldots \text{ for } -1 < x < 1$$

Answer to On Your Own 3

$$e^x = 1 + x + \frac{x^2}{2!} + \frac{x^3}{3!} + \frac{x^4}{4!} + \ldots$$

$$e^{\frac{1}{3}} = 1 + \left(\frac{1}{3}\right) + \frac{\left(\frac{1}{3}\right)^2}{2!} + \frac{\left(\frac{1}{3}\right)^3}{3!} + \frac{\left(\frac{1}{3}\right)^4}{4!} + \ldots$$

n	1	2	3	4	5	6	7
S_n	1.333333	1.388889	1.3950617	1.3955761	1.3956104	1.3956123	1.3956124

We can hope that we have the approximation correct to 5 decimals here, but with the skills developed so far, we have no way of being certain.

10.4 THE ERROR IN TAYLOR POLYNOMIAL APPROXIMATIONS

Objective: To estimate the magnitude of error when using an nth degree Taylor polynomial to approximate a function $f(x)$.

Preparations: Read and take notes on section 10.4.
Suggested practice problems: 2,4,6,8,14,15.

✠ Key Ideas of the Section

You should be able to:

- Find the upper bound of the magnitude of the error in approximation.
- See the effect on the error by changing the degree of the polynomial.
- Understand that the error of approximation decreases the closer x is to x = a, the center of the Taylor polynomial.

Key Questions to Ask and Answer

1. ***What factors affect the error in approximating a function with a Taylor polynomial of degree n centered at x = a?***

✍ ____

____ *(See text p.454)*

In Section 10.1, we discussed that the higher the degree the polynomial, the better the approximation of a Taylor polynomial is to the function. We also saw that when using Taylor polynomials of degree n, centered at $x = a$, the approximation is better the closer you are to $x = a$. (see **Example 10.1.3** in this Study Guide).

Example 10.4.1

Recall the Taylor series for $f(x) = \dfrac{1}{1-x} = 1 + x + x^2 + x^3 + \ldots$, centered at $x = 0$. Compute the errors in using second degree Taylor polynomials to approximate $f(0.1) = \dfrac{1}{1-0.1} \approx 1 + 0.1 + 0.1^2 = 1.11000$ and $f(0.05) = \dfrac{1}{1-0.05} \approx 1 + 0.05 + 0.05^2 = 1.05250$, where the second approximation is for x closer to $x = 0$ (and, as a matter of fact, it is decreased by a factor of 2) than the first.

$f(0.1) - \left(1 + 0.1 + 0.1^2\right) = 1.111111 - 1.11 = 0.001111$

$f(0.05) - \left(1 + 0.05 + 0.05^2\right) = 1.052632 - 1.052500 = 0.000132$. Notice that $\dfrac{0.001111}{0.000132} \approx 8.4$ which is the factor by which the error has decreased ($\approx 2^{2+1}$).

✎ On Your Own 1

Compare the errors in the approximations of $y = ln(1.1)$ and $y = ln(1.05)$ using third degree polynomial approximations $ln(1+x) \approx P_3(x) = x - \dfrac{x^2}{2} + \dfrac{x^3}{3}$.

(The answer appears at the end of this section.)

---

2. ***How do you calculate the upper bound of the error in using $P_n(x)$?***

✍ ____

____ *(See text p.454)*

The function f and its derivatives are continuous and $P_n(x)$ is the nth degree Taylor polynomial approximation to $f(x)$ about a and $En(x)$ is the error of approximation at x. Then

$|E_n(x)| = |f(x) - P_n(x)| \leq \dfrac{M}{(n+1)!}|x-a|^{n+1}$ where $M = |\max f^{(n+1)}|$ on the interval between a and x.

You should read the proof of this theorem carefully. Notice how the size of the error depends on n, on how close x is to a and on the $(n+1)$st derivative on the interval.

Example 10.4.2

Give a bound on the error, E_2, when $f(x)=\dfrac{1}{1-x}$ is approximated by a second degree polynomial for $-0.1 \le x \le 0.1$. $f^{(3)}(x)=\dfrac{3!}{(1-x)^4}$ is an increasing function on $-0.1 \le x \le 0.1$.

So, $\left|f^{(3)}(x)\right| \le \dfrac{3!}{(1-0.1)^4} = 9.145 < 10$ on $-0.1 \le x \le 0.1$.

$|E_2| = |f(x)-P_2(x)| \le \dfrac{10}{3!}|x|^3$ or $|E_2| \le \dfrac{10}{3!}|0.1|^3 = 0.00167$.

On Your Own 2

Give a bound on the error E_4 when the function $y = \ln(1+x)$ is approximated by a fourth degree Taylor polynomial on $-0.5 \le x \le 0.5$.

(The answer appears at the end of this section.)

Key Notation

- $f^{(n+1)}$ is the $(n+1)^{\text{st}}$ derivative.

Key Study Skills

- Choose two problems that represent this section and put each on a 3 x 5 note card with the solution on the back.
- Use the ✓ ? ★ system on homework problems and get answers to all ★ problems.
- Learn to use the technology preferred by your instructor. The text assumes you have a graphing calculator available or a computer algebra system.

Answer to On Your Own 1

$ln(1+x) \approx x - \dfrac{x^2}{2} + \dfrac{x^3}{3}$ so $ln(1+0.1) \approx 0.1 - \dfrac{0.1^2}{2} + \dfrac{0.1^3}{3}$ and $ln(1+0.05) \approx 0.05 - \dfrac{0.05^2}{2} + \dfrac{0.05^3}{3}$

$ln(1+0.1) - \left(0.1 - \dfrac{0.1^2}{2} + \dfrac{0.1^3}{3}\right) = 0.095310 - 0.09533 = -0.000020$ and

$ln(1+0.05) - \left(0.05 - \dfrac{0.05^2}{2} + \dfrac{0.05^3}{3}\right) = 0.048790 - 0.048792 = -0.000002$

Answer to On Your Own 2

$f(x) = ln(1+x) \approx x - \dfrac{x^2}{2} + \dfrac{x^3}{3} - \dfrac{x^4}{4}$. The fifth derivative, $f^{(5)}(x) = \dfrac{4!}{(1+x)^5}$ is decreasing and positive on $-0.5 \le x \le 0.5$ so $\left|f^{(5)}(x)\right| \le \dfrac{4!}{(1+(-0.5))^5} = 768 < 800$ on $-0.5 \le x \le 0.5$.

$|E_4| = |f(x) - P_4(x)| \le \dfrac{800}{5!}|0.5|^5 = 0.20833$

10.5 FOURIER SERIES

Objective: To construct Fourier approximations to a function.

Preparations: Read and take notes on section 10.5.
Suggested practice problems: 1, 3, 5, 8, 14, 15.

✠ Key Ideas of the Section

You should be able to:

- Calculate Fourier coefficients.
- Construct Fourier polynomials and series to approximate a periodic function of any period.
- Use Fourier series to analyze a function and to find a good global approximation.
- Compute and graph the energy spectrum for a function.

Key Questions to Ask and Answer

1. How does a Fourier polynomial compare to a Taylor polynomial as an approximation for a function?

✍ ____

____ *(See text p 457-458.)*

Taylor polynomials provide good approximations locally, whereas Fourier polynomials (note that they're not really polynomials in the true sense) give good approximations globally. Since Fourier polynomials are periodic functions, they give good approximations to periodic functions.

Example 10.5.1

The square wave function in Example 1 in the text (p. 459) is $f(x) = \begin{cases} 0 & -\pi \le x < 0 \\ 1 & 0 \le x < \pi \end{cases}$

The Taylor polynomial centered at $x \neq 0$ has derivatives equal to 0. So the Taylor polynomial is either going to be $f(x) = 0$ or $f(x) = 1$ depending on where the polynomial is centered. Although this gives a perfect approximation locally, it is not a good global approximation.

2. How do you construct a Fourier polynomial of degree n as an approximation for a periodic function?

✍ ____

____ *(See text p 459.)*

A *Fourier polynomial of degree n* is: $f(x) \approx F_n(x) = a_0 + \sum_{k=1}^{n} a_k \cos(kx) + \sum_{k=1}^{n} b_k \sin(kx)$

The *Fourier Coefficients* for a periodic function f of period 2π are:

$$a_0 = \frac{1}{2\pi}\int_{-\pi}^{\pi} f(x)\,dx$$

$$a_k = \frac{1}{\pi}\int_{-\pi}^{\pi} f(x)\cos(kx)\,dx,\ k>0$$

$$b_k = \frac{1}{\pi}\int_{-\pi}^{\pi} f(x)\sin(kx)\,dx,\ k>0$$

Example 10.5.2

Construct Fourier approximations for the function of period 2π and graph the Fourier polynomials $F_1, \ldots, F_4$.

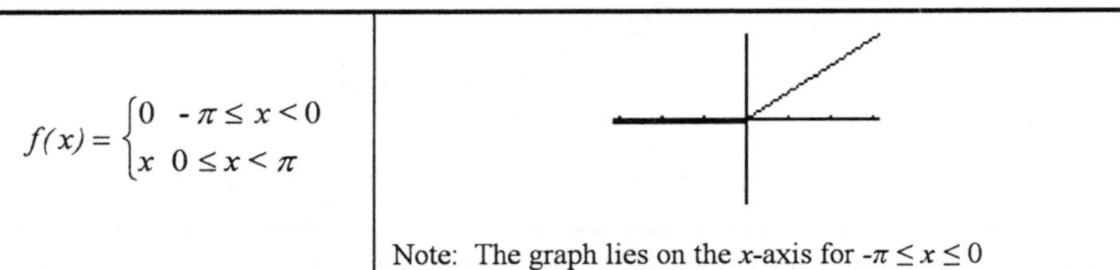

$f(x) = \begin{cases} 0 & -\pi \le x < 0 \\ x & 0 \le x < \pi \end{cases}$	Note: The graph lies on the x-axis for $-\pi \le x \le 0$

First write down the Fourier coefficients:

$$a_0 = \frac{1}{2\pi}\int_{-\pi}^{\pi} f(x)\,dx = \frac{1}{2\pi}\int_{0}^{\pi} x\,dx = \frac{\pi}{4}$$

To construct the coefficients, you get the following (using a Table of Integrals)

$$a_n = \frac{1}{\pi}\int_{-\pi}^{\pi} f(x)\cos(nx)\,dx = \frac{1}{\pi}\int_{0}^{\pi} x\cos(nx)\,dx$$

$$= \frac{1}{\pi}\left(\frac{x}{n}\sin(nx) + \frac{1}{n^2}\cos(nx)\right)\Bigg|_0^{\pi} = \frac{1}{n^2\pi}\left(\cos(n\pi) - 1\right)$$

which gives $a_n = 0,\ n$ even, $a_n = \dfrac{-2}{n^2\pi},\ n$ odd , $n \ge 1$.

$$b_n = \frac{1}{\pi}\int_{-\pi}^{\pi} f(x)\sin(nx)\,dx = \frac{1}{\pi}\int_{0}^{\pi} x\sin(nx)\,dx$$

$$= \frac{1}{\pi}\left(\frac{-x}{n}\cos(nx) + \frac{1}{n^2}\sin(nx)\right)\Bigg|_0^{\pi} = \frac{1}{\pi}\left(\frac{-\pi}{n}\cos(n\pi)\right)$$

$$= \frac{-1}{n}\cos(n\pi)$$

which gives $b_n = \dfrac{(-1)^{n+1}}{n}$ if $n \ge 1$.

k	a_k	b_k	$F_n(x)$	Fourier Polynomials
0	$a_0 = \frac{\pi}{4}$			
1	$a_1 = \frac{-2}{\pi}$	$b_1 = 1$	$F_1(x) = \frac{\pi}{4} - \frac{2}{\pi}\cos(x) + \sin(x)$	
2	$a_2 = 0$	$b_2 = \frac{-1}{2}$	$F_2(x) = \frac{\pi}{4} - \frac{2}{\pi}\cos(x) + \sin(x) - \frac{1}{2}\sin(2x)$	

3	$a_3 = \dfrac{-2}{9\pi}$	$b_3 = \dfrac{1}{3}$	$F_3(x) = \dfrac{\pi}{4} + \dfrac{-2}{\pi}cos(x) + sin(x) + \dfrac{-1}{2}sin(2x)$ $+\dfrac{-2}{9\pi}cos(3x) + \dfrac{1}{3}sin(3x)$	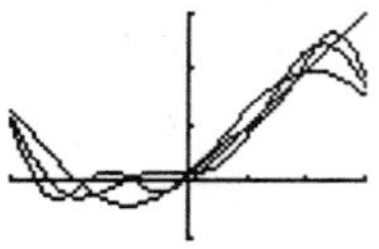
4	$a_4 = 0$	$b_4 = \dfrac{-1}{4}$	$F_4(x) = \dfrac{\pi}{4} + \dfrac{-2}{\pi}cos(x) + sin(x) + \dfrac{-1}{2}sin(2x)$ $+\dfrac{-2}{9\pi}cos(3x) + \dfrac{1}{3}sin(3x) + \dfrac{-1}{4}sin(4x)$	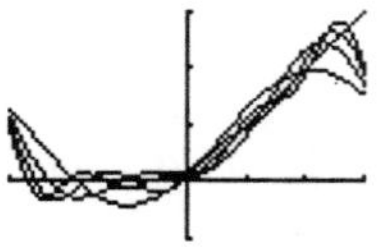

The Fourier Series for this function f on $[-\pi,\pi]$ is

$$f(x) = \frac{\pi}{4} + \frac{-2}{\pi}cos(x) + sin(x) + \frac{-1}{2}sin(2x)$$
$$+\frac{-2}{9\pi}cos(3x) + \frac{1}{3}sin(3x) + \frac{-1}{4}sin(4x) + \ldots.$$

3. *What if the period is not 2π?*

✍ ____

____ *(See text p 464.)*

Example 10.5.3

Suppose the function in the example above had period $b = 4$ and is defined as follows:

$f(x) = \begin{cases} 0 & -2 \le x < 0 \\ x & 0 \le x < 2 \end{cases}$	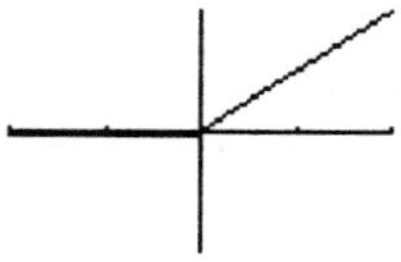 *The graph lies on the x*-axis on $-2 \le x < 0$

We simply make a substitution to get a function that is of period 2π, find the Fourier approximation to this function, and then substitute back.

Let $x = \dfrac{4t}{2\pi} = \dfrac{2t}{\pi}$. x varies over the interval $[-2,2]$ when t varies over $[-\pi,\pi]$.

$$g(t) = f(\frac{2t}{\pi}) = \begin{cases} 0 & -\pi \le t < 0 \\ t & 0 \le t < \pi \end{cases}$$

g has period 2π. $g(t) \approx \dfrac{\pi}{4} - \dfrac{2}{\pi}cos(t) + sin(t) - \dfrac{1}{2}sin(2t)$ and since $t = \dfrac{\pi x}{2}$,

$$f(x) = g\left(\frac{\pi x}{2}\right) \approx \frac{\pi}{4} - \frac{2}{\pi}cos\left(\frac{\pi x}{2}\right) + sin\left(\frac{\pi x}{2}\right) - \frac{1}{2}sin(\pi x) + \ldots.$$

The graph of the first degree polynomial looks like this:

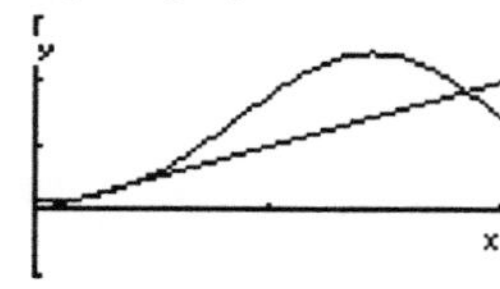

window: $0 < x < \pi$, $0 < y < 2$

4. *What does the energy spectrum tell us about a periodic function?*

✍ ____

____ *(See text p 462.)*

The *energy* E of a periodic function f of period 2π is the number :

$$E = \frac{1}{\pi}\int_{-\pi}^{\pi} (f(x))^2 dx = A_0^2 + A_1^2 + A_2^3 + \ldots$$

where $A_0 = \sqrt{2}a_0$ and $A_k = \sqrt{a_k^2 + b_k^2}$ (for all integer $k \geq 1$),
which turns out to be the sum of the energy of its harmonics.

✍Your Turn

Complete the following table and graph the energy spectrum for the function from **Example 10.5.2**, $f(x) = \begin{cases} 0 & -\pi \leq x < 0 \\ x & 0 \leq x < \pi \end{cases}$ and give the fraction of the energy carried by the constant term and the first four harmonics.

Calculate the Energy of f	*Answer*
$E = ?$	$E = \frac{1}{\pi}\int_{-\pi}^{\pi} (f(x))^2 dx = \frac{1}{\pi}\int_{0}^{\pi} (x)^2 dx = \frac{\pi^2}{3} = 3.290$

The fraction of energy is:

a_k	b_k	*Find* A_n^2	*Answers*
$a_0 = \frac{\pi}{4}$			$A_0^2 = 2a_0^2 = 2\left(\frac{\pi}{4}\right)^2 = \frac{\pi^2}{8} = 1.2337$
$a_1 = \frac{-2}{\pi}$	$b_1 = 1$		$A_1^2 = a_1^2 + b_1^2 = \left(\frac{-2}{\pi}\right)^2 + 1 = 1.4053$
$a_2 = 0$	$b_2 = \frac{-1}{2}$		$A_2^2 = a_2^2 + b_2^2 = \left(\frac{-1}{2}\right)^2 = 0.2500$
$a_3 = \frac{-2}{9\pi}$	$b_3 = \frac{1}{3}$		$A_3^2 = a_3^2 + b_3^2 = \left(\frac{-2}{9\pi}\right)^2 + \left(\frac{1}{3}\right)^2 = 0.1161$
$a_4 = 0$	$b_4 = \frac{-1}{4}$		$A_4^2 = a_4^2 + b_4^2 = \left(\frac{-1}{4}\right)^2 = 0.0625$
Fraction of energy carried by the constant term and the first 4 harmonics is:			$1.2337 + 1.4053 + 0.25 + 0.1161 + 0.0625 = 3.0676$ Fraction = 3.0676/3.290 =0.93.2 or 93.2%

🗁 Key Study Skills

- Review Trigonometric functions (Section 3.5).
- Review using Tables of Integrals (Section 7.3).
- Choose at least two problems that represent this section; put each on a 3 x 5 note card with the solution on the back. Review note card problems from 10.1-10.5.
- Use the ✓ ? ★ system on homework problems and get answers to all ★ problems.

CUMULATIVE REVIEW CHAPTER TEN

Key Study Skills

- Review all of the **Key Ideas.**
- Choose from Review Exercises and Problems from the text, p. 470-473.
- Do Check Your Understanding from text, p. 473-474.
- Review the note card problems from 10.1 -10.5.
- Make up your own test.
- Review homework problems and get answers to all ★ (starred) problems.
- Take the Chapter Ten Practice Test.

Chapter Ten Practice Test

1. Construct the Taylor polynomial approximation of degree 3 to the *function* $f(x) = \cos x$ about $x = -\pi/4$. How does the approximation compare to the exact answer at $x = 0$?

2. Find $\lim\limits_{x\to 0} \dfrac{(\sin x - x)^3}{x(1-\cos x)^4}$ by approximating *sin x* and *cos x* with suitable Taylor polynomials.

3. Find the Taylor series about $x = 0$ for $\cos\sqrt{x}$, $x \geq 0$ without computing the derivatives. That is, modify a familiar Taylor series by integrating it, differentiating it, making a substitution in it, or some combination of these. Use that series to approximate $\cos\sqrt{2}$.

4. Find the Taylor series expansion for $f(x) = -\ln(1-4x)$ by substituting *t* into the series for *ln(1+x)*. Plot *f(x)* and its Taylor polynomials up to degree 4. Use the graphs to guess the interval of convergence.

5. a. Find the Taylor series for $f(t) = te^t$ about $t = 0$.

   b. Use your answer to part a to find a Taylor series expansion about $x = 0$ for $g(x) = \int_0^x te^t dt$.

   c. Use your answer to part b to show that $\dfrac{1}{2} + \dfrac{1}{3(1!)} + \dfrac{1}{4(2!)} + \dfrac{1}{5(3!)} + \dfrac{1}{6(4!)} + \ldots = 1$

(The answers appear at the end of this study guide.)

CHAPTER ELEVEN

KEY CONCEPT: DIFFERENTIAL EQUATIONS

Differential equations model some complex problems which have functions as solutions rather than numbers. These models include exponential growth or decay, logistic population growth, or laws of motion such as oscillations of a mass on a spring. General first order differential equations can be seen using slope fields, using numerical methods, or analytically. Finding antiderivatives is solving a differential equation of a particular type.

11.1 WHAT IS A DIFFERENTIAL EQUATION?

Objective: To know what a differential equation and its family of solutions look like.

Preparations: Read and take notes on section 11.1.
Suggested practice problems: 1,5,10,12,14,15.

✠ Key Ideas of the Section

You should be able to:

- Recognize a differential equation as a model.
- Demonstrate that a family of functions is a solution to a differential equation.
- Demonstrate that a specific function is a solution to an initial-value problem.
- Distinguish between a first-order and a second-order differential equation.

Key Questions to Ask and Answer

1. ***What is a differential equation and what is a solution?***

✍ ____

____ *(See text p. 478-479.)*

Algebraic equations model situations for which *numbers* are solutions. Differential equations model situations for which *functions* are solutions.

Example 11.1.1

Suppose you place an amount of money, y, into a savings account paying a continuous annual rate of 8% interest. The growth rate of your money is proportional to the amount of money itself. Then the differential equation $\frac{dy}{dt} = 0.08y$ describes your bank account balance over time. The *general solution* to this differential equation is $y = Ce^{0.08t}$. This is an infinite *family of functions*, several of which look like:

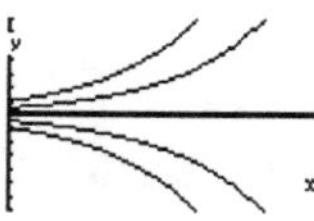

This is a *first-order* differential equation because it involves only the first derivative of the function y.

2. *How do you verify that a function is a solution to a differential equation ?*

✍ ____

____ *(See text p. 479.)*

Example 11.1.2

Show that the family of functions $y = Ce^{0.08t}$ satisfies the differential equation $\frac{dy}{dt} = 0.08y$ from **Example 11.1.1** above.

Using $y = Ce^{0.08t}$ Left side: $\frac{dy}{dt} = \frac{d\left(Ce^{0.08t}\right)}{dt} = C(0.08e^{0.08t})$

Using $\frac{dy}{dt} = 0.08y$ Right side: $0.08y = 0.08(Ce^{0.08t})$

Since the two sides are the same, $y = Ce^{0.08t}$ is the general solution to the differential equation $\frac{dy}{dt} = 0.08y$.

On Your Own 1

a. Verify that the specific function $y = x - 1$ is a solution to the differential equation $\frac{dy}{dx} = x - y$

b. Verify that both functions $y = 2t^2 + t - 3$ and $y = 2t^2 + 5t + 6$ are solutions to the differential equation $\frac{d^2y}{dt^2} = 4$.

(The answer appears at the end of this section.)

3. *What is an initial value problem and what does its solution look like?*

✍ ____

____ *(See text p. 479.)*

Requiring an *initial condition* or a specific value of the function at a specific value of the independent variable gives an additional requirement on the family of solutions to a differential equation. This usually allows you to pick out one function that satisfies both the differential equation and the initial condition, called the *initial value problem.*

Example 11.1.3

Suppose you placed \$300 in the savings account in **Example 11.1.1**. This means that $y = \$300$ when $t = 0$. Substituting into the general solution $y = Ce^{0.08t}$ yields $300 = Ce^{0.08(0)}$ or $C = \$300$. The specific function $y = 300e^{0.08t}$ is the solution to the differential equation $\frac{dy}{dt} = 0.08y$ with initial condition that $y = \$300$ when $t = 0$. The initial value here was the initial amount placed into the savings account, picking out that specific solution function from the general solution.

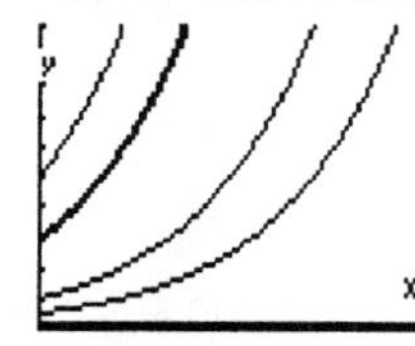

window: $0 \le x \le 50$, $0 \le y \le 1000$

On Your Own 2

a. Show that the function $y = 2t^2 + t - 3$ satisfies the differential equation $\frac{d^2y}{dt^2} = 4$ with initial conditions $y(0) = -3$ and $y'(0) = 1$.

b. Show that the function $y = 2t^2 + 5t + 6$ does not satisfy the initial value problem in part a.

(The answer appears at the end of this section.)

4. What is the relationship between finding antiderivatives and solving differential equations?

____

____ *(See text p. 480)*

Finding solutions to a differential equation of the form $\frac{dy}{dx} = f(x)$ is the same as finding the antiderivative of $f(x)$.

Example 11.1.4

Finding solutions to the differential equation $\frac{dy}{dx} = \sin x$ is the same as finding the anti-derivative of $\sin x$, which you know is $y = -\cos x + C$. Notice here, the solution involves one arbitrary constant.

Example 11.1.5

Find the solutions to $\frac{d^2y}{dx^2} = \sin x$.

$\frac{dy}{dx} = -\cos x + C_1$, and $y = -\sin x + C_1x + C_2$. Notice that the solution involves two arbitrary constants because there were two antidifferentiations.

Key Study Skills

- Review antiderivatives and differential equations in Section 6.3 of the text.
- Choose at least two problems that represent this section; put each on a 3 x 5 note card with the solution on the back.
- Use the ✓ ? ★ system on homework problems and get answers to all ★ problems.

Answer to On Your Own 1

a. Left side: $\frac{dy}{dx} = \frac{d(x-1)}{dx} = 1$ and the Right side: $x - y = x - (x-1) = 1$

b. $\frac{d^2y}{dt^2} = \frac{d}{dt}\left(\frac{d(2t^2+t-3)}{dt}\right) = \frac{d}{dt}(4t+1) = 4$ and $\frac{d^2y}{dt^2} = \frac{d}{dt}\left(\frac{d(2t^2+5t+6)}{dt}\right) = \frac{d}{dt}(4t+5) = 4$.

Answer to On Your Own 2

a. $\frac{d^2y}{dt^2} = \frac{d}{dt}\left(\frac{d(2t^2+t-3)}{dt}\right) = \frac{d}{dt}(4t+1) = 4$, $y(0) = 2(0)^2 + 0 - 3 = -3$, and $y'(t) = 4t + 1$, $y'(0) = 4(0) + 1 = 1$.

b. You showed in **On Your Own 1** that $y = 2t^2 + 5t + 6$ is a solution to $\frac{d^2y}{dt^2} = 4$. However the function does not satisfy the initial condition since $y(0) = 6$ and $y'(0) = 5$.

11.2 SLOPE FIELDS

Objective: To visualize first order differential equations graphically.

Preparations: Read and take notes on section 11.2.
Suggested practice problems: 1,3,5,6,8,10.

✠ Key Ideas of the Section

You should be able to:

Visualize a solution to a first order differential equation by looking at the slope field.
Construct a slope field for a differential equation.

Key Questions to Ask and Answer

1. What is a slope field for a first order differential equation?

✍ ____

____ *(See text p. 482.)*

A slope field for a first order differential equation is a set of signposts in the plane. Wherever you are, there is a signpost to tell you which direction to move. After moving in that direction for a small amount, there is another signpost to tell you which direction to move from there.

Example 11.2.1

Section 11.1 of this Study Guide showed that the general solution for the differential equation $\frac{dy}{dt} = 0.08y$ modeling your bank account earning 8% interest was $y = Ce^{0.08t}$. The slope field for this differential equation looks like:

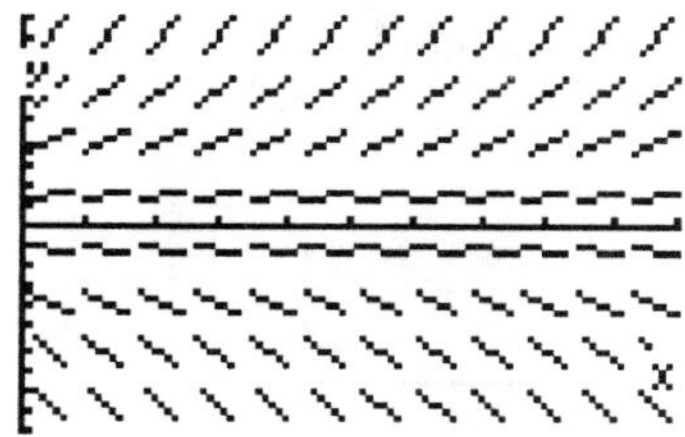

Note: window: $0 \le x \le 100$, xscl=10, $-200 \le y \le 200$, yscl=10

For example, the slope at the point (0,$0) is $0/year; the slope at the point (0,$100) is $8/year. (an initial investment of $100 is increasing initially at a rate of $8/year); if you are at the point (0,$300) the slope there is $24/year.

Example 11.2.2

Illustrated to the right is the specific solution $y = 300e^{0.08t}$ to the initial value problem for the example above where y = $300 when $t = 0$.

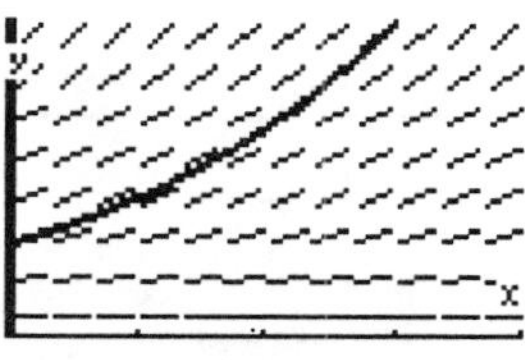

window: $0<x<20$, $0<y<1000$

On Your Own 1

Find the general solution to the differential equation $y' = x^2$, sketch the slope field, and find the specific solution with initial condition $y = 2$ if $x = 0$.

(The answer appears at the end of this section.)

2. *Do the solutions to initial value problems always exist and, if so, are they unique?*

____

____ *(See text p. 484-485.)*

Initial value problems (a differential equation and an initial condition) almost always have solutions. The slope field helps visualize this, that with a given initial condition, you can trace out a unique curve.

Example 11.2.3

From Section 11.1 **On Your Own 1** in this Study Guide, you showed that $y = x - 1$ satisfied the differential equation $\frac{dy}{dx} = x - y$. Specifically, it is the solution to the initial value problem with initial condition $y(0) = -1$. Show this solution on the slope field. Notice that solutions with different initial conditions look very different from the solution to this particular initial value problem.

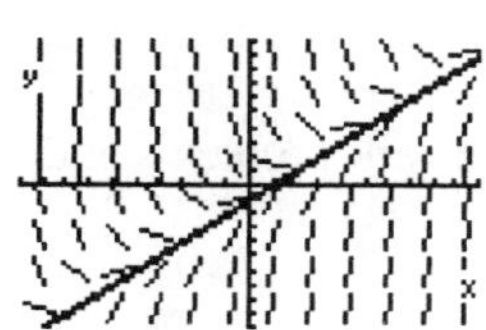

window: $-10 \le x \le 10$, xscl=1, $-10 \le y \le 10$, yscl=1

On Your Own 2

Sketch the slope field for the differential equation $y' = x^2$ and find the solution with initial condition a) $y = 4$ if $x = 0$ and b) with initial condition $y = 2$ if $x = 3$. Use the following grid.

(The answer appears at the end of this section.)

Your Turn

Match the following slope fields with their differential equations:

Differential equation	Slope field Note: windows: $-10 \le x \le 10$, xscl=1, $-10 \le y \le 10$, yscl=1	Answers:
a) $y' = x$	(i) 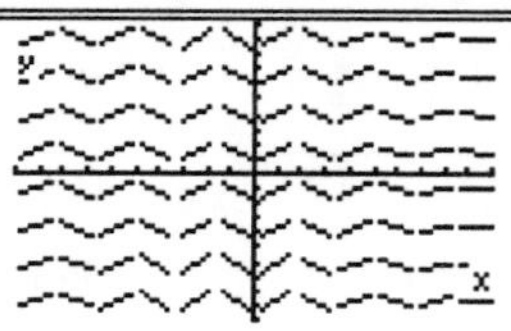	(a) matches (iii). Notice that at $x = 0$, the slope is 0; at $x = 1$, the slope is 1, at $x = -1$, the slope is -1, etc.
b) $y' = sin(2x)$	(ii) 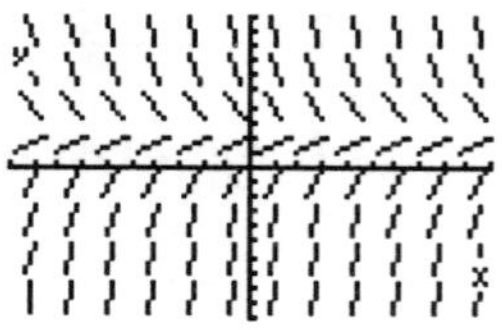	(b) matches (i). Notice that the slopes behave as a periodic function.
c) $y' = 2 - y$	(iii) 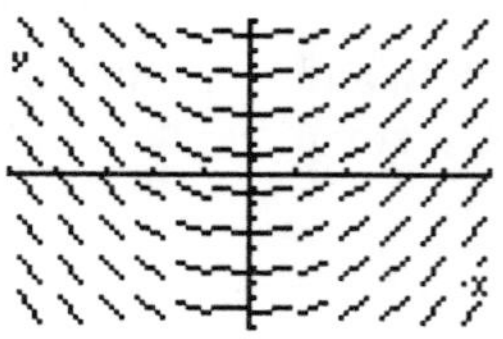	(c) matches (ii). Notice that for $x = 0$, if the value of $y > 2$, the slopes are negative; if $y < 2$, the slopes are positive.

Key Study Skills

- If you are using a graphing calculator, you should add the program for slope fields and practice recognizing slope fields for various types of first order differential equations.
- Choose at least two problems that represent this section; put each on a 3 x 5 card with the solution on the back.
- Use the ✓ ? ★ system on homework problems and get answers to all ★ problems.

Answer to On Your Own 1

Finding the antiderivative $\int x^2 dx = \frac{x^3}{3} + C$ gives the general solution with the following slope field. The specific solution is the function passing through the point (0, 2).

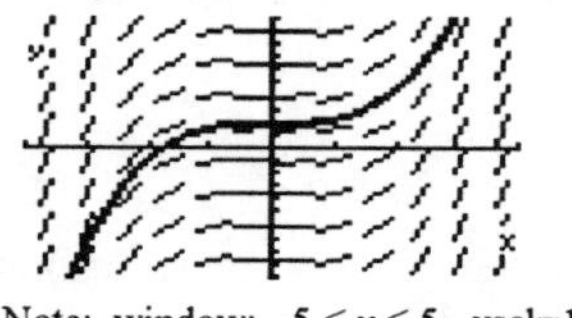

Note: window: $-5 \le x \le 5$, xscl=1, $-20 \le y \le 2\ 0$, yscl=2

At (0,2) since $y = \frac{x^3}{3} + C$, then $2 = \frac{0^3}{3} + C$, so $C = 2$.

The solution we want is $y = \frac{x^3}{3} + 2$

Answer to On Your Own 2

	Solution	***Slope Field*** windows: $5 \le x \le 5$, xscl=1, $-20 \le y \le 20$, yscl=2
Initial conditions $y = 4\ if\ x = 0$	$y = \frac{x^3}{3} + 4$	
$y = 2\ if\ x = 3$	$y = \frac{x^3}{3} - 7$	

11.3 EULER'S METHOD

Objective: To solve first order differential equations numerically

Preparations: Read and take notes on section 11.3.
Suggested practice problems: 2,6,7,9,10.

✠ Key Ideas of the Section

You should be able to:

- Compute numerical values of the solution to a first order differential equation.
- Determine whether the estimate is an overestimate or an underestimate.
- Have an approximation of the error in using Euler's method.

Key Questions to Ask and Answer

1. ***How do you use Euler's Method to construct a solution for a differential equation passing through a given point?***

✍ ____

____ *(See text p. 488-489.)*

Given a point on the solution curve, Euler's method uses the slope at that point (the signpost of the slope field) to approximate the next value of y. To get from point P_{n-1} to P_n, there will need to be an increase of $\Delta y = (slope\ \ at\ \ P_{n-1})\Delta x$.

The next y-value at the point P_n = (the y- value at P_{n-1}) + Δy . It helps to understand this by looking at it graphically in Figure 11.25 on page 489 in the text.

Example 11.3.1

In Section 11.2 **On Your Own 2** of this Study Guide, you found that $y = \frac{x^3}{3} - 7$ is the solution to the initial value problem, $\frac{dy}{dx} = x^2$ which passes through the point (3,2). Use Euler's Method to approximate the value of y at x=3.5 and compare with the true y-value. Let $\Delta x = 0.1$.

	x	y	$\Delta y = (slope)\Delta x$
P_0	3.0	2.0	$\Delta y = (3^2)(.1) = 0.9$
P_1	3.1	2.0 + 0.9 = 2.9	$\Delta y = (3.1^2)(.1) = 0.961$
P_2	3.2	2.9 + 0.961 = 3.861	$\Delta y = (3.2^2)(.1) = 1.024$
P_3	3.3	3.861 + 1.024 = 4.885	$\Delta y = (3.3^2)(.1) = 1.089$
P_4	3.4	4.885 + 1.089 = 5.974	$\Delta y = (3.4^2)(.1) = 1.156$
P_5	3.5	5.974 + 1.156 = 7.13	True value of y at 3.5 = $\frac{(3.5)^3}{3} - 7 = 7.2916$
		Comparison of estimated points (dots) and the true value (line) on the slope field: window: $3 < x < 3.5,\ 2 < y < 8$	

On Your Own 1

The solution to the differential equation $\frac{dy}{dx} = \frac{1}{x}$ for x > 0 is $y = ln\,|x| + C$. Pick the solution that passes through the point (1,0) and use Euler's Method to estimate $y = ln(2)$. Let $\Delta x = 0.5$, that is $n = 2$.

(The answer appears at the end of this section.)

2. *When does Euler's Method give an underestimate and when does it give an overestimate?*

✍ ____

____ *(See text p. 489)*

The concavity determines whether the approximations are overestimates or underestimates. Curves which are concave up have estimates which are too small; curves which are concave down have estimates which are too large.

Example 11.3.2

As you can see from **Example 11.3.1**, the curve is concave up in this interval [3,3.5], and the slopes will give underestimates for the change in y. Hence, the estimates will be too small.

$\frac{dy}{dx} = x^2$ passing through (3,2)

window: $3 < x < 3.5$, $2 < y < 8$

x	**Approx. y-value**	**True y-value**
3.0	2.0	2.0
3.1	2.9	2.9303
3.2	3.861	3.9227
3.3	4.885	4.979
3.4	5.794	6.1013
3.5	7.13	7.2917

Example 11.3.3

In **On Your Own 1**, the curve is concave down, and exactly the opposite occurs from the previous example – the slopes give overestimates and the approximations will be too large.

$\frac{dy}{dx} = \frac{1}{x}$ passing through (1,0)

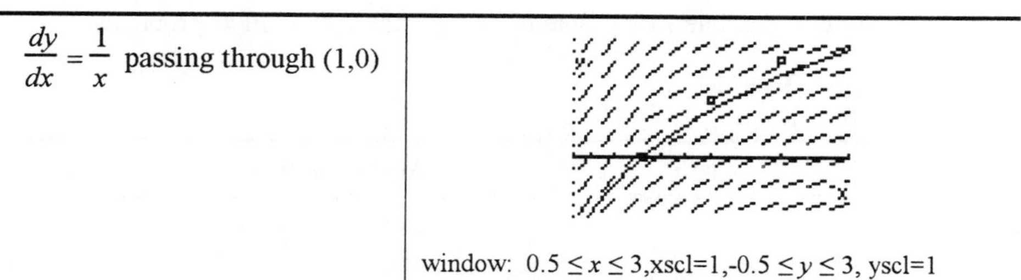

window: $0.5 \le x \le 3$,xscl=1,$-0.5 \le y \le 3$, yscl=1

x	y	**True y-value**
1	0	0
1.5	0.5	0.4054
2.0	0.8333	0.69314

3. *How accurate is Euler's Method?*

✍ ____

____ *(See text p. 490-491.)*

Smaller increments of Δx give more accurate estimates.

✎ Your Turn

Use Euler's method to approximate *ln* (*2*) starting at the point (1,0) with $\Delta x = 0.2$, that is $n = 5$. Compare the error for $n = 5$ with the error for $n = 2$ used in On Your Own 1.

Answers

	x	y	Δy=(*slope*)Δx	**Approx. *Y*-value**
P_0	1	0		0
P_1	1.2			0.2
P_2	1.4			.3666
P_3	1.6			0.5095
P_4	1.8			0.6345
P_5	2.0			0.7456
	Error(n = 5)			Error (n = 5)=0.7456 – 0.69314 = 0.05246
	Error(n = 2)			Error(n = 2)=0.8333 – 0.69314 = 0.14016

Key Algebra Skills

- Review linear functions.

Key Study Skills

- Choose at least two problems that represent this section; put each on a 3 x 5 card with the solution on the back.
- Use the ✓ ? ★ system on homework problems and get answers to all ★ problems.

Answer to On Your Own 1

The solution we want is $y = ln\ x$.

	x	y	Δy = (*slope*)Δx
P_0	1	0	Δy = (1)(.5) = 0.5
P_1	1.5	0 + 0.5 = 0.5	Δy = (1/1.5)(.5) = 0.333
P_2	2.0	0.5 + 0.333 = 0.8333	*ln(2)* = 0.693
		Comparison of estimated points (dots) and the true value (line) on the slope field.	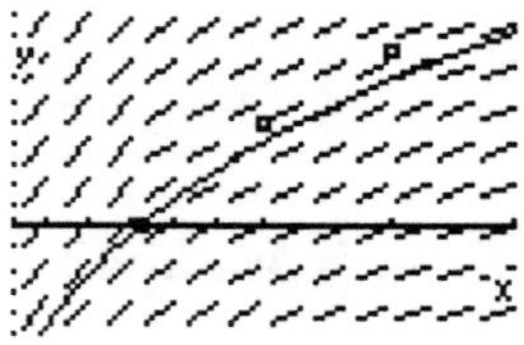 Note: window: $0.5 \le x \le 3$, xscl = 1, $-0.5 \le y \le 3$, yscl = 1

11.4 SEPARATION OF VARIABLES

Objective: To solve certain first order differential equations analytically.

Preparations: Read and take notes on section 11.4.
Suggested practice problems: 2,6,8,11,14,26,32,35,40,42,46.

✠ Key Ideas of the Section

You should be able to:

- Recognize which first order differential equations can be solved by separating the variables.
- Find the solution to the differential equation by separation of variables.
- Find the specific constants, given initial conditions.
- Relate the solution you find analytically with its graphical representation.

Key Questions to Ask and Answer

1. What types of first order differential equations can be solved using separation of variables?

✍ ____

____________________________________ *(See text p. 492.)*

A first order differential can be solved using separation of variables if dy/dx can be factored into a function of x, say $f(x)$, times a function of y, say $g(y)$. Even then, you may or may not be able to solve for y directly, depending on whether or not there are explicit antiderivatives for $f(x)$ and $g(y)$.

✎ Your Turn

Which of the following differential equations can be solved by separation of variables?

Equation	*Your Response*
$\frac{dy}{dx} = xy$	
$\frac{dy}{dx} = x - y$	
$\frac{dP}{dt} + 3P = 90$	
$\frac{dy}{dt} = ty^2 \cos(t^2)$	
$\frac{dy}{dt} = 0.08y$	

Answers:
Yes, $\frac{1}{y} dy = x dx$
No
Yes, $\frac{1}{90 - 3P} dP = dt$
Yes, $\frac{1}{y^2} dy = t \cos(t^2) dt$
Yes, $\frac{1}{y} dy = 0.08 dt$

2. *What are the steps to solve first order differential equations of this type using separation of variables?*

✍ ____

____ *(See text p. 492-493.)*

To solve a differential equation:

- separate the variables,
- integrate both sides,
- solve for y.

Example 11.4.1

Solve the following initial value problem using separation of variables and illustrate the solution graphically.

Initial Value Problem	*Analytical solution*	*Graph*
$\frac{dP}{dt} + 3P = 90, \quad P(0) = 10$	Solve for dt, then integrate: $\int \frac{1}{90-3P}\,dP = \int dt$ $\frac{-1}{3} \ln\lvert 90-3P\rvert = t + C$ $\ln\lvert 90-3P\rvert = -3t + C'$ $\lvert 90-3P\rvert = e^{-3t+C'}$ $90-3P = Ae^{-3t}, \quad A = \pm e^{C'}$ $P(0) = 10$ gives $90-30 = Ae^{0}$, so $A = 60$ $90-3P = 60e^{-3t}$ $P = 30 - 20e^{-3t}$	window: $0 < x < 5,\ 0 < y < 4.0$

Example 11.4.2

Suppose you leave a cup of coffee in a room of 70° F. Let T (degrees Fahrenheit) be the temperature of the coffee at time t (minutes). The rate at which the temperature of the coffee decreases is proportional to the difference between the temperature of the coffee and the temperature of the room. This gives the following differential equation: $\frac{dT}{dt} = k(T-70)$. Suppose you also know that when the temperature of the coffee is 150°, its temperature is decreasing at 4° per minute. $-4\,{}^{\circ}\!/_{\text{min}} = k(150-70)$ or $k = -0.05$. Solve the differential equation $\frac{dT}{dt} = -0.05(T-70)$ if you know at time $t = 0$, $T = 150°$.

Steps	*Solution*
• The initial value problem	$\frac{dT}{dt} = -0.05(T-70), \quad T(0)=150°$
• Separate variables	$\frac{1}{T-70}\,dT = -0.05dt$
• Integrate both sides	$\int \frac{1}{T-70}\,dT = \int -0.05dt$
• Find antiderivatives	$\ln\lvert T-70\rvert = -0.05t + C$

• Take exponential of each side	$\|T-70\| = e^{-0.05t+C} = e^{C}e^{-0.05t}$
• Solve for T	$T = 70 + Ae^{-0.05t}$, $A = \pm e^{C}$
• Apply initial conditions	$150 = 70 + Ae^{-0.05(0)}$ so $A = 80$
	$T = 70 + 80e^{-0.05t}$
• Illustrate graphically window: $0 \le x \le 50$, $50 \le y \le 175$	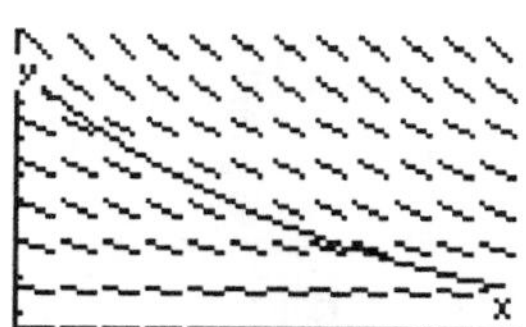

On Your Own 1

Use separation of variables to find the solution to the following initial value problems.

a. $\dfrac{dy}{dx} = xy,\ \ y(0) = 1$

b. $\dfrac{dy}{dx} = \dfrac{2y}{x},\ \ y(1) = 5$

c. $\dfrac{dy}{dt} = ty^2\cos(t^2),\ \ y(0) = 1$

d. $\dfrac{dw}{dz} = w + wz^2,\ \ w(0) = 3$

(The answer appears at the end of this section.)

Key Algebra Skills

- Review solving logarithmic equations.
- Review integration.

Key Study Skills

- Choose at least two problems that represent this section; put each on a 3 x 5 card with the solution on the back.
- Use the ✓ ? ★ system on homework problems and get answers to all ★ problems.

Answer to On Your Own 1

a. $\dfrac{dy}{dx} = xy$ 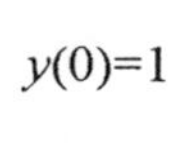 $y(0)=1$	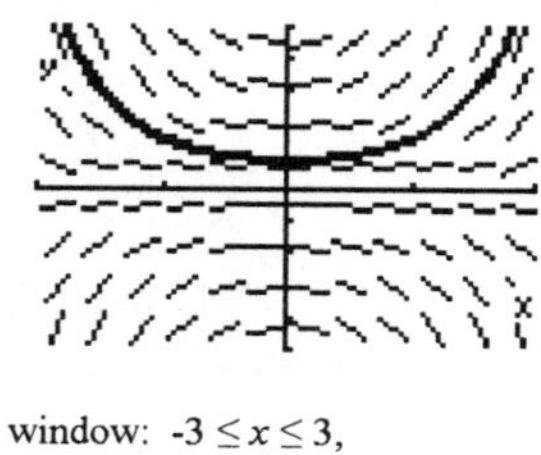 window: $-3 \le x \le 3$, $-10 \le y \le 10$	$\dfrac{1}{y}dy = xdx$ $\displaystyle\int \frac{1}{y}dy = \int xdx$ $ln\|y\| = \dfrac{x^2}{2} + C$ $\|y\| = e^{\frac{x^2}{2}+C} = Ae^{\frac{x^2}{2}}$, $A = e^{C}$ $y(0) = 1$ gives A $= 1$ $y = e^{\frac{x^2}{2}}$

b. $\frac{dy}{dx}=\frac{2y}{x}$ $y(1)=5$	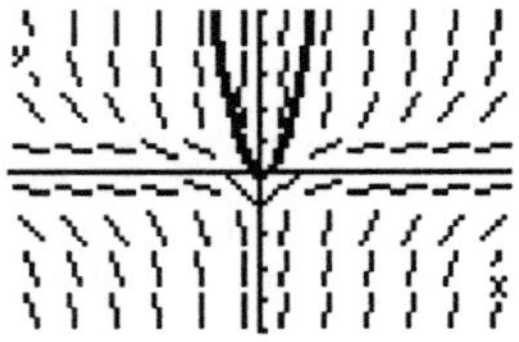 window: $-5 \le x \le 5$, $-10 \le y \le 10$	$\frac{1}{y}\,dy=\frac{2}{x}\,dx$ $\int \frac{1}{y}\,dy=\int \frac{2}{x}dx$ $ln\lvert y\rvert = 2\,ln\lvert x\rvert + C$ $\lvert y\rvert = e^{2\,ln\lvert x\rvert + C}$ $y = Ax^2,\ \ A=\pm e^C$ $y=5x^2$
c. $\frac{dy}{dt}=ty^2\,cos(t^2)$ $y(0)=1$	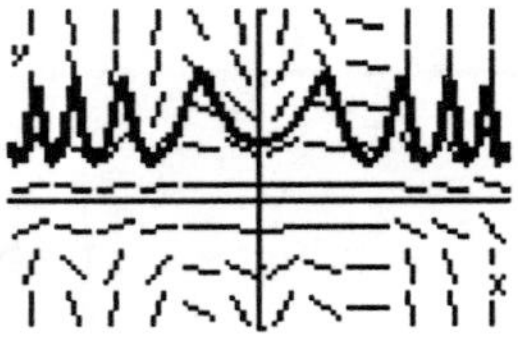 window: $-5 \le x \le 5$, $-3 \le y \le 3$	$\frac{1}{y^2}\,dy = t\,cos(t^2)dt$ $\int \frac{1}{y^2}\,dy = \int t\,cos(t^2)dt$ $-\frac{1}{y}=\frac{1}{2}\,sin\,t^2 + C$ $y=\frac{-2}{sin\,t^2 + C'}$ $1=\frac{-2}{0+C'}$ $y=\frac{-2}{sin\,t^2-2}$
d. $\frac{dw}{dz}=w+wz^2$ $w(0)=3$	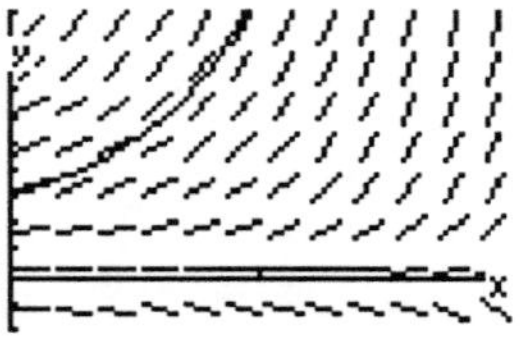 window: $-3 \le x \le 3$, $-10 \le y \le 10$	$\frac{1}{w}\,dw=\left(1+z^2\right)dz$ $\int \frac{1}{w}\,dw = \int (1+z^2)dz$ $ln\ \lvert w\rvert = z+\frac{z^3}{3}+C$ $e^{(z+\frac{z^3}{3}+C)} = \lvert w\rvert$ $w=Ae^{(z+\frac{z^3}{3})}$ let $A=\pm e^C$ $3=Ae^0 \Rightarrow A=3$ $w=3e^{(z+\frac{z^3}{3})}$

11.5 GROWTH AND DECAY

Objective: To model exponential growth and decay situations.

Preparations: Read and take notes on section 11.5.
Suggested practice problems: 1,2,5,9,10,18,21.

✠ Key Ideas of the Section

You should be able to:

- Distinguish between relative and absolute growth rates.
- Create a differential equation from a verbal statement of an exponential growth or decay problem.
- Recognize the existence of and find an equilibrium solution.
- Recognize when an equilibrium solution is stable or unstable.

Key Questions to Ask and Answer

1. What is the difference between relative and absolute growth rates?

✍ ____

____ (See text p. 497)

The *absolute growth rate* of a population over time is the rate of change of the population, or $\frac{dP}{dt}$; the units for absolute growth rate are: units (people, $, etc.)/ time unit.

The *relative growth rate* is the percentage change over time, or $\frac{1}{P}\frac{dP}{dt}$; the units for relative growth rate are: percent/ time unit;

Example 11.5.1

A population of 500 rabbits is increasing at a continuous growth rate of 1% per month.

a) The *relative growth rate* = $\frac{1}{P}\frac{dP}{dt}$ % per unit time =1% per month, where t is measured in months. The rate of growth of the population of rabbits = 0.01(current population) or, if P is the population of rabbits at time t, the differential equation $\frac{dP}{dt}=0.01P$, models the rabbit population growth. The units are rabbits/month. This equation has solution $P = P_0e^{0.01t}$, or specifically $P = 500e^{0.01t}$, where 500 is the initial population.

b) After 1 month there will be 505.025 rabbits or about 5 additional rabbits, and the *absolute growth rate* for the first month = rate of change of population = $\frac{dP}{dt}$ rabbits per month or 5.025 rabbits per month.

2. How do you translate a percentage growth (or decay) problem into a statement about derivatives?

✍ ____

____ (See text p. 497-502.)

Example 11.5.2

Write each problem as a differential equation with its solution.

Situation	*Differential Equation*	*Solution*
From **Example 11.5.1**: A population of rabbits is increasing at a continuous growth rate of 1% per month. The rate of growth of the population of rabbits = 0.01(current population)	$\frac{dP}{dt}=0.01P$	$P=P_0e^{0.01t}$
From Section 11.1 in the Study Guide: Suppose you place an amount of money, y_0, into a savings account paying a continuous rate of interest of 8% per year. The growth rate of your money at any time is proportional to the amount of money itself.	$\frac{dy}{dt}=0.08y$	$y=y_0e^{0.08t}$
From Section 11.4 in the Study Guide: A cup of coffee whose temperature is left in a room of 70° F. The rate at which the temperature of the coffee decreases is proportional to the difference between the temperature of the coffee and the temperature of the room.	$\frac{dT}{dt}=k(T-70)$	$T=70+Ae^{kt}$

On Your Own 1

Complete the following table:

Situation	*Differential Equation*	*Solution*
The rate at which a drug is eliminated from the body is proportional to the amount present. Let Q be the amount present at time *t*.		

(The answer appears at the end of this section.)

Example 11.5.3

The rate at which barometric pressure decreases with altitude is proportional to the barometric pressure at that altitude. If the barometric pressure, *P*, is measured in inches of mercury, and the altitude, *h*, in feet, then the constant of proportionality is -3.7×10^{-5}. Suppose the barometric pressure at sea level is 29.92 inches of mercury. Calculate the barometric pressure at the top of Mount Kilimanjaro, 19,590 feet; at San Francisco, 65 feet.

Situation	*Differential equation*	*Solution*
General equation:	$\frac{dP}{dh}=-3.7\times10^{-5}P$ $h=0$ ft., $P_0=29.92$ inches	$P=P_0e^{-3.7\times10^{-5}h}$ $P=29.92e^{-3.7\times10^{-5}h}$
Mount Kilimanjaro		$P=29.92e^{-3.7\times10^{-5}(19,590)}=14.49$ inches of mercury
San Francisco		$P=29.92e^{-3.7\times10^{-5}(65)}=29.85$ inches of mercury

On Your Own 2

Use **Example 11.5.3** to answer the following:

a. Calculate the barometric pressure at Denver (5280 feet).

b. Evidently, people cannot easily survive at a pressure below 15 inches of mercury. What is the highest altitude that you can safely go?

(The answer appears at the end of this section.)

3. What are equilibrium solutions and what do they look like?

____

____ *(See text p. 502.)*

An *equilibrium solution* is constant for all values of the independent variable; that is, its graph is a horizontal line. An *equilibrium solution* is *stable* if, as the independent variable tends towards ∞, the near by solution curves get closer and closer to that line. An *equilibrium solution* is *unstable* if as the independent variable tends toward ∞, the nearby solution curves move further away from this line.

Example 11.5.4

Determine the equilibrium solution to each of the following and whether the solution is stable or unstable:

Situation	*Solution Curves*	*Graphs*
Recall the problem of a cup of coffee at 150° left in a room of 70° F. • Place a second cup of coffee at 100° F, and, third, a glass of ice water at 45° F. • The differential equation is: $\frac{dT}{dt} = -k(T-70)$, for some constant of proportionality k. (Assume the same k). Solve $\frac{dT}{dt} = 0$	$T = 70 + Ae^{-kt}$	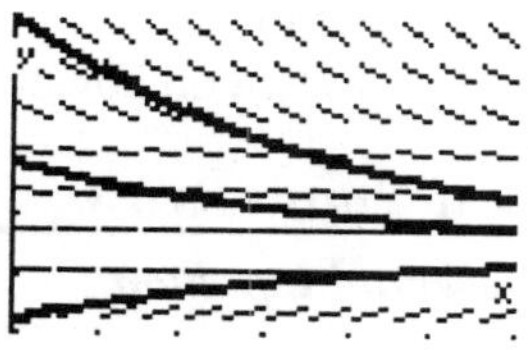 window: $0<x<50$, $40<y<150$ Stable equilibrium solution: T=70. The curves are approaching the line.
Recall the interest problem from Section 11.1 in the Study Guide where an amount of money is placed into a savings account paying a continuous rate of interest of 8% per year. • Place first \$300, or \$200, or \$100. • The differential equation is: $\frac{dy}{dt} = 0.08y$ • Solve $\frac{dy}{dt} = 0$.	$y = y_0 e^{0.08t}$	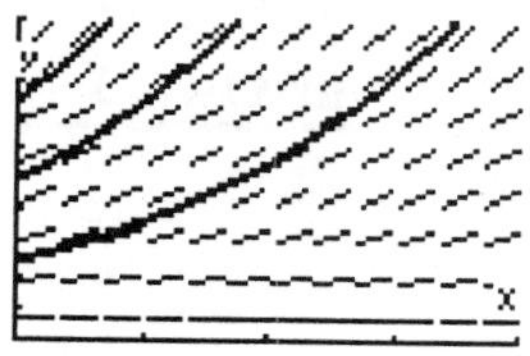 window: $0<x<20$, $0<y<400$ Unstable equilibrium solution: y=0. The curves are moving away from the line.

On Your Own 3

The slope fields for various differential equations are given below. Estimate all equilibrium solutions for each and indicate whether or not each is stable or unstable.

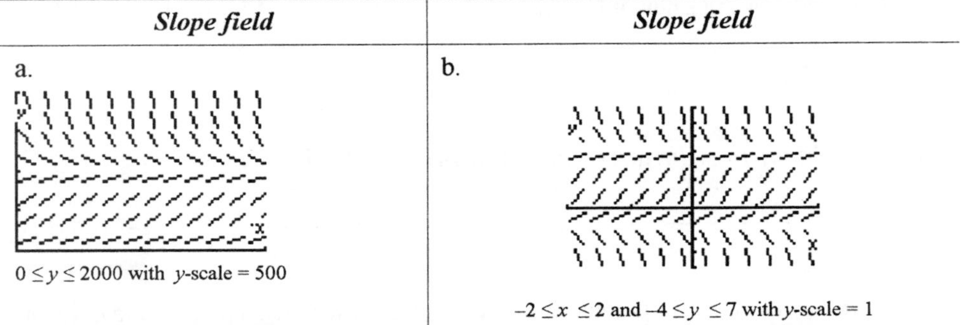

Slope field	*Slope field*
a. $0 \le y \le 2000$ with y-scale = 500	b. $-2 \le x \le 2$ and $-4 \le y \le 7$ with y-scale = 1

(The answer appears at the end of this section.)

Key Algebra Skills

- Review Chapter 1.2 from text on exponential functions.
- Review solving exponential and logarithmic equations.

Key Study Skills

- Choose at least two problems that represent this section; put each on a 3 x 5 card with the solution on the back. Review problems 11.1-11.5.
- Read the examples in this section of the text carefully.
- Use the ✓ ? ★ system on homework problems and get answers to all ★ problems.

Answer to On Your Own 1

Differential Equation	*Solution*
$\frac{dQ}{dt} = -kQ \qquad k > 0$	$Q = Q_0 e^{-kt} \qquad k > 0$

Answer to On Your Own 2:

a. $P = 29.92e^{-3.7\times10^{-5}(5280)} = 24.61$ inches of mercury.

b. $15 = 29.92e^{-3.7\times10^{-5}h}$

$$h = \frac{-1}{-3.7 \times 10^{-5}} \ln\left(\frac{15}{29.92}\right) = 18{,}661.5 \textit{ feet}$$

Answer to On Your Own 3

Answers	
a. $y = 1000$ is stable solution, $y = 0$ is unstable.	b. Stable solution $y = 4$, Unstable solution $y = -1$

11.6 APPLICATIONS AND MODELING

Objective: To create a differential equation from a verbal description.

Preparations: Read and take notes on section 11.6.
Suggested practice problems: 1,9,14,18,21,22.

✠ Key Ideas of the Section

You should be able to:

- Express the relationships between rates of change and variables in symbolic form.
- Create the differential equation which represents the situation.
- Find the solution to the differential equation.
- Recognize an equilibrium solution for differential equations of the form $\frac{dy}{dt} = f(y)$ and whether it is stable or unstable.
- Decide whether or not the model and the numbers make sense.

Key Questions to Ask and Answer

1. What are the steps in creating a differential equation from a verbal description?

✍ ____

____ *(See text p. 506-511)*

The steps in creating a differential equation are:

- Identify all variables (and units) in words.
- Establish the relationship between variables. Ask, "Which are independent?" and "Which are dependent?".
- Write down how you expect them to behave, are they increasing? decreasing?
- Write the equations in words.
- Check the units.
- Create the resulting differential equation.
- Establish any given initial conditions.
- Find the solution and interpret in your own words.
- Ask does the solution make sense?

Example 11.6.1

Refer to Problem 6 in Section 11.6 of the text (p. 512). According to a simple physiological model, an athletic adult needs 20 calories per day per pound of body weight to maintain his weight. If he consumes more or fewer calories than those required to maintain his weight, his weight will change at a rate proportional to the difference between the number of calories consumed and the number needed to maintain his current weight; the constant of proportionality is 1/3500 pounds per calorie. Suppose that a particular person has a constant caloric intake of I calories per day. Let $W(t)$ be the person's weight in pounds at time t, measured in *days*.

a) Begin the modeling process:

- Identify all variables (and units) in words.
- Establish the relationship between variables: which are independent?, which are dependent?
- Write down how you expect them to behave: increasing? or decreasing?

Variables/constants	*Units*	*Information*
W = weight	pounds	A function of time; the dependent variable being examined changes over time.
t = time	days	The independent variable.
I = calories consumed	calories/day	Changes in I affect the change in weight over time; the larger I, the larger W over time.
k = 1/3500	pounds/calorie	The constant of proportionality.
20 = calories needed to maintain body weight	(calories/pound)/day	A constant.

b) Form the equation and solve:

- Write the equations in words
  rate of change in weight over time
  = k(calories consumed – calories needed to maintain body weight)
- Create the resulting differential equation
  $$\frac{dW}{dt} = \frac{1}{3500}(I - 20W)$$
- Check units
  The units on the left are *pounds/day* and so must be the units on the right.
  $$\frac{dW}{dt}\frac{pounds}{day} = \frac{1}{3500}\frac{pounds}{calories}\left(I\frac{calories}{day} - 20\frac{calories}{pound - day}Wpounds\right)$$
- Find the solution and interpret in your own words
  $$\frac{dW}{dt} = \frac{1}{3500}(I - 20W)$$
  $$\frac{dW}{dt} = \frac{-20}{3500}\left(W - \frac{I}{20}\right)$$
  Separating variables: $\dfrac{1}{\left(W - \frac{I}{20}\right)}dW = \dfrac{-2}{350}dt$

  Integrating: $\displaystyle\int \frac{1}{\left(W - \frac{I}{20}\right)}dW = \int \frac{-2}{350}dt$ giving $ln\left|W - \frac{I}{20}\right| = \frac{-2}{350}t + C$

  Taking the exponential: $W - \dfrac{I}{20} = Ae^{\frac{-2}{350}t}$ or $W = \dfrac{I}{20} + Ae^{\frac{-2}{350}t}$

- What is the equilibrium solution for a person consuming 3000 calories per day?
  It is given that I = 3000 calories per day, so $\dfrac{dW}{dt} = \dfrac{1}{3500}(3000 - 20W)$. Setting $\dfrac{dw}{dt} = 0$, gives the equilibrium solution W = 150 pounds. This means that if you eat 3000 calories per day, your weight over time will tend toward 150 pounds.

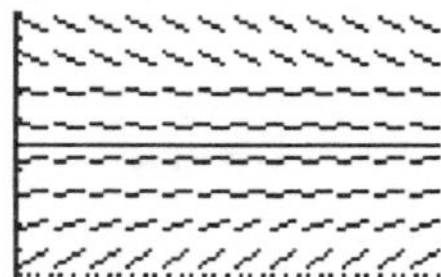

window: $0 \le t \le 360$ days, $100 \le W \le 200$ pounds

- Establish any given initial conditions

If $W(0)= W_0$, then $A = W_0 - \frac{I}{20}$ and $W = \frac{I}{20} + \left(W_0 - \frac{I}{20}\right)e^{-\frac{2t}{350}}$

Analyze for: W_0=160 pounds ; W_0=120 pounds; W_0=200 pounds, W_0=150 pounds.

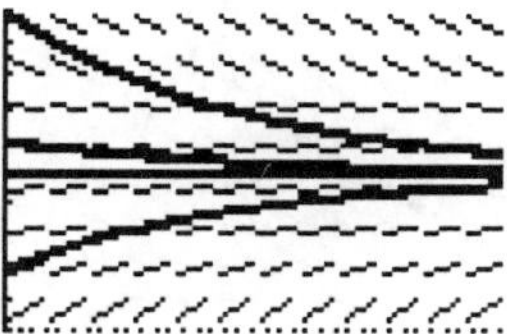

window: $0 \le t \le 360$ days, $100 \le W \le 200$ pounds

Example 11.6.2

Let's re-visit the savings account problem from Section 11.1 of this Study Guide. You invested an amount of money into an account paying a continuous rate of 8% per year. y is the amount of money at a given time t in years, or $y = f(t)$, and clearly, y is increasing.

Recall that the rate at which the money is increasing = 0.08 (the amount of money) which yields the differential equation $\frac{dy}{dt} = 0.08y$ with solution $y = y_0 e^{0.08t}$.

Situation 1:

You receive a constant amount $\$M$ each year as a bonus and will invest this in the same savings account each year.

Rate at which the money is increasing = 0.08 (the amount of money) + rate of investments

$$\frac{dy}{dt} = 0.08y + M \text{ or } \frac{dy}{dt} = 0.08\left(y + \frac{M}{0.08}\right)$$

Separating variables: $\frac{1}{y + \frac{M}{0.08}}\,dy = 0.08dt$

Integrating:

$$\int \frac{1}{y + \frac{M}{0.08}}\,dy = \int 0.08dt$$

$$ln\left|y + \frac{M}{0.08}\right| = 0.08t + C$$

Taking the exponential: $e^{0.08t+C} = \left|y + \frac{M}{0.08}\right|$

$$y + \frac{M}{0.08} = Ae^{0.08t}, \quad \text{where } A = \pm e^{C}$$

$$y = Ae^{0.08t} - \frac{M}{0.08}$$

To find A, use the initial condition, $y(0)= y_0$.

$$A = y_0 + \frac{M}{0.08} \text{ and } y = \left(y_0 + \frac{M}{0.08}\right)e^{0.08t} - \frac{M}{0.08}$$

Suppose $\$M$=\$1000

If y_0= \$5000, $y = \left(5000 + \frac{1000}{0.08}\right)e^{0.08t} - \frac{1000}{0.08}$

$y_0 = \$8000 \quad y = \left(8000 + \frac{1000}{0.08}\right)e^{0.08t} - \frac{1000}{0.08}$

$y_0 = \$10{,}000 \quad y = \left(10,000 + \frac{1000}{0.08}\right)e^{0.08t} - \frac{1000}{0.08}$

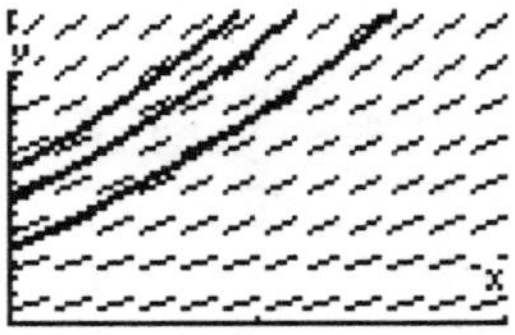

window: 0 < x < 10, 0 < y < 20,000

On Your Own 1:

Continue **Example 11.6.2** for **Situation 2:**
You want to *draw out* a constant amount $\$M$=\$1000 each year. Find the solution and analyze for y_0 = \$12,500, y_0 = \$15,000, and y_0 = \$10,000.

(The answer appears at the end of this section.)

Key Study Skills

- Read the examples in this section of the text carefully.
- Choose at least two problems that represent this section; put each on a 3 x 5 card with the solution on the back. Review note card problems 11.1-11.6.
- Use the ✓ ? ★ system on homework problems and get answers to all ★ problems.

Answer to On Your Own 1

Rate at which the money is changing = 0.08(the amount of money) – (rate of withdrawals). The question is: How much would your initial investment need to be in order for your earnings to be exactly the same as the withdrawals?

Rate at which money is increasing = (rate of withdrawals).

$0.08\, y_0 = \$M$

$y_0 = \$M/0.08$

If M = \$1000, then y_0 = \$12,500. That is, if you initially invest \$12,500 the rate at which the money is increasing is exactly the same as the rate at which you are withdrawing. This is the equilibrium solution.

If the initial amount is larger than \$12,500, you are earning more than you are withdrawing; if the initial amount is less than \$12,500, you are withdrawing more than you are earning.

$\frac{dy}{dt} = 0.08y - M$ or $y = Ae^{.08t} + \frac{M}{0.08}$, *where* $A = \pm e^C$

To find A, use the initial condition, y(0)= y_0. $A = y_0 - \frac{M}{0.08}$ and therefore

$$y = \left(y_0 - \frac{M}{0.08}\right)e^{0.08t} + \frac{M}{0.08}$$

If $y_0 = \$12{,}500,\ y = \left(12,500 - \frac{1000}{0.08}\right)e^{0.08t} + \frac{1000}{0.08} = \$12{,}500$

$y_0 = \$15{,}000, \quad y = \left(15{,}000 - \dfrac{1000}{0.08}\right)e^{0.08t} + \dfrac{1000}{0.08} = 2500e^{0.08t} + 12{,}500$

$y_0 = \$10{,}000, \quad y = \left(10{,}000 - \dfrac{1000}{0.08}\right)e^{0.08t} + \dfrac{1000}{0.08} = -2500e^{0.08t} + 12{,}500$

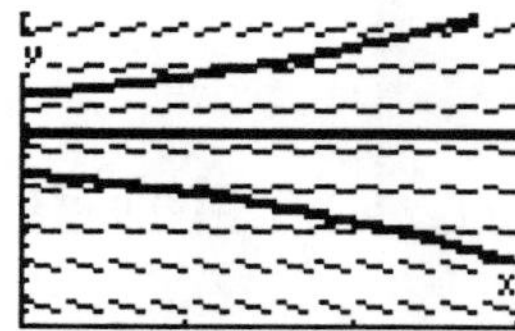

window: $0 < x < 15$, $0 < y < 20{,}000$

The initial investment of \$12,500 is an equilibrium solution, but an unstable one.

11.7 MODELS OF POPULATION GROWTH

Objective: To create exponential and logistic models of population growth.

Preparations: Read and take notes on section 11.7.
Suggested practice problems: 2,3,6,8,9,13.

✠ Key Ideas of the Section

You should be able to:

- Recognize when an exponential model is appropriate for population growth.
- Recognize when a logistic model is appropriate for population growth.
- Recognize the general form of a logistic differential equation.
- Find and interpret the equilibrium solutions.

Key Questions to Ask and Answer

1. When is a logistic model appropriate for population growth?

✍ ____

____ *(See text p. 518-520)*

The logistic model, unlike the exponential model, has an upper bound or limiting value L; that is, growth can't continue indefinitely.

The *exponential model* assumes the relative growth rate is constant, or

$$\frac{1}{P}\frac{dP}{dt} = k, \text{ for which the solution is } P = P_0 e^{kt}$$

The *logistic model* assumes the relative growth rate is a linearly decreasing function of P, or

$$\frac{1}{P}\frac{dP}{dt} = k - aP, \text{ where } a = \frac{k}{L}, \text{ or } \frac{1}{P}\frac{dP}{dt} = k\left(1 - \frac{P}{L}\right).$$

The solution for the logistic model is:

$$P = \frac{L}{1 + Ae^{-kt}} \text{ where } A = \frac{L - P_0}{P_0}.$$

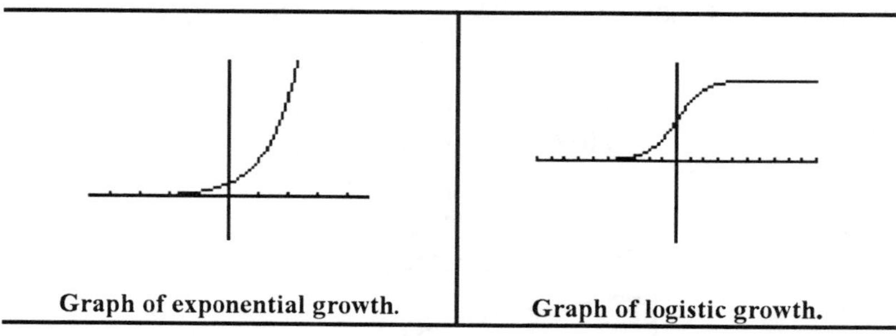

Graph of exponential growth. | Graph of logistic growth.

Example 11.7.1

The rumor of a dramatic increase in tuition is spreading throughout a small private college with student body size 1000. Let $P(t)$ = number of students who have heard the rumor by day t. Suppose you have the following:

T (days)	0	1	2
P (students)	10	14	22

a) Assume an *exponential model.* $\frac{1}{P}\frac{dP}{dt} = k$, where k is the continuous growth rate, with solution $P(t) = P_0 e^{kt}$

T (days)	0	1
Relative growth rate	$\dfrac{\frac{14-10}{1}}{10} = 0.4$	$\dfrac{\frac{22-14}{1}}{14} = 0.571$

These growth rates aren't constant. The question is, what is k? To find the relative growth rate for the P_0 to $P(2)$:

$$\frac{P(2)}{P_0} = e^{k \cdot 2} \quad or \quad \frac{22}{10} = e^{2k} \quad \text{giving} \quad k = \frac{1}{2} ln\, 2.2 = 0.394 \quad \text{giving the solution}$$

$P = 10e^{0.394t}$.

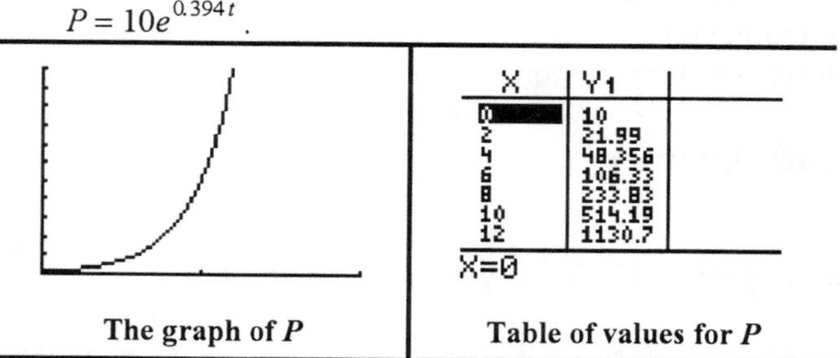

X | Y1
0 | 10
2 | 21.99
4 | 48.356
6 | 106.33
8 | 233.83
10 | 514.19
12 | 1130.7
X=0

The graph of P | Table of values for P

Notice that 12 days after the rumor started the number of students who have heard exceeds the student body of 1000. The model may have been appropriate at the beginning, but as P is a larger proportion of the student body, the rumor can not continue spreading at the same rate and so the relative growth rates should decrease.

b) Assume a *logistic model*: $\frac{1}{P}\frac{dP}{dt} = k\left(1 - \frac{P}{L}\right)$ *or* $\frac{dP}{dt} = kP\left(1 - \frac{P}{L}\right)$

This model implies that the rate at which the rumor is spread is proportional to the product of the number of students who have heard the rumor and the number of students who have not.

The model behaves like the exponential model above at the very early stages. Assume $k = 0.394$. The carrying capacity, $L = 1000$, gives $\frac{dP}{dt} = 0.394P\left(1 - \frac{P}{1000}\right)$

The solution is $P = \dfrac{1000}{1+99e^{-0.394t}}$ where $A = \dfrac{1000-10}{10} = 99$.

The slope field and solution is:	window: 0<x<20,0<y<1100
Comparison of the exponential model and the logistic model. The logistic model is a better model for this situation.	window: 0<x<20,0<y<1100

On Your Own 1

For **Example 11.5.1**, find and graph the solution for the cases when 5% of the student body heard the rumor initially; 20%; 50%, assume that $k = 0.394$ in all cases. Compare with the solution above (initially 1% heard the rumor).

2. What are the characteristics of the logistic curve?

____

____ *(See text p. 519.)*

- Shape is *sigmoid*, or S-shaped

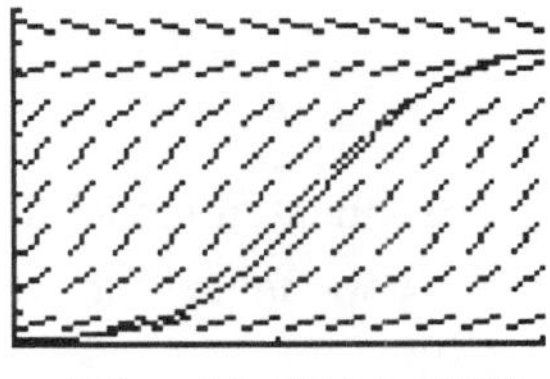

window: 0<x<20,0<y<1100

- Concave up for $0 < P < L/2$ and concave down for $L/2 < P < L$
- Inflection point when $P = L/2$

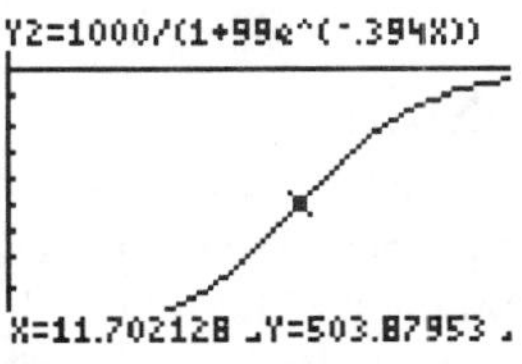

- $P = 0$ unstable equilibrium; $P = L$ stable equilibrium

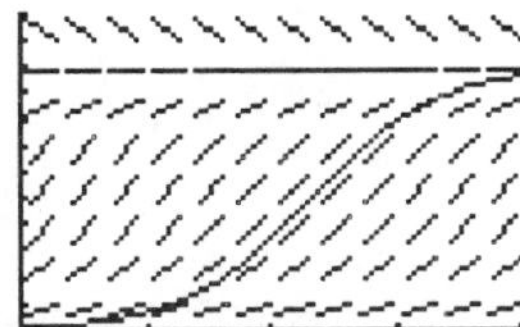

Example 11.7.2

The following table gives the Consumer price indices for expenditures, E, for Medical care services[1]. Determine the behavior of the relative growth rates and how that might lead to a logistic model.

Year	*1970*	*1980*	*1990*	*1996*	*1997*
Price Index	32.3	74.8	162.7	229.3	236.3
Relative growth rates	0.08	0.07	0.057	0.03	

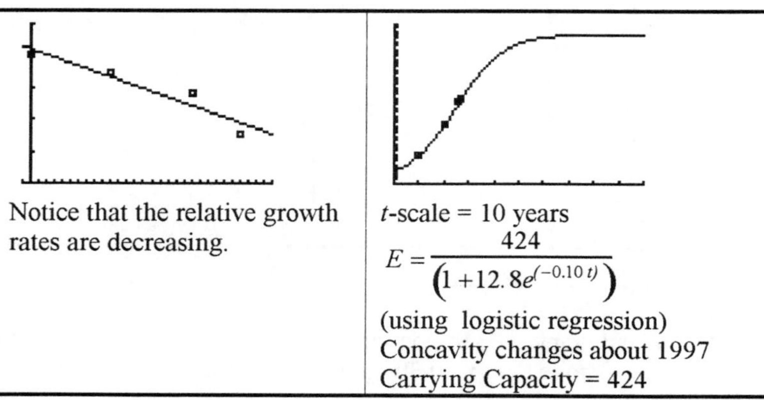

Notice that the relative growth rates are decreasing.
t-scale = 10 years
$E = \frac{424}{\left(1 + 12.8e^{(-0.10\,t)}\right)}$
(using logistic regression)
Concavity changes about 1997
Carrying Capacity = 424

On Your Own 2

The following table gives the population of San Francisco. The total area of San Francisco is 7 square miles, and since San Francisco is nearly surrounded by water, it is assumed that there is limit to the population growth.

Year	*1980*	*1990*	*1994*
Population in thousands	679	724	735

- Write the exponential model for the population growth and find the solution.
- Write the logistic model for the population, assuming that the maximum population is 750,000 people.

Key Study Skills

- Read carefully the examples in this section of the text.
- Choose at least two problems that represent this section; put each on a 3 x 5 card with the solution on the back. Review problems 11.1-11.7.
- Use the ✓ ? ★ system on homework problems and get answers to all ★ problems.

[1] Department of Labor, Bureau of Labor Statistics

Answers to On Your Own 1

1% or 10 students	$P = \dfrac{1000}{1+99e^{-.394t}}$	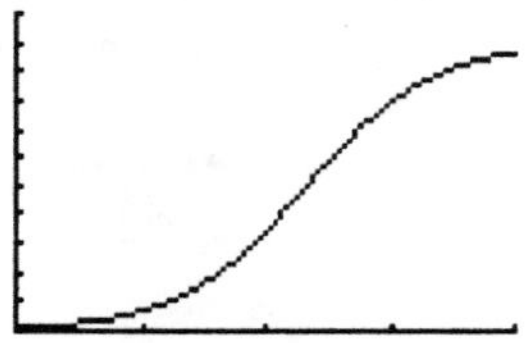 window: 0 < x < 20, 0 < y < 1100
5% or 50 students	$P = \dfrac{1000}{1+19e^{-.394t}}$	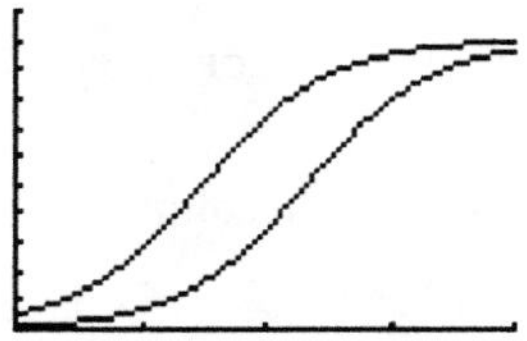
20% or 200 students	$P = \dfrac{1000}{1+4e^{-.394t}}$	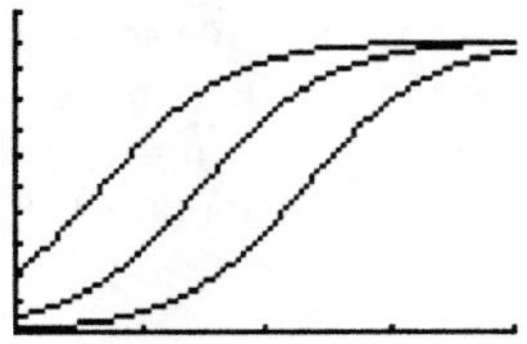
50% or 500 students	$P = \dfrac{1000}{1+1e^{-.394t}}$	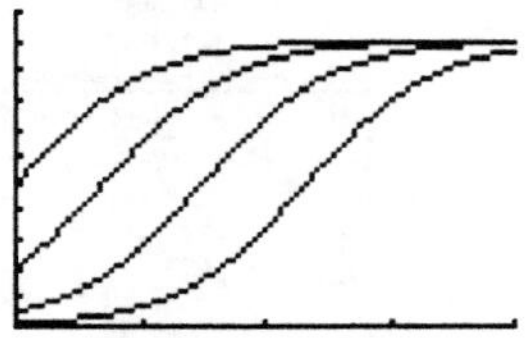

Answer to On Your Own 2

a.

Year	1980	1990
Relative growth rates	$\frac{1}{679}\left(\frac{724-679}{10}\right) = 0.7\%$	$\frac{1}{724}\left(\frac{735-724}{4}\right) = 0.4\%$ per year

$\frac{dP}{dt} = kP$ where $k = 0.006$ since $\frac{P(14)}{P_0} = e^{14k}$ or $\frac{735}{679} = e^{14k}$

and (1/14)(ln(735/679) = 0.006.

The solution of the differential equation $\frac{dy}{dt} = 0.006t$ is $y(t) = 679e^{0.006t}$

The predicted population in the year 2000 is 765,570 people which is larger than the assumed maximum.

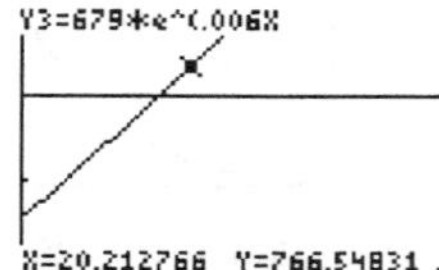

b. Assume the same relative growth rate for P near 1980, or $k = 0.006$. The model is

$$\frac{dP}{dt} = 0.006P\left(1 - \frac{P}{750}\right) \text{ with solution } P = \frac{750}{1 + 0.105e^{-.006t}}$$

The predicted population in the year 2000 is 686,110 people, which is too low, because it is below the population in 1990. We really needed to have some additional earlier data before the relative growth rates had begun to decrease. The constant k is the relative growth rate when the population is just a small fraction of the carrying capacity. Our mistake was to estimate it from the growth rate of San Francisco when population was already 679,000 or about or about 90% of the limiting population of 750,000.

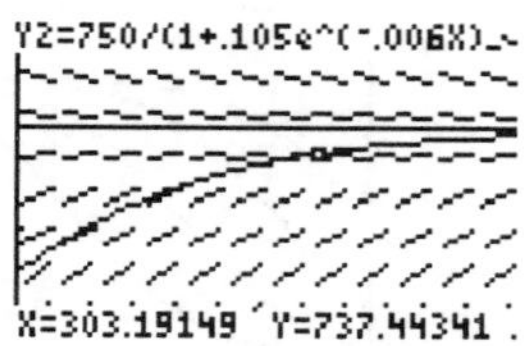

Note: With this model the maximum would not be reached for at least 300 years!

11.8 SYSTEMS OF DIFFERENTIAL EQUATIONS

Objective: To model situations with two populations with systems of differential equations.

Preparations: Read and take notes on section 11.8.
Suggested practice problems: 1,3,9,11-14,16,18.

✠ Key Ideas of the Section

You should be able to:

- Interpret a system of differential equations modeling interaction between two populations.
- Draw the slope field and phase plane for a system of differential equations.
- Describe how two interacting populations are changing with time.
- Determine if interactions are symbiosis, competition, or predator-prey.

Key Questions to Ask and Answer

1. What does a system of differential equations tell you about the behavior of two populations.

✍ ____

____ *(See text p. 524,527.)*

The system of differential equations gives the relationship between two interacting populations.

Example 11.8.1

Suppose F represents a population of foxes in thousands (the predator) and R represents a population of rabbits in thousands (the prey). Describe the interaction between the two populations modeled by the following system of differential equations: $\frac{dR}{dt} = 0.4R - 0.01RF$, $\frac{dF}{dt} = -0.3F + 0.005RF$

This is a predator-prey model where we observe the following:

- in the absence of foxes, rabbits increase
- in the absence of rabbits, foxes decrease
- in the presence of both, the two encounter each at a rate proportional to both populations.

✎ On Your Own 1

Consider each of the following systems for two populations x and y. Describe the behavior of each:

1. $$\frac{dx}{dt} = x$$
   $$\frac{dy}{dt} = -y$$

2. What do solutions to a system of equation look like?

✍ ____

____ *(See text p.524,527-529.)*

From the system of equations: $\frac{dx}{dt} = f(x, y)$ and $\frac{dy}{dt} = g(x, y)$ we can use the chain rule to get $\frac{dy}{dx} = h(x, y)$. The slope field for this differential equation gives the phase trajectory or orbit which is the path along which the point (x,y) moves over time. The equilibrium point is that point for which $\frac{dx}{dt} = 0$ and $\frac{dy}{dt} = 0$.

Example 11.8.2

Find the slope field and describe the behavior of foxes and rabbits from **Example 11.8.1** if $R = 80$ and $F = 30$ at time $t = 0$. Put the predator F on the vertical axis and the prey R on the horizontal axis.

Using the chain rule, to get $\frac{dF}{dt} = \frac{dF}{dR} \cdot \frac{dR}{dt}$ or $\frac{dF}{dR} = \frac{dF/dt}{dR/dt} = \frac{-0.3F + 0.005RF}{0.4R - 0.01RF}$.

The slope field looks like this:

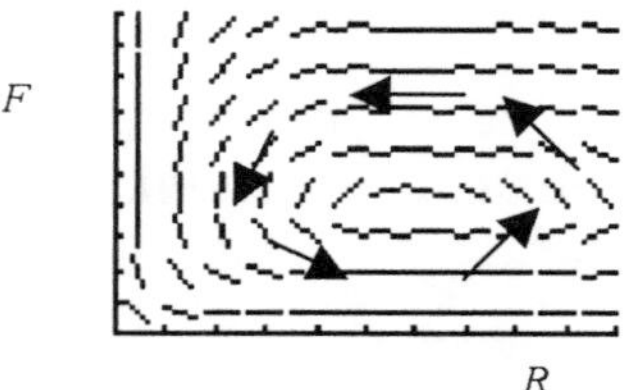

window : $0 < R < 100$; $0 < F < 100$) scales = 10

Notice that at $t = 0$, you begin at the point (80,30) and travel a counter-clockwise orbit. Initially as rabbits increase, foxes increase; then rabbits begin to decrease as foxes continue to increase (although more slowly) and then decrease; then rabbits begin to increase again, completing the orbit. **Note:** This happens in the presence of foxes.

On Your Own 2

Find the slope field and describe the behavior of x and y from **On Your Own 1a** if $x= 4$ and $y = 4$ at time $t = 0$; for $x= 4$ and $y = -4$ at time $t = 0$. Find the equilibrium point.

$$\frac{dx}{dt} = x, \quad \frac{dy}{dt} = -y$$

(The answer appears at the end of this section.)

Example 11.8.3

Find the equilibrium point for **Example 11.8.1** and interpret what it means with respect to foxes and rabbits.

We need to solve the following equations: $\frac{dR}{dt} = R(0.4 - 0.01F) = 0$ and $\frac{dF}{dt} = F(-0.3 + 0.005R) = 0$. The first equation has solutions $R = 0$ or $F = 40$. The second has solutions $F = 0$ or $R = 60$. Thus the equilibrium points are (0,0) and (60,40). This means that if at some time, there are 60 rabbits and 40 foxes, the populations will remain constant. Obviously if there are 0 rabbits and 0 foxes, neither population will grow!

3. *How do the models differ for populations with the following interaction: symbiosis, competition, predator-prey)?*

✍ ____

____ (See text p.530)

Symbiosis is when the interaction of two species benefits both. Both populations decrease if alone, but are helped by the presence of the other. If the interaction is one of competition, both populations grow when alone, but are harmed by the presence of the other. With predator-prey, the predator decreases when alone, but is helped by the prey; however, the prey increases when alone, but is harmed by the predator.

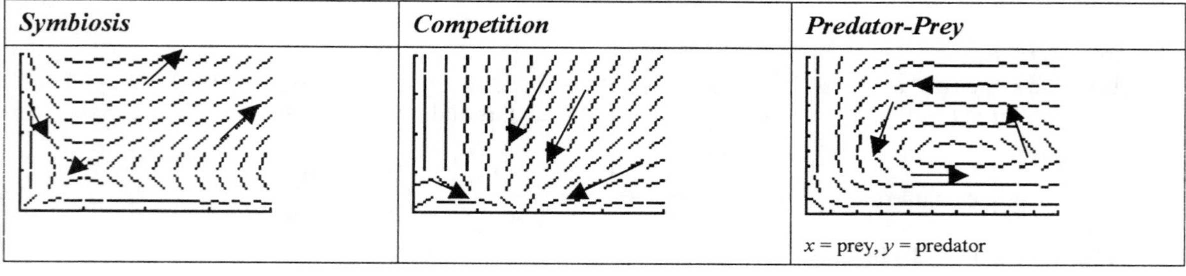

Symbiosis	*Competition*	*Predator-Prey*
		x = prey, y = predator

Key Study Skills

- Choose two problems that represent this section and put each on a 3 x 5 note card with the solution on the back.
- Use the ✓ ? ★ system on homework problems and get answers to all ★ problems.
- Learn to use the technology preferred by your instructor. The text assumes you have a graphing calculator available or a computer algebra system.

Key Notation

- The phase plane gives the graph of *y* against *x* both of which are functions of time. Time is represented by the arrow showing the direction a point moves along that trajectory.

Answers to On Your Own 1

a) $\dfrac{dx}{dt} = x$ $\dfrac{dy}{dt} = -y$	Over time, x is increasing if $x > 0$; x is decreasing if $x < 0$ regardless of y; y is decreasing if $y > 0$; y is increasing if $y < 0$ regardless of x.

Answers to On Your Own 2

Use the chain rule to get: $\dfrac{dy}{dx} = \dfrac{-y}{x}$ which has the following slope field:

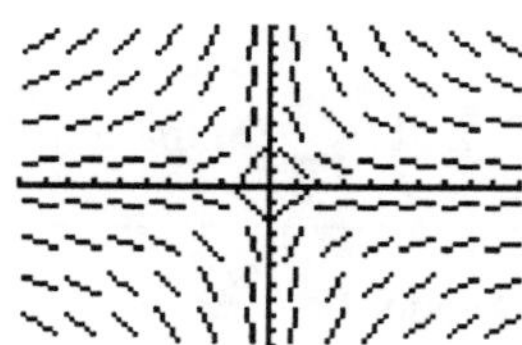

window: $-10 < x < 10$. $-1 < y < 10$

At the point (4,4), as x increases y decreases; at (4,–4), as x increases, y increases. The equilibrium point is (0,0). None of the trajectories pass through this point.

11.9 ANALYZING THE PHASE PLANE

Objective: To analyze a system of differential equations using nullclines.

Preparations: Read and take notes on section 11.9.
Suggested practice problems: 1,2,4,13.

✠ Key Ideas of the Section

You should be able to:

- Find the nullclines and equilibrium point(s) for a system of differential equations.
- Divide the phase plane into regions using nullclines.
- Interpret the behavior of the populations between the nullclines.

Key Questions to Ask and Answer

1. ***How does a system of differential equations for two populations having similar niches in competition for food and space behave?***

✍ ____

____ *(See text p. 532.)*

Each of the two populations grow logistically when alone, but the presence of the other population harms the growth of the each.

Example 11.9.1

Suppose you have two competing populations, P_1 and P_2. Compare the behavior of two models given below:

Model I	Model II
$\frac{dP_1}{dt} = 0.4P_1\left(1-\frac{P_1}{1000}\right)$	$\frac{dP_1}{dt} = 0.4P_1\left(1-\frac{P_1}{1000}\right) - 0.01P_1P_2$
$\frac{dP_2}{dt} = 0.2P_2\left(1-\frac{P_2}{500}\right)$	$\frac{dP_2}{dt} = 0.2P_2\left(1-\frac{P_2}{500}\right) - 0.005P_1P_2$

Model I shows both populations growing logistically (refer to Section 11.7 of text) with the following equations (assume $P_1(0)$=80 and $P_2(0)$=30): $P_1 = \frac{1000}{1+11.5e^{-0.4t}}$ and $P_2 = \frac{500}{1+15.7e^{-0.2t}}$ with maximum capacity of 1000 and 500, respectively. The graphs are:

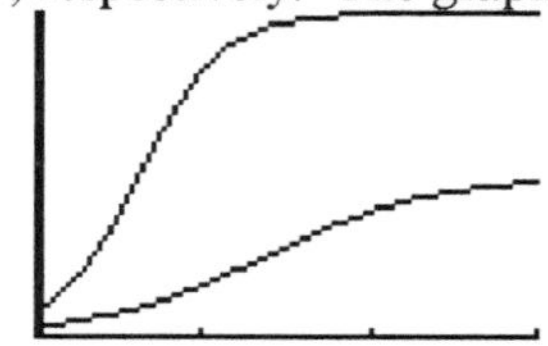

window: 0 < x < 30 and 0 < y < 1000

Model II assumes that the presence of the other reduces the growth for each. Note that the growth of population P_1 is being more affected by the interaction than population P_2.

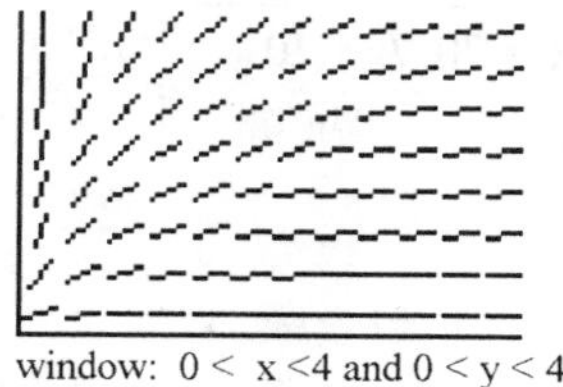
window: $0 < x < 4$ and $0 < y < 4$

2. *What do nullclines and equilibrium points tell me about the phase plane (P_1, P_2)?*

✍ ____

____________________________________ *(See text p.533.)*

We saw in Section 11.8 that an equilibrium point is a point at which neither population changes, or $\frac{dP_1}{dt} = 0$ and $\frac{dP_2}{dt} = 0$. Nullclines are curves at which $\frac{dP_1}{dt} = 0$ and the trajectory is vertical (that is, population P_1 is momentarily constant) or $\frac{dP_2}{dt} = 0$ and the trajectory is horizontal (that is, population P_2 is momentarily constant).

Example 11.9.2

Recall the behavior of foxes and rabbits from **Example 11.8.1** described by the system of differential equations: $\frac{dR}{dt} = 0.4R - 0.01RF$ and $\frac{dF}{dt} = -0.3F + 0.005RF$. We found the equilibrium points to be (0,0) and (60,40) (R is on the horizontal axis, F on the vertical).

To find the nullclines, we need $\frac{dR}{dt} = 0.4R - 0.01RF = 0.1R(4 - 0.1F) = 0$ or where R=0 or 4–0.1F=0. This gives the vertical line R=0 (the vertical axis) and the horizontal line $F = 40$. On each of these lines the rabbit population stays constant, the trajectory is vertical (put little vertical lines on these). Similarly, $\frac{dF}{dt} = -0.3F + 0.005RF = -0.1F(3 - 0.05R) = 0$ when $F = 0$ (the horizontal axis) and $R = 60$ (a vertical line). On these lines, the trajectory is horizontal (put little horizontal lines on these). This divides the phase plane into 4 regions:

Regions	***Sample points***		***Trajectories***
Region I: $R < 60$ and $F > 40$	$R = 50$, $F = 50$	$\frac{dR}{dt} < 0$ $\frac{dF}{dt} < 0$	
Region II: $R > 60$ and $F > 40$	$R = 70$, $F = 50$	$\frac{dR}{dt} < 0$ $\frac{dF}{dt} > 0$	
Region III: $R < 60$ and $F < 40$	$R = 20$, $F = 20$	$\frac{dR}{dt} > 0$ $\frac{dF}{dt} < 0$	

Region IV: $R > 60$ and $F < 40$	$R = 70, F = 20$	$\frac{dR}{dt} > 0$ $\frac{dF}{dt} > 0$	

Notice that this is consistent with the slope field generated in **Example 11.8.2**.

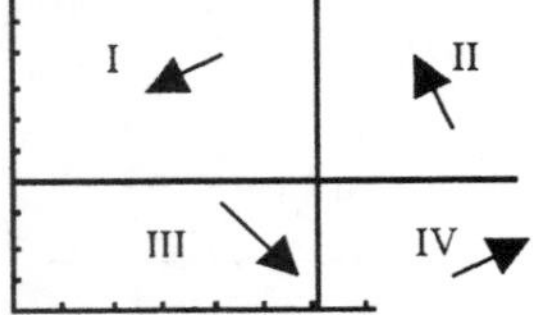

✎ On Your Own 1

Analyze the phase plane of the differential equations for x, y ≥0. Show the nullclines and equilibrium points, and sketch the direction of the trajectories in each region.

$$\frac{dx}{dt} = 0.1x(1 - 0.01x + 0.005y), \quad \frac{dy}{dt} = 0.03y(1 + 0.04x - 0.1y)$$

Key Study Skills

- Choose two problems that represent this section and put each on a 3 x 5 note card with the solution on the back.
- Use the ✓ ? ★ system on homework problems and get answers to all ★ problems.
- Learn to use the technology preferred by your instructor. The text assumes you have a graphing calculator available or a computer algebra system.
- Review logistic models in Section 11.7 of the text.

Key Notation

- The phase plane gives the graph of y against x both of which are functions of time. Time is represented by the arrow showing the direction a point moves along that trajectory.

Answers to On Your Own 1

• To find the vertical trajectories:	• $\frac{dx}{dt} = 0.1x(1 - 0.01x + 0.005y) = 0$ if $x = 0$ (the vertical axis) or $1 - 0.01x + 0.005y = 0$ (the heavy black line below, say y_1)
• To find the horizontal trajectories:	• $\frac{dy}{dt} = 0.03y(1 + 0.04x - 0.1y) = 0$ if $y = 0$ (the horizontal axis) or $1 + 0.04x - 0.1y = 0$ (the finer black line below, say y_2)
• The equilibrium points are:	• (0,0) and (131.25,62.5) (the intersection of the lines)

• Region I: above y_1 and above y_2	• (50,80)	• $\frac{dx}{dt} > 0, \frac{dy}{dt} < 0$	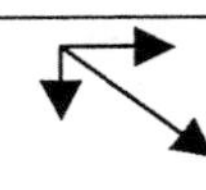
• Region II: above y_1 and below y_2	• (50,20)	• $\frac{dx}{dt} > 0, \frac{dy}{dt} > 0$	

• Region III: below y_1 and below y_2	• (150,40)	• $\frac{dx}{dt} < 0$, $\frac{dy}{dt} > 0$	
• Region IV: below y_1 and above y_2	• (150,80)	• $\frac{dx}{dt} < 0$, $\frac{dy}{dt} < 0$	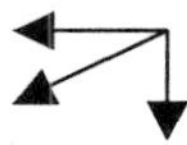

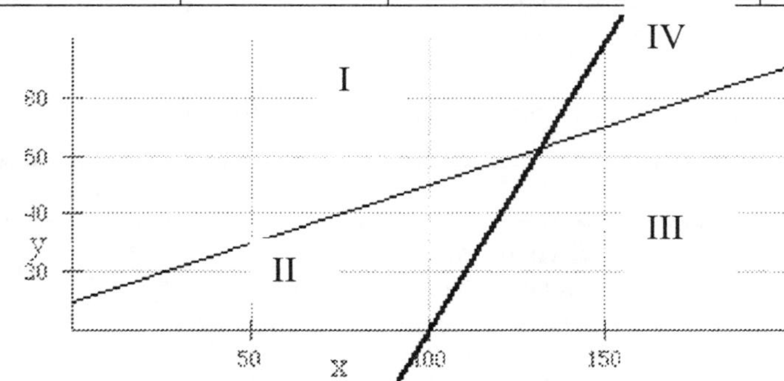

Slope field:

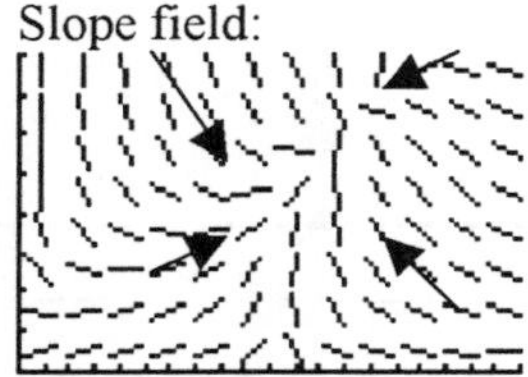

11.10 SECOND-ORDER DIFFERENTIAL EQUATIONS: OSCILLATIONS

Objective: To derive differential equations from laws of motion.

Preparations: Read and take notes on section 11.10.
Suggested practice problems: 1,4,8,13,17,19,20.

✠ Key Ideas of the Section

You should be able to:

- Recognize a second-order differential equation.
- Verify solutions to second-order differential equations.
- Interpret the coefficients and initial conditions with respect to the motion.
- Distinguish between initial value problems and boundary value problems.

Key Questions to Ask and Answer

1. *What is a second-order differential equation?*

✍ ____

____ *(See text p. 537.)*

Second-order differential equations involve, at most, the second derivative. The general solution will involve two constants, which, when determined by two given initial values or boundary values, pick out the specific solution from the family of solutions.

Example 11.10.1

Which of the following are second-order differential equations? Find the solution(s).

Equation		*Solution*
$\frac{d^2s}{dt^2} = -g,\ s(0) = 10,\ s'(0) = 20$	Second-order initial value problem	$s(t) = -\frac{1}{2}gt^2 + 20t + 10$
$y'' + y = 0,\ y(0)=1,\ y(\pi/2)=1$	Second-order boundary value problem	$y = \cos t + \sin t$
$\frac{dy}{dt} = ky$	First-order differential equation	$y = Ce^{kt}$

On Your Own 1

Show that $y = 5\cos 2t$ is a solution to $y'' + 4y = 0$

(The answer appears at the end of this section.)

2. *What is the solution to the General Spring equation and what does it look like?*

✍ ____

____ *(See text p. 538.)*

The differential equation for oscillations of a mass on a spring is the second order differential equation $\frac{d^2s}{dt^2} = -\frac{k}{m}s$, where $k > 0$ is the spring constant, s the displacement from equilibrium and m the mass.

The general solution is of the form $s(t) = C_1\cos\omega t + C_2\sin\omega t$, where $\omega = \sqrt{\frac{k}{m}}$. The period of this oscillation (called simple harmonic motion) is $T = \frac{2\pi}{\omega}$.

Example 11.10.2

Find the solution to the general spring equation $y'' + 4y = 0$.

Solution:

Here, $\omega^2 = 4,\ \omega = 2$ giving $y(t) = C_1\cos 2t + C_2\sin 2t$ as the general solution with period $T = \frac{2\pi}{\omega} = \pi$

Graphs of several solutions:

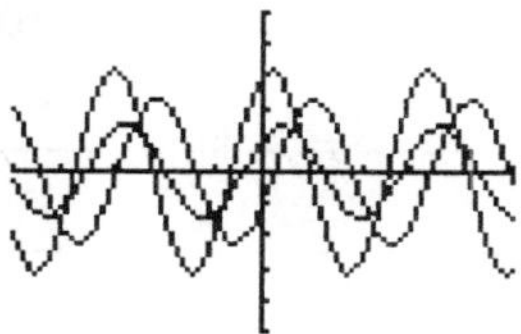

Example 11.10.3

Find and graph the solutions to the following *initial value* or *boundary value* problems for $y'' + 4y = 0$.

Initial or boundary values	*General Solution* $y(t) = C_1 \cos 2t + C_2 \sin 2t$	*Graph*
$y(0) = 5$ $y'(0) = -2$	$y(0) = C_1 \cos 0 + C_2 \sin 0 = 5$ or $C_1 = 5$ $y'(t) = -2C_1 \sin 2t + 2C_2 \cos 2t$ $y'(0) = -2C_1 \sin 0 + 2C_2 \cos 0 = -2$ or $C_2 = -1$ $y(t) = 5\cos 2t - \sin 2t$	
$y(0) = 1$ $y'(0) = 3$	$y(0) = C_1 \cos 0 + C_2 \sin 0 = 1$ or $C_1 = 1$ $y'(0) = -2C_1 \sin 0 + 2C_2 \cos 0 = 3$ or $C_2 = \frac{3}{2}$ $y(t) = \cos 2t + \frac{3}{2}\sin 2t$	
$y(0) = 0$ $y(3\pi/4) = -2$	$y(0) = C_1 \cos 0 + C_2 \sin 0 = 0$ or $C_1 = 0$ $y(\frac{3\pi}{4}) = C_2 \sin\frac{3\pi}{2} = -2$ or $C_2 = 2$ $y(t) = 2\sin 2t$	

On Your Own 2

The functions below describe the motion of a mass on a spring satisfying the differential equation $s'' + 4s = 0$, where s is the displacement of the mass from the equilibrium point at time t, with upwards as positive. Describe how the motion starts when $t = 0$.

$s(t) = 3\cos(2t)$

$s(t) = -\sin(2t)$

$s(t) = -0.5\cos(2t)$

(The answer appears at the end of this section.)

3. *What are the two forms of the solution to the undamped spring equation?*

✍ ____

____ *(See text p. 541-542.)*

$y = C_1 \cos \omega t + C_2 \sin \omega t = A \sin(\omega t + \phi)$, where the amplitude $A = \sqrt{C_1^2 + C_2^2}$ and ϕ is the phase shift such that:

$$\tan \phi = \frac{C_1}{C_2}, -\pi < \phi \leq \pi, \text{ and if } C_1 > 0, \text{ then } \phi > 0 \text{ and if } C_1 < 0, \text{ then } \phi < 0 \cdot$$

Thus $\phi = \arctan\left(\frac{C_1}{C_2}\right)$ or $\phi = \arctan\left(\frac{C_1}{C_2}\right) \pm \pi$.

Example 11.10.4

Write $y = 4\cos(2t) - \sin(2t)$ in the form $y = A\sin(\omega t + \phi)$ and give the phase shift and amplitude.

Here $C_1 = 4,\ C_2 = -1,\ \omega = 2$ and the graph of $y = 4\cos(2t) - \sin(2t)$ is:

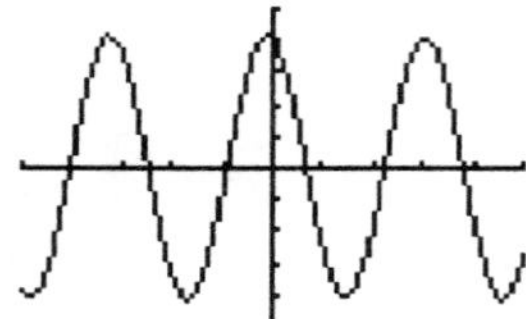

The *amplitude* $A = \sqrt{C_1{}^2 + C_2{}^2} = \sqrt{4^2 + (-1)^2} = \sqrt{17} \approx 4.12$.

The *phase shift* ϕ in $(-\pi, \pi]$ is such that

$$\tan \phi = \frac{C_1}{C_2} = \frac{4}{-1} \text{ or}$$

$\arctan(-4) \approx -1.3258$

Since $C_1 > 0, \phi > 0$ then $\phi = -1.3258 + \pi = 1.8158$

then, $y = \sqrt{17}\sin(2t + 1.805)$ which has the graph:

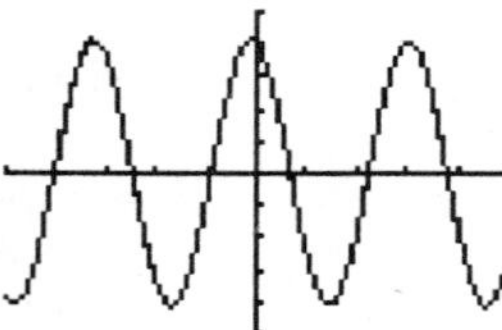

Note: The phase shift is a translation, but by a different amount: $\sin(2t + 1.805) = \sin(2(t + 0.9025))$

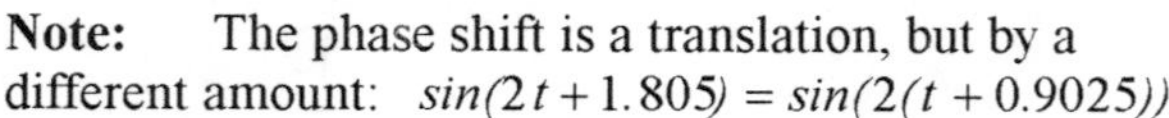

🗁 Key Study Skills

- Read carefully the examples in this section of the text.
- Choose at least two problems that represent this section; put each on a 3 x 5 card with the solution on the back. Review problems 11.1-11.10.
- Use the ✓ ? ★ system on homework problems and get answers to all ★ problems.

Answer to On Your Own 1

$y' = -10\sin 2t$

$y'' = -20\cos 2t$

$y'' + 4y = (-20\cos 2t) + 4(5\cos 2t) = 0$

Answer to On Your Own 2

Solution		*Graph*
$s(t) = 3\,cos(2t)$	At $t = 0$, $s = 3$ which is the highest point; since $s'(t) = -6\,sin(2t),\ \ s'(0) = 0$, the object is at rest.	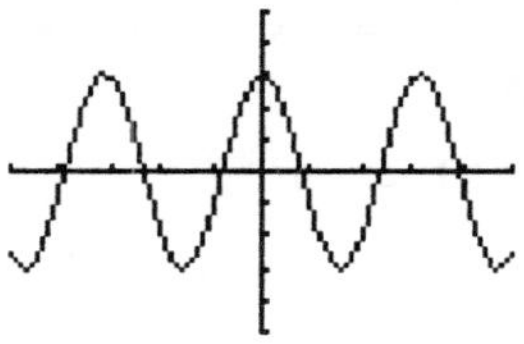
$s(t) = -\,sin(2t)$	At $t = 0$, $s = 0$, the object is starting at the equilibrium point, since $s'(t) = -2\,cos(2t),\ \ s'(0) = -2$ the object is moving downward.	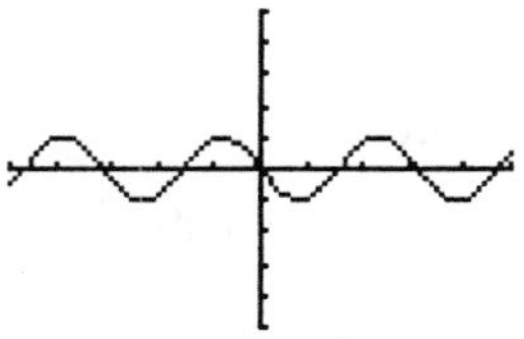
$s(t) = -0.5\,cos(2t)$	At $t = 0$, $s = -0.5$ which is the lowest point; since $-1 \le cos(2t) \le 1$; and since $s'(t) = sin(2t),\ \ s'(0) = 0$ the object is at rest.	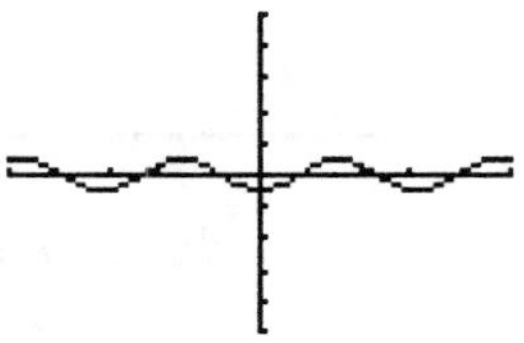

11.11 LINEAR SECOND-ORDER DIFFERENTIAL EQUATIONS

Objective: To find and interpret solutions to linear second-order differential equations.

Preparations: Read and take notes on section 11.11.
Suggested practice problems: 2,5,8,14,19,22,25,34.

✠ Key Ideas of the Section

You should be able to:

- Recognize a linear second-order differential equation.
- Find solutions to linear second-order differential equations.
- Interpret the solution with respect to damped oscillations of a spring.
- Distinguish among over damped, critically damped, and under damped motions.

Key Questions to Ask and Answer

1. What is the difference between undamped and damped oscillations of a spring?

✍ ____

____ *(See text p. 544.)*

Damped oscillation of the spring adds the effect of friction to the model describing the motion, whereas the undamped oscillation model in Section 11.10 disregards friction as a factor.

Example 11.11.1

	Equation	*Examples of Solutions*
Undamped oscillation • no friction • oscillates indefinitely	$\frac{d^2s}{dt^2}+\omega^2 s=0$	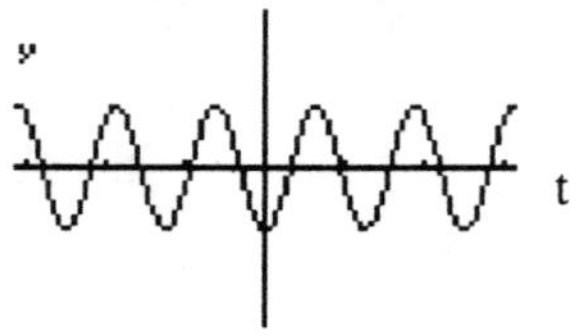 y, t
Damped oscillation • friction as a factor • motion eventually stops and solution dies away over time	$\frac{d^2s}{dt^2}+\frac{a}{m}\frac{ds}{dt}+\frac{k}{m}s=0$	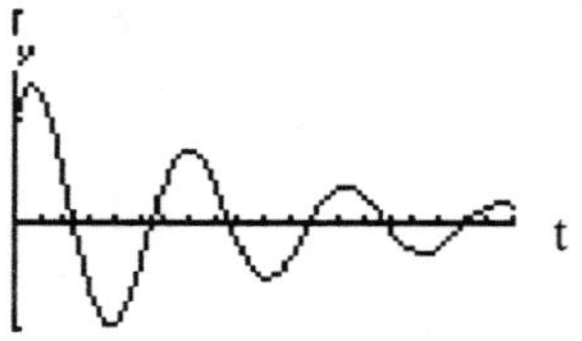 y, t

2. ***How do you find the general solution to the linear second-order differential equation*** $\frac{d^2y}{dt^2}+b\frac{dy}{dt}+cy=0$***?***

✍ ____

____ *(See text p. 545-546.)*

There are three different types of solutions to the differential equation, depending on whether or not the solutions to the corresponding characteristic equation are real and unequal, real and equal, or complex.

Examine the discriminant b^2-4ac, of the characteristic equation $ar^2+br+c=0$, where $a = 1$ or examine b^2-4c:

Linear second-order differential equation with constant coefficients	***Characteristic equation:*** $r^2+br+c=0$	***Solutions***
$\frac{d^2y}{dt^2}+b\frac{dy}{dt}+cy=0$	$b^2-4c>0$	$y=C_1e^{r_1t}+C_2e^{r_2t}$, $r=r_1$ or r_2
	$b^2-4c=0$	$y=(C_1t+C_2)e^{rt}$, $r=\frac{-b}{2}$
	$b^2-4c<0$	$y=C_1e^{\alpha t}\cos\beta t+C_2e^{\alpha t}\sin\beta t$ $r=\alpha\pm i\beta$

Example 11.11.2

Find the general solution to each of the following equations.

Equation	*Characteristic equation*	*General Solution*
1. $y'' + 6y = 0$	$r^2 + 6 = 0 \Rightarrow r = \pm\sqrt{6}i$ • $b^2 - 4ac < 0$	As in Section 11.10, the solution can be found to be: $y(t) = C_1 \cos(\sqrt{6}t) + C_2 \sin(\sqrt{6}t)$
2. $y'' - 4y' + 3y = 0$	$r^2 - 4r + 3 = 0$ $r_1 = 3,\ r_2 = 1$ • $b^2 - 4ac > 0$	$y = C_1 e^{3t} + C_2 e^{t}$
3. $y'' + 5y' + 6y = 0$	$r^2 + 5r + 6 = 0$ $r_1 = -2, r_2 = -3$ • $b^2 - 4ac > 0$	$y = C_1 e^{-2t} + C_2 e^{-3t}$
4. $y'' + 6y' + 9y = 0$	$r^2 + 6r + 9 = 0$ $r_1 = r_2 = -3$ • $b^2 - 4ac = 0$	$y = (C_1 t + C_2) e^{\frac{-6t}{2}}$
5. $y'' + 4y' + 5y = 0$	$r^2 + 4r + 5 = 0$ $r = -2 \pm i$ • $b^2 - 4ac < 0$	$y = C_1 e^{-2t} \cos t + C_2 e^{-2t} \sin t$

✎ On Your Own 1

Use the results of **Example 11.11.2** to solve the following initial value problems. Graph the specific solution.

Initial or boundary values
a. $y'' - 4y' + 3y = 0$ $y(0) = 2$ $y'(0) = 0$
b. $y'' + 5y' + 6y = 0$ $y(0) = -1$ $y'(0) = 4$
c. $y'' + 6y' + 9y = 0$ $y(0) = -1$ $y'(0) = 4$
d. $y'' + 4y' + 5y = 0$ $y(0) = -1$ $y'(0) = 4$

(The answer appears at the end of this section.)

3. *How do you interpret the solution to the problem of damped oscillation of a spring?*

✍ ____

____ *(See text p. 546-548)*

Motion is *over-damped, critically damped* or *under-damped.*

Example 11.11.3

Initial Value Problem:		*Solution and Graph*	
$y'' + 5y' + 6y = 0$ $y(0) = -1$ $y'(0) = 3$	$b^2 - 4ac > 0$ $r^2 + 5r + 6$ $r_1 = -2,\ r_2 = -3$	$y = e^{-2t} - 2e^{-3t}$  scale: $0 \le x \le 6,\ -1 \le y \le 0.1$	$r_1, r_2 < 0$, the motion is overdamped. The mass is slowed by the friction and the motion is damped out.
$y'' + 6y' + 9y = 0$ $y(0) = -1$ $y'(0) = 4$	$b^2 - 4ac = 0$ $r^2 + 6r + 9 = 0$ $r_1 = r_2 = -3$	$y = (t-1)e^{-3t}$  scale: $0 \le x \le 2,\ -0.1 \le y \le 0.05$	$b > 0$, the motion is critically damped. The object returns to equilibrium as slowly as possible. There is no oscillation.
$y'' + 4y' + 5y = 0$ $y(0) = -1$ $y'(0) = 4$	$b^2 - 4ac < 0$ $r^2 + 4r + 5$ $r = -2 \pm i$ $\alpha = -2 < 0$	$y = -e^{-2t}\cos t + 2e^{-2t}\sin t$ 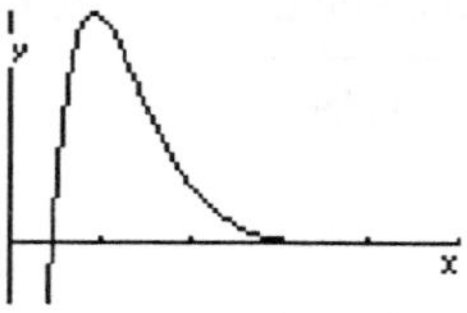 scale: $0 \le x \le 5,\ -0.05 \le y \le 0.2$	$\alpha < 0$, the motion is called *underdamped* as the motion still oscillates, but is dampened very quickly.

Key Algebra Skills

- Review complex numbers.

Key Study Skills

- Read carefully the examples in this section of the text.
- Choose at least two problems that represent this section; put each on a 3 x 5 card with the solution on the back. Review note card problems 11.1-11.11.
- Use the ✓ ? ★ system on homework problems and get answers to all ★ problems.

Answers to On Your Own 1

Initial or boundary values	*Solution*	*Graph*
a. $y'' - 4y' + 3y = 0$ $y(0) = 2$ $y'(0) = 0$	$C_1 = -1$ $C_2 = 3$ $y = -e^{3t} + 3e^{t}$	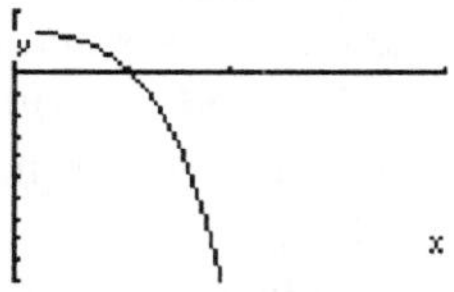
b. $y'' + 5y' + 6y = 0$ $y(0) = -1$ $y'(0) = 4$	$C_1 = 1$ $C_2 = -2$ $y = e^{-2t} - 2e^{-3t}$	
c. $y'' + 6y' + 9y = 0$ $y(0) = -1$ $y'(0) = 4$	$C_1 = 1$ $C_2 = -1$ $y = (t-1)e^{-3t}$	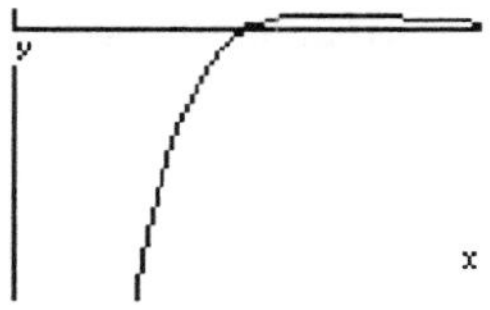 $0 \le x \le 2$, $-0.1 \le y \le 0.007$
d. $y'' + 4y' + 5y = 0$ $y(0) = -1$ $y'(0) = 4$	$C_1 = -1$ $C_2 = 2$ $y = -e^{-2t} \cos t + 2e^{-2t} \sin t$	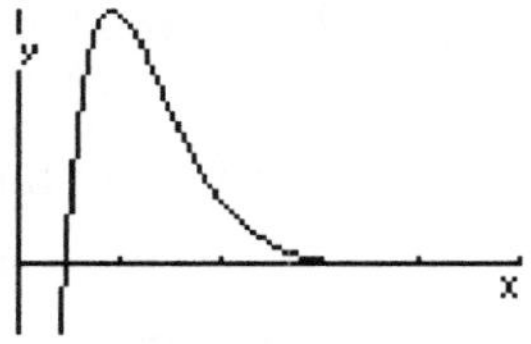 $0 \le x \le 5$, $-0.04 \le y \le 0.015$ 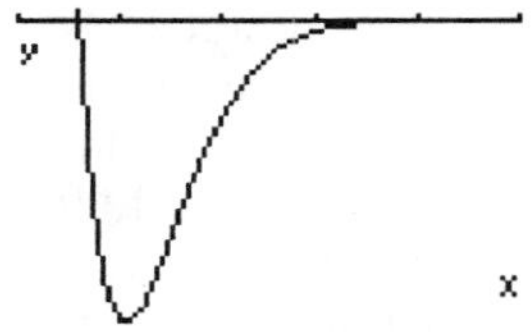 $3 \le x \le 8$, $-0.0003 \le y \le 0.00001$

CUMULATIVE REVIEW CHAPTER ELEVEN

Key Study Skills

- Review all of the **Key Ideas.**
- Choose from Review Exercises and Problems from the text, p. 552-554.
- Do Check Your Understanding from text, p. 555-556.
- Review the note card problems from 11.1 –11.11.
- Make up your own test.
- Review homework problems and get answers to all ★ (starred) problems.
- Take the Chapter Eleven Practice Test.

Chapter 11 Practice Exam

1. Using what you know about functions like $e^x, \ln x,$ *and* $\sin x$ find functions which satisfy the following differential equations:

   a. $\frac{dy}{dt} - y = 0$

   b. $\frac{dy}{dt} + y = 0$

   c. $\frac{d^2y}{dx^2} + y = 0$ (find two functions)

   d. $\frac{d^2y}{dt^2} - y = 0$ (find two functions)

2. Water is being pumped continuously from a lake at a rate proportional to the amount of water left. Let y = number of gallons of water left at any time t (months). Initially there are 2,000,000 gallons of water and 8 months later 1,000,000 gallons remain. The authorities will stop pumping if there is less than 100,000 gallons of water.

   a. Write a differential equation that models this situation.

   b. Find the solution to the differential equation.

   c. When there are 1,000,000 gallons of water remaining, what is the rate at which the water is decreasing?

   d. When will they stop pumping?

3. Given the following slope field for $\frac{dy}{dx} = \frac{-x}{y}$.

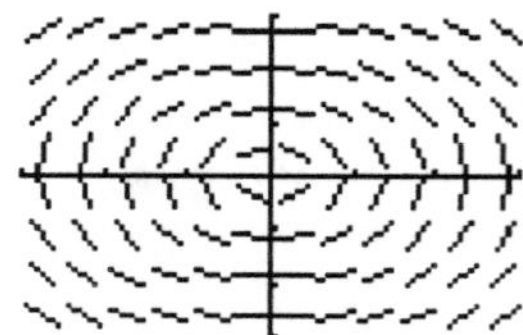

x-scale:[-10,10]; y-scale [-10,10]

   a) Sketch the solution curve that passes through the point $x = 0, y = 2$.

b) Use Euler's method with $n = 4$ to approximate y when $x = 1$. Is this an overestimate or an underestimate? Explain.

c) Solution curves satisfy the equation $x^2 + y^2 = C$. Find the solution that passes through $x = 0$ and $y = 2$ and show that it satisfies the differential equation.

4. Find the solution to the following initial value problem: $\frac{dy}{dx} = \frac{1}{\sqrt{xy}}$, $x = 1, y = 4$.

5. The rates of change of the populations for four different species are given by the differential equations below. (These species do not interact, and so the differential equations are all independent of one another). P measures the population in thousands at time t months. Each species starts with 1000 members ($P = 1$) at time $t = 0$. Assume that each of these differential equations holds indefinitely.

(i) $\frac{dP}{dt} = 0.05$ (ii) $\frac{dP}{dt} = -0.05P$ (iii) $\frac{dP}{dt} = 0.05(1000 - P)$

(iv) $\frac{dP}{dt} = 0.0005P(1000 - P)$

a. For each of the four species, describe the model. Find the solution to the differential equation with initial condition $t = 0$, $P = 1$.

b. For each of the four species, find all equilibrium values of the population and state whether each equilibrium value is stable or unstable.

c. For each of the four species, sketch a graph of the population as a function of time given the initial condition $P(0) = 1$.

6. Consider the following system of differential equations:

$$\frac{dx}{dt} = x(1 - 0.5x - 0.5y)$$

$$\frac{dy}{dt} = y(1 - x - y)$$

a. Find a differential equation describing the relationship between x and y.

b. Find the slope field.

c. If $x = 3$ and $y = 3$, do the values of x and y increase or decrease?

d. Show the nullclines and equilibrium points, and sketch the direction of the trajectories in each region.

ANSWERS TO PRACTICE TESTS

Possible Answers To Chapter Practice Tests

Chapter One Practice Test

1. $f(x) = 7.00(2.5)^x$ (an exponential growth function)
   $g(x) = 7.00 - 3.5x$ (a linear function, decreasing)
   $h(x) = 20.0(0.9)^x$ (an exponential decay function)
   $k(x) = x^3$ (a power function)

2. a. Linear model: $A_l(t) = 27{,}049 + 2500t$

   where $m = \dfrac{29{,}549 - 27{,}049 \; new\, cases}{1 - 0 \; year}$

   b. Exponential model: $A_e(t) = 27{,}049(1.092)^t$

   where the growth factor $a = \dfrac{29{,}549}{27{,}049} = 1.092$, or the growth rate is 9.2% per year.

   c. In 1992 $t=4$.

   $$A_l(4) = 27{,}049 + 2{,}500(4) = 37{,}049 \;\; new\ cases$$

   $$A_e(4) = 27{,}049(1.092)^4 = 38{,}463 \;\; new\ \ cases$$

   The exponential model is a better predictor because the number of reported new cases of males with AIDS is increasing at an increasing rate.

   d. In 2002 $t=14$.

   $$A_l(14) = 27{,}049 + 2{,}500(14) = 62{,}049 \;\; new\ \ cases$$

   $$A_e(14) = 27{,}049(1.092)^{14} = 92{,}740 \;\; new\ \ cases$$

3.

r(cm)	V(cubic cm)
0	0
1	4.189
2	33.510
3	113.097
4	268.083
5	523.599

If the radius doubles from $r = 2$ to $r = 4$, the volume is 8 times as large, or, in general:

$$V(r) = \frac{4}{3}\pi r^3 \;\; to \;\; V(2r) = \frac{4}{3}\pi(2r)^3 = 8V(r)$$

4.

Function	Graph	Some Possible Answers
$y_1 = 2^x$		• exponential function • domain set of real numbers • increasing and concave up • horizontal asymptote at $y = 0$ as $x \to -\infty$
$y_2 = x^2$		• power function with positive integer exponent • domain: the set of real numbers • increasing for $x > 0$; decreasing for $x < 0$ • concave up

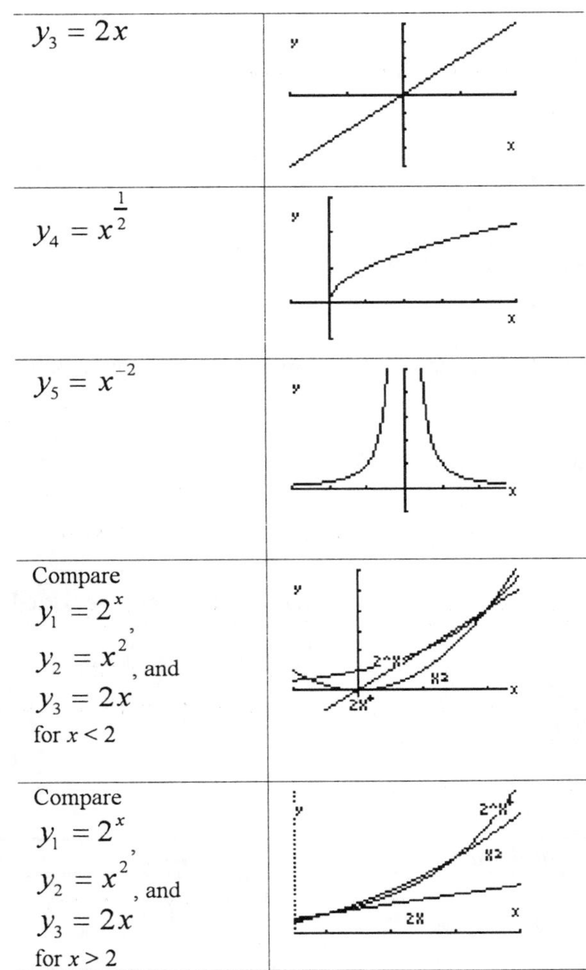

$y_3 = 2x$
$y_4 = x^{\frac{1}{2}}$
$y_5 = x^{-2}$
Compare $y_1 = 2^x$, $y_2 = x^2$, and $y_3 = 2x$ for $x < 2$
Compare $y_1 = 2^x$, $y_2 = x^2$, and $y_3 = 2x$ for $x > 2$

- linear function
- domain: the set of real numbers
- always increasing
- constant rate of change $m = 2$

- power function with positive fractional exponent
- domain: the set of non-negative real numbers
- always increasing
- concave down

- power function with negative integer exponent
- domain: the set of real numbers $\neq 0$
- increasing and concave up for $x < 0$; decreasing and concave up for $x > 0$

- $y_3 = 2x$ dominates for $1 < x < 2$
- $y_2 = x^2$ dominates for $x < -.7667$
- $y_1 = 2^x$ dominates $-.7 < x < 2$

- $y_2 = x^2$ dominates for $2 < x < 4$
- $y_3 = 2^x$ dominates for $x > 4$

5. Let t=0 , the number of years since 1990.

a. $R_L(t) = 2840t + 39,982$ (*millions of dollars*)

$$\text{where } m = \frac{45,662 - 39,982}{2} = 2840 \frac{(millions\ \ of\ \ dollars)}{year}$$

b. $R_E(t) = 39,982(1.068)^t$, where t = the number of years from 1990.

$$39,982a^2 = 45,662$$

$$a = 1.068$$

c. $t = 5$ in 1995

$$R_L(5) = 2840(5) + 39,982 \ (millions\ \ of\ \ dollars)$$

$$= \$54,182(millions\ \ of\ \ dollars)$$

$$R_E(5) = 39,982(1.068)^5 \ (millions\ \ of\ \ dollars)$$

$$= \$55,554(millions\ \ of\ \ dollars)$$

The exponential model is closer since, evidently, the annual receipts are increasing at an increasing rate.

d. $2840t + 39,982 = 100,000$

$t = 21$ years, or year 2011

$$39{,}982(1.068)^t = 100{,}000$$

$$(1.068)^t = \frac{100{,}000}{39{,}982}$$

$$t\ln(1.068) = \ln(2.5011)$$

e. $t = 14$ years or year 2004

6. $S_1(t)$ is an exponential function, $S_2(t)$ is a power function.

a. They are the same, since $S_1(1) = S_2(1)$.

b. $S_2(t) > S_1(t)\ \ for\ \ 2 < t < 4$

c. $S_1(t)$ dominates $S_{21}(t)$ for large t. An exponential function always will dominate a power function as $x \to \infty$.

7. There may be several answers to this problem.

a. Amplitude = $\dfrac{4.35 - 0.03}{2} = 2.16$

b. Period is 12 months.

c. This looks like a *cosine* curve (you could also use *sine*), which is stretched, the period adjusted and then translated up.
2π/B= 12 or B = 2π/12 adjusts the period
The stretch is the amplitude, or 2.16.
The graph is lifted vertically 2.19 units
Or, $p(t) = 2.16\cos(\frac{2\pi}{12}t) + 2.19$, where t=0 in January

8.

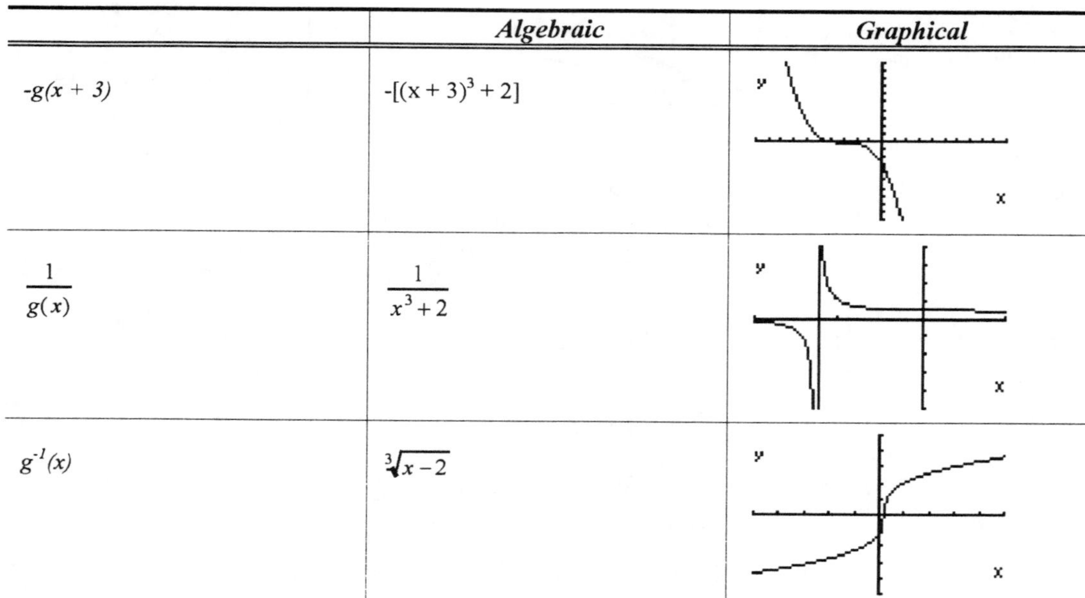

	Algebraic	*Graphical*
$-g(x+3)$	$-[(x+3)^3+2]$	
$\dfrac{1}{g(x)}$	$\dfrac{1}{x^3+2}$	
$g^{-1}(x)$	$\sqrt[3]{x-2}$	

9.

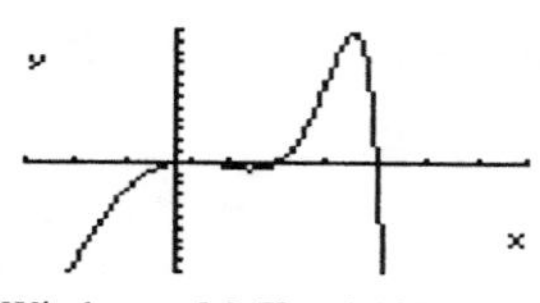

Window: x:[-3,7], y:[-100,120]

b. Estimates are: f increases $-3 < x < 0.3$ and then again for $1.5 < x < 3.5$; otherwise f decreases.

c. Estimates are: f changes concavity twice over this interval. f is concave down for $-3 < x < 1$ and then again for $3 < x < 7$

10.

$y_1 = 3\sin(4x-2)$	• Stretch factor 3 • Period = $\pi/2$ • Shifted right 1/2 units • No asymptotes • Domain =$(-\infty,+\infty)$ • Many Zeros $\sin(4x-2)=0$ $4x-2 = n\pi$ $x=0.5 \pm n(\pi/4),\ \ n = 0, \pm 1, \pm 2\ldots$	
$y_2 = \ln(x+1)$	• *natural log* function translated 1 unit left. • Vertical asymptote: $x=-1$ • Domain: $(-1,+\infty)$ • 1 Zero: $x=0$	
$y_3 = \dfrac{2x-3}{x^2-4}$	• Rational function • Vertical asymptotes: $x=2$ *and* $x=-2$ • Horizontal asymptote: $y=0$ • 1 Zero: $x=1.5$	

11. $f(x) = -3(x-2)(x+1)(x-1)x^2$ since

$$f(x) = k(x-2)(x+1)(x-1)x^2$$
$$-216 = k(3-2)(3+1)(3-1)3^2$$
$$k = -3$$

Chapter Two Practice Test

1a.

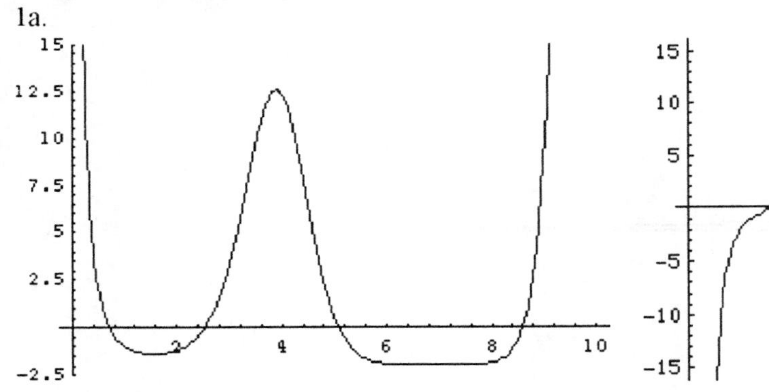

The graph of $f(x) = x^2\cos(x-30) - 2$

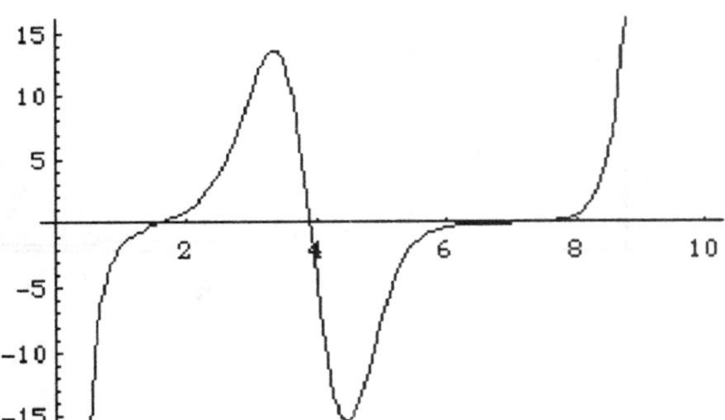

The graph of $f'(x)$

1b. $f'(3) > 0,\ f'(5) < 0, f'(7) \approx 0,$ therefore $f'(5) < f'(7) < f'(3).$

1c. $f''(2) > 0$ because $f(x)$ is concave up at $x = 2$, $f''(4) < 0$ because $f(x)$ is concave down at $x = 4$, therefore $f''(4) < f''(2)$.

2.

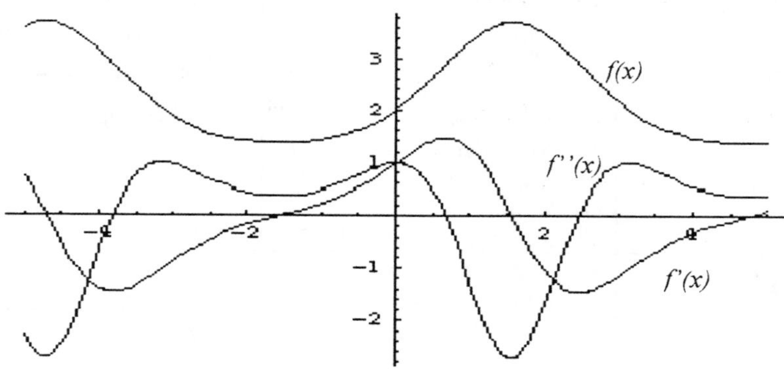

Label *f(x), f'(x), and f''(x)*

Explanation:

Where *f(x)* has turning points , $f'(x) = 0$.
Where *f(x)* is increasing, $f'(x) > 0$ and where *f(x)* is decreasing $f'(x) < 0$.
Where *f(x)* is concave up, $f''(x) > 0$ and where *f(x)* is concave down $f''(x) < 0$.
Where $f'(x)$ has a turning point , $f''(x) = 0$.

3a. $$f'(a) = \lim_{h \to 0} \frac{[(\log(a+h-1))/\log 2] - [(\log(a-1))/\log 2]}{h},$$

$f'(a)$ is the instantaneous rate of change of f at the point $x = a$, and also the slope of the tangent line to the graph of f at that point.

3b. The slope of the line tangent to the graph of f at $x = 2$ has the value 1.4427 At $x=2$, *f(x)* is changing at a rate of 1.4427 units per unit change in x.

3c. Use the tangent line to approximate the value:
$y_{\tan} = 1.4427(x-2) + \log(2-1)/\log 2 = 1.4427(x - 2) + 0$,
therefore $f(2.5) \approx 1.4427(2.5 - 2) = .72135$, this will be an over approximation because the the graph of f is concave down and the tangent line will lie above *f(x)*.

3d.

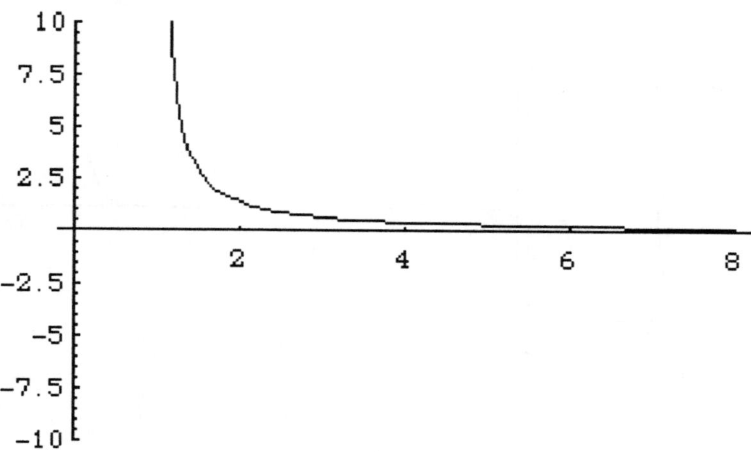

The domain for *f(x)* is $x > 1$, so $f'(1)$ is undefined and f has a vertical asymptote at $x = 1$.

4. Let $f(x) = x^{2/3}$

4a. $f'(0) = \lim_{x \to 0} \frac{(0+h)^{\frac{2}{3}} - (0)^{\frac{2}{3}}}{h} = \frac{h^{\frac{2}{3}}}{h} = \frac{1}{h^{\frac{1}{3}}}$ = does not exist . Since as $h \to 0$, if $h > 0$ the denominator becomes small so the fraction grows without bound so $f' \to +\infty$, if $h < 0$ the denominator becomes small so the fraction decreases without bound so $f' \to -\infty$.

4b. Is *f(x)* continuous at $x = 0$? YES

- $f(0)=0$
- $\lim_{x \to 0^-} f(x) = \lim_{x \to 0^+} f(x) = 0$
- $\lim_{x \to 0} f(x) = f(0) = 0$

4c. Is *f(x)* differentiable at $x = 0$? NO based on 4A above.

Chapter Three Practice Test

1a. $f'(x) = \dfrac{(14x + \frac{1}{2}x^{-\frac{1}{2}})(x^2+8) - (7x^2 + \sqrt{x} + 1)(2x)}{(x^2+8)^2}$

1b. $y' = (\ln a)(a^x)(5x)^{\frac{1}{3}} + a^x(\frac{1}{3})(5x)^{-\frac{2}{3}}(5)$

1c. $P' = 3(6t^2 - 14t + 100)^2(12t - 14)$

1d. $s' = 4(\sin(5w - \pi))^3(\cos(5w - \pi))(5)$

1e. $D' = 2e^{\tan x} + (2x)e^{\tan x}(\dfrac{1}{\cos^2 x})$

2.

$$f(x) = \ln(x+2) \Rightarrow f'(x) = \frac{1}{x+2}$$

$$f'(0) = \frac{1}{2} \text{ and } f(0) = \ln(0+2) = \ln 2$$

$$\text{Tangent line equation: } f(x) \approx f(0) + f'(0)(x-0) = \ln 2 + \frac{1}{2}x$$

which will be an overestimate for x near 0, since f is concave down.

$$f(x) \approx \ln 2 + \frac{1}{2}(x)$$

$$f(0.1) \approx \ln 2 + \frac{1}{2}(0.1) = 0.7178$$

$$f(0.2) \approx \ln 2 + \frac{1}{2}(0.2) = 0.7419$$

$$f(0.5) \approx \ln 2 + \frac{1}{2}(0.5) = 0.8109$$

3.

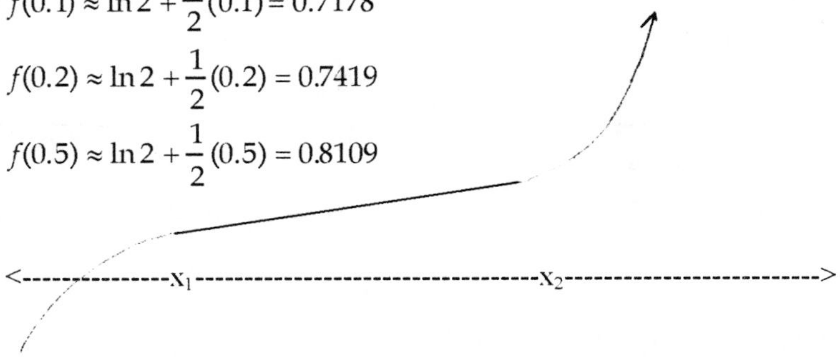

<---------------x_1-----------------------------------x_2------------------------->

4. Area of top and bottom = 2 (area of circle) = $2\pi r^2$
area of side = (circumference)(height of can) = $2\pi rh$
Volume of can = $\pi r^2 h = 40$, so $h = 40/(\pi r^2)$

surface area = $SA(r) = 2\pi r^2 + 2\pi rh = 2\pi r^2 + 2\pi r(40/(\pi r^2)) = 2\pi r^2 + 80/r$

$SA'(r) = 4\pi r - 80/r^2 = (4\pi r^3 - 80)/r^2 = 0$

$r = 1.853$, $h = 40/(\pi r^2) = 3.707$. The can with minimum surface area that has a volume of 40 in^3 will have a radius of 1.853 inches and a height of 3.707 inches.

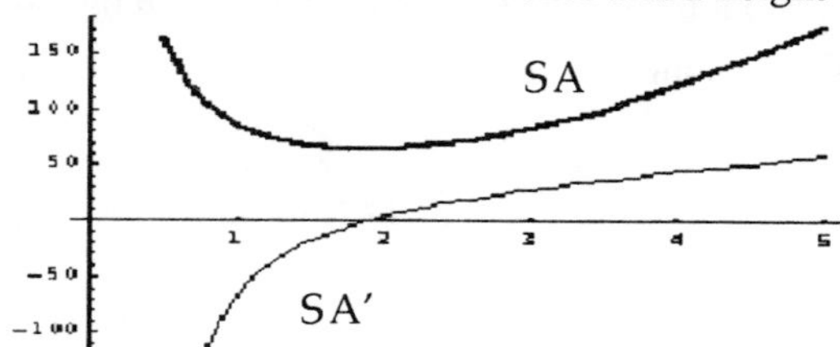

At r= 1.853, SA' = 0 and the SA is at a minimum.

5.

a) $\lim\limits_{x\to o}\frac{sin(5x)}{3x} = \lim\limits_{x\to o}\frac{\frac{d}{dx}sin(5x)}{\frac{d}{dx}3x} = \lim\limits_{x\to o}\frac{5\,cos5x}{3} = \frac{5}{3}$

b) $\lim\limits_{t\to\infty}\frac{t^2}{e^t} = \lim\limits_{t\to\infty}\frac{\frac{d}{dt}t^2}{\frac{d}{dt}e^t} = \lim\limits_{t\to\infty}\frac{2t}{e^t} = \lim\limits_{t\to\infty}\frac{\frac{d}{dt}2t}{\frac{d}{dt}e^t} = \lim\limits_{t\to\infty}\frac{2}{e^t} = 0$

c)

$\lim\limits_{x\to 0^+} 3x\ln x^2 = \lim\limits_{x\to 0^+}\frac{\ln x^2}{\frac{1}{3x}}$ and $\lim\limits_{x\to 0^+}\ln x^2 = -\infty$, $\lim\limits_{x\to 0^+}\frac{1}{3x} = +\infty$

So L'Hospital's rule applies.

$$\lim_{x\to 0^+}\frac{\ln x^2}{\frac{1}{3x}} = \lim_{x\to 0^+}\frac{\frac{1}{x^2}2x}{\frac{-1}{3x^2}} = \lim_{x\to 0^+} -6x = 0$$

6.

a)

X1T=cos(3T) Y1T=sin(T)

T=.78539816
X=-.7071068 Y=.70710678

b)

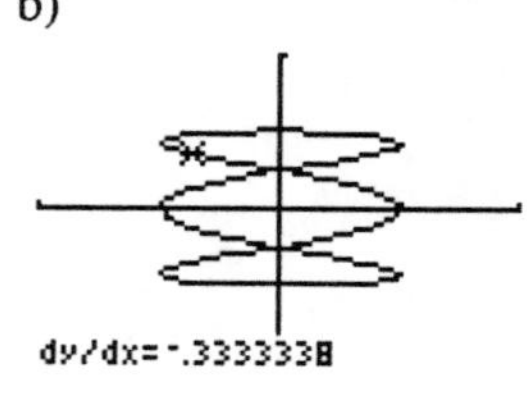

$\frac{dx}{dt} = -3\sin 3t \Rightarrow \left.\frac{dw}{dt}\right|_{t=\frac{\pi}{4}} = -3\sin(3\cdot\pi/4) = -3/\sqrt{2}$

$\frac{dy}{dt} = \cos t \Rightarrow \left.\frac{dy}{dt}\right|_{t=\frac{\pi}{4}} \cos(\pi/4) = 1/\sqrt{2}$

$\frac{dy}{dx} = \frac{1/\sqrt{2}}{-3/\sqrt{2}} = -\frac{1}{3}$

c)The tangent line: $x = \frac{-1}{\sqrt{2}} - \frac{3}{\sqrt{2}}t$, $y = \frac{1}{\sqrt{2}} + \frac{1}{\sqrt{2}}t$

Chapter Four Practice Test

1. a. f is increasing on [2,7] and [9.5,10]; otherwise f is decreasing
   b. f is concave up on [0,4] and [8.5,10]; otherwise f is concave down.
   c. f has a local maxima when $x = 7$; f has local mimima when x=2 and x=9.5.
   d. The global maximum occurs at x=7

e. $f(2) = 1 + \int_0^2 f'(x)dx$

f. $f(2) \approx f(0) +$ (area between the graph of f ' and the x-axis) $\approx 1 + (-4) = -3$.

2.

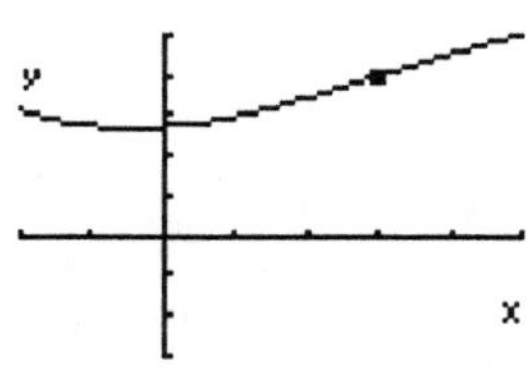

3. A possible graph is

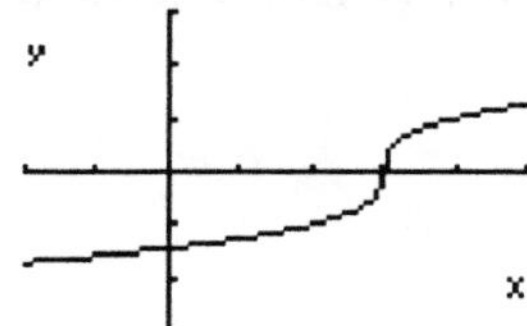

4.

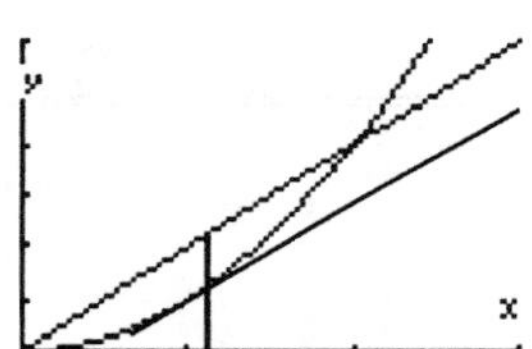

This is where the distance between R and C is the greatest and where the slopes of the two curves are equal. Mathematically

$$\pi'(q) = R'(q) - C'(q) = 2 - C'(q)$$

$$0 = 2 - C'(q)$$

which gives a critical point at q_0 when $C'(q_0) = 2$. This is a local maximum since

$$\pi''(q_0) = -C''(q_0) < 0$$

5. Let r =televisions/hour and the hourly costs $H(r) = kr^2$. Then $400 = k(10)^2$ or $H(r) = 4r^2$. The number of hours to produce the total order of T televisions is $\dfrac{T \text{ televisions}}{r \text{ televisions/hour}} = \dfrac{T}{r} \text{ hours}$. The total cost to produce the order is the hourly cost times the number of hours plus the fixed cost per hour times the number of hours, or $C(r) = 4r^2 \dfrac{T}{r} + 1600\dfrac{T}{r} = 4rT + \dfrac{1600T}{r}$.

$C'(r) = 4T - \dfrac{1600T}{r^2}$, which has a critical point when $r = 20$ televisions/hour. The total cost is a minimum there since $C''(20) > 0$. (Note that the minimum cost does not depend on the size of the order, T).

6.

a) The domain: $x \neq 0$.

b) $f'(x) = \dfrac{(1 - \cosh x)\sinh x - (1 + \cosh x)(-\sinh x)}{(1 - \cosh x)^2} = \dfrac{2\sinh x}{(1 - \cosh x)^2} = \dfrac{4(e^x - e^{-x})}{(2 - e^x - e^{-x})^2}$

c) 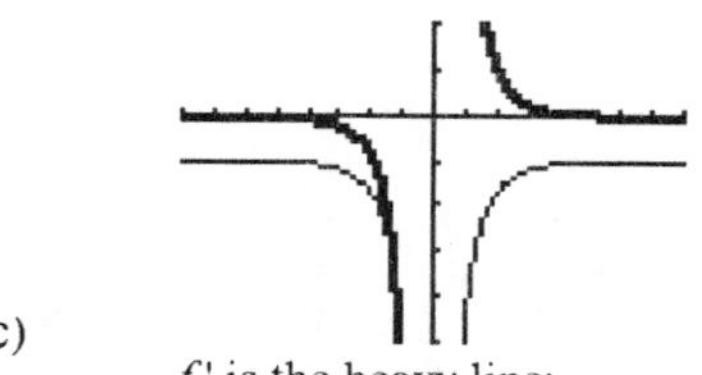

f' is the heavy line:
window: -8 < x < 8,-5 < y < 2

7. The function is continuous on the closed interval and differentiable on the open interval, and therefore the Mean Value Theorem applies. $\dfrac{f(4)-f(0)}{4-0}=\dfrac{0-1}{4}=-0.25$. We need a number c such that $f'(c)=2^c \ln 2-2c=-0.25$ and $c\approx 0.681$ (found graphically) which is in the interval [0.4].

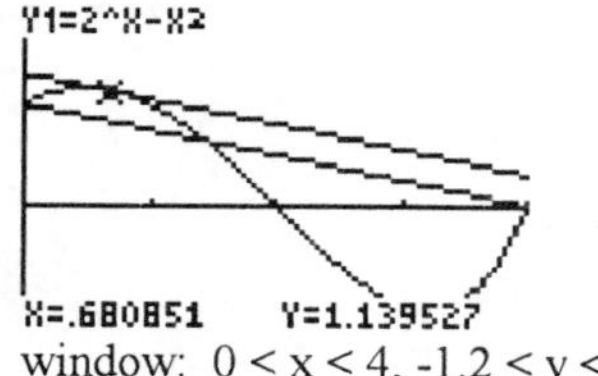

window: 0 < x < 4, -1.2 < y < 2

Chapter Five Practice Test

1a. $h(t)=\int_0^t r(t)dt$

1b. Using geometry to find the area under the rate curve $r(t)$, by summing the areas.

$h(0) = 0$	$h(30) = 60 + 20 = 80$
$h(10) = 0 + 20 = 20$	$h(40) = 80 + (-30) = 50$
$h(20) = 20 + 40 = 60$	$h(50) = 50 +(-60) = -10$

1c. The sketch of the graph of the height of the river above flood stage, $h(t)$, over the 50 hour time period:

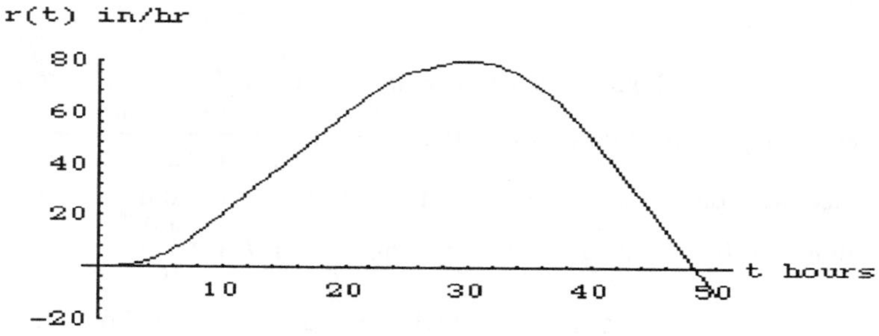

1d. The maximum height occurs at t= 30, where the rate of change graph equals zero. The height is 80 inches above flood stage.

1e. At t = 50, the river is 10 inches below flood stage. The area of $r(t)$ below the x-axis between $30 \le t \le 50$ is greater than the area above the x-axis between $0 \le t \le 30$.

2a. The area between the x-axis and the graph of $f(x)$ over [-1,0] is part below and part above the x-axis and it appears that there is more area below the x-axis (thus the signed area is negative). The area between the x-axis and the curve over [0,1] is all above the x-axis, therefore $\int_{-1}^{0} f(x)dx < \int_{0}^{1} f(x)dx$

2b. $\Delta x = \frac{b-a}{n} = \frac{5-1}{4} = 1$

$$\int_1^5 f(x)dx \approx leftsum = \Delta x(f(1) + f(2) + f(3) + f(4)) = 20.9037$$

$$\int_1^5 f(x)dx \approx rightsum = \Delta x(f(2) + f(3) + f(4) + f(5)) = 20.2282$$

2c. Since *f(x)* is decreasing in the interval [1,5], the true value if the interval must lie between its right and left sum:

$$20.2282 \le \int_1^5 f(x)dx \le 20.9037$$

The *Error* $\le \; | f(5) - f(1)) | \cdot \Delta x = | 5.0011 - 5.6767 | (1) = .6755$ or

20.9037-20.2282 =.6755, so ½(.6755) would be the estimate.

Chapter Six Practice Test

1.

2a. The bounded region of $f(x) = -x^3 + 7x^2 - 10x$ is:

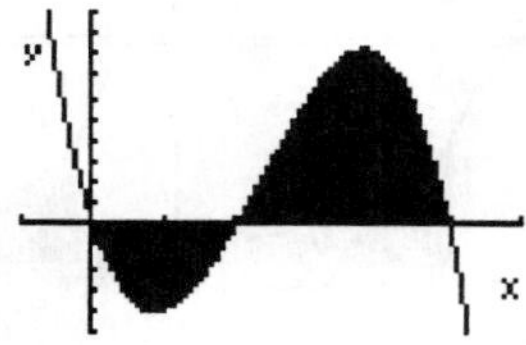

Since $f(x) = -x^3 + 7x^2 - 10x = -x(x-5)(x-2)$ the zeros are $x = 0$, $x = 2$, and $x = 5$.

For $0 \le x \le 2$, $f(x) \le 0$, the area of the region =

$$-\int_0^2 (-x^3 + 7x^2 - 10x)dx = -(\frac{-x^4}{4} + \frac{7x^3}{3} - \frac{10x^2}{2})\Big|_0^2 = \frac{16}{3} = 5.33$$

For $2 \le x \le 5$, $f(x) \ge 0$ the area of the region =

$$\int_2^5 (-x^3 + 7x^2 - 10x)dx = (\frac{-x^4}{4} + \frac{7x^3}{3} - \frac{10x^2}{2})\Big|_2^5 = \frac{63}{4} = 15.75$$

So total area, without regard to sign, $A_1 + A_2$ is about 5.33 + 15.75 = 21.08

2b. the accumulated <u>signed</u> area is:

$$\int_0^5 (-x^3 + 7x^2 - 10x)dx = (\frac{-x^4}{4} + \frac{7x^3}{3} - \frac{10x^2}{2})\Big|_0^5 = \frac{125}{12} \approx 10.42$$ or using the results from above,

$-A_1 + A_2$. or $-5.33 + 15.75 = 10.42$

3a. No, because

$$y'(t) = \frac{d}{dt}\left(\frac{1}{3}\cos^3 t\right) = \cos^2 t(-\sin t) \ne \cos^2 t$$

3b. The graph of $f(t)=\dfrac{dy}{dt}=\cos^2 t,\ y(0)=0$

An antiderivative is shown below:

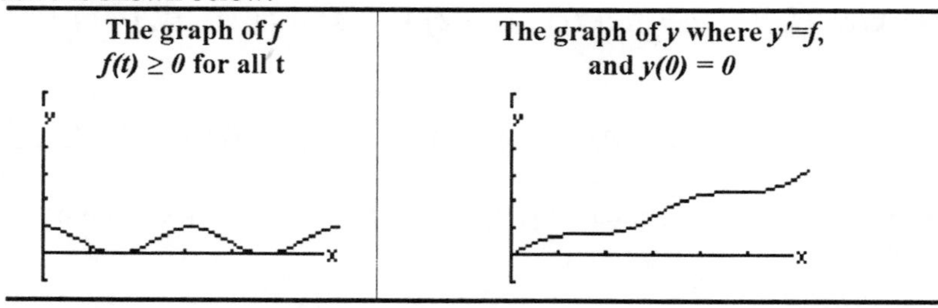

3c.

$$\frac{dy}{dt}=\cos^2 t=\frac{1}{2}(1+\cos 2t)$$

has general solution $y=\dfrac{1}{2}(t+\dfrac{1}{2}\sin 2t)+C$

y(0) = 0 gives $C=0$ so the solution to the initial value problem is $y=\dfrac{1}{2}(t+\dfrac{1}{2}\sin 2t)$

To show this, find the derivative: $\dfrac{dy}{dt}=\dfrac{d}{dt}\left[\dfrac{1}{2}(t+\dfrac{1}{2}\sin(2t)\right]=\dfrac{1}{2}(1+\cos(2t))=\cos^2 t$

4. a) $h(t)=-32t^2-50t+300$ in feet

b) $v(t)=h'(t)=-32t-50$, $v(1)=-82$ ft/sec

c) $h(t)=0$ when $t=3.014$ sec.

5.

x	1	1.5	2.0	2.5	3
F(x)	∫f(x)dx=0 *F(1) = 0*	∫f(x)dx=.40546511 *F(1.5) = 0.405*	∫f(x)dx=.69314718 *F(2)= 0.693*	∫f(x)dx=.91629073 *F(2.5) =. 916*	∫f(x)dx=1.0986123 *F(3) = 1.099*
Graph of *F* *F* is increasing at a decreasing rate.			y x		

6. a) iv b) i c) iii d)ii

Chapter Seven Practice Test

1. Complete the square:

$$\left(x^2+6x+12\right)=\left(x^2+6x+9\right)+3$$

$$=(x+3)^2+\sqrt{3}^2$$

$\displaystyle\int\frac{dx}{x^2+6x+12}=\int\frac{dx}{(x+3)^2+3}=\frac{1}{\sqrt{3}}\arctan\frac{x+3}{\sqrt{3}}+C$, using Table V-24.

2. Use integration by parts, or Table II-8.

Let $u = e^{3x}$, $v' = \sin 3x$, $so\ u' = 3e^{3x}$, $v = \frac{-1}{3}\cos 3x$

$$\int e^{3x}\sin 3x dx = e^{3x}\left(\frac{-1}{3}\cos 3x\right) - \int\left(\frac{-1}{3}\cos 3x\right)\left(3e^{3x}\right)dx$$

$$= \frac{-1}{3}e^{3x}\cos 3x + \int e^{3x}\cos 3x dx$$

Use parts again: $u = e^{3x}$, $v' = \cos 3x$, $so\ u' = 3e^{3x}$, $v = \frac{1}{3}\sin 3x$

$$\int e^{3x}\sin 3x dx = \frac{-1}{3}e^{3x}\cos 3x + \int e^{3x}\cos 3x dx =$$

$$\int e^{3x}\sin 3x dx = \frac{-1}{3}e^{3x}\cos 3x + \frac{1}{3}e^{3x}\sin 3x - \int e^{3x}\sin 3x dx$$

$$\int e^{3x}\sin 3x dx = \frac{1}{2}\left[\frac{-1}{3}e^{3x}\cos 3x + \frac{1}{3}e^{3x}\sin 3x\right] + C$$

3. Use substitution: let $u = x - 3$, $du = dx$, $u = u + 3$.

$$\int_4^{12}\frac{x}{\sqrt{x-3}}dx = \int_1^9 \frac{u+3}{\sqrt{u}}du = \int_1^9 u^{\frac{1}{2}} + 3u^{\frac{-1}{2}}du = \frac{2}{3}u^{\frac{3}{2}} + 6u^{\frac{1}{2}}\Big|_1^9 = 29.\overline{33}$$

4. $n =$ 50 subdivisions:
Left-rule = 0.3019
Right-rule = 0.3187
Trapezoid rule = 0.3103
Midpoint rule = 0.31025
Simpson's rule = 0.310268

5. Since

$$\int \tan^2\theta d\theta = \int\frac{\sin^2\theta}{\cos^2\theta}d\theta = \int\frac{1-\cos^2\theta}{\cos^2\theta}d\theta$$

$$= \int\left(\sec^2\theta - 1\right)d\theta = \tan\theta - \theta + C$$

then,

$$\int_0^{\frac{\pi}{2}}\tan^2\theta d\theta = \lim_{b\to\frac{\pi}{2}^-}\int_0^b \tan^2\theta d\theta$$

$= \lim_{b\to\frac{\pi}{2}^-}\left[\tan\theta - \theta\Big|_0^b\right]$ which does not exist. Therefore, $\int_0^{\frac{\pi}{2}}\tan^2\theta d\theta$ diverges.

6. $$\frac{x+1}{x^2-x} = \frac{x+1}{x(x-1)} = \frac{A}{x} + \frac{B}{x-1}$$

$x + 1 = A(x-1) + Bx$

or A = -1 and B = 2, so $\int\frac{x+1}{x^2-x}dx = \int\frac{-1}{x}dx + \int\frac{2}{x-1}dx = -ln|x| + 2\,ln|x-1| + C$

7. Use $x = 2\sin\theta$, or $\theta = \arcsin(x/2)$, for $\frac{-\pi}{2} \le \theta \le \frac{\pi}{2}$

so $dx = 2\cos\theta\, d\theta$.

$$\int \frac{1}{\sqrt{4-x^2}}dx = \int \frac{\cos\theta}{\sqrt{1-\sin^2\theta}}d\theta = \int \frac{\cos\theta}{\sqrt{\cos^2\theta}}d\theta = \int d\theta = \theta + C = \arcsin\frac{x}{2} + C$$

Note that we use the fact that for $\frac{-\pi}{2} \le \theta \le \frac{\pi}{2}$, $\cos\theta \ge 0$ so $\sqrt{\cos^2\theta} = \cos\theta$.

Chapter Eight Practice Test

1. The base of the pyramid is 25 feet square. Take n horizontal slices of thickness Δy a height y_i from the base with sides of length s_i. The volume of the slice is $\approx s_i^2\,\Delta y$. Since the side length s_i is 1/4 the perimeter which is given as (100 - y), we have $s = \frac{1}{4}(100 - y_i)$.

   The total volume $V \approx \sum_{i=0}^{n-1} s_i^2 \Delta y = \sum_{0}^{n-1}\left(\frac{1}{4}(100 - y_i\right)^2 \Delta y$ ft^3. As n gets large,

   $$V = \int_0^{100}\left(\frac{1}{4}(100 - y\right)^2 dy = \frac{-1}{48}(100-y)^3\Big|_0^{100} = 20{,}833.3 \text{ ft}^3$$

2. Divide the highway into n pieces of length $\Delta x = 50/n$ miles each. For each segment, the approximate number of automobiles is $\rho(x_i)\Delta x$. The approximate total number of automobiles is

   $T \approx \sum_{i=0}^{n-1} \rho(x_i)\Delta x$. As n gets large,

   $$T = \int_0^{50} \rho(x)dx = \int_0^{50}(100 + 50\sin(\pi x))dx$$

   $$= 100x - \frac{50}{\pi}\cos(\pi x)\Big|_0^{50} = 5000$$

   The total number is 5000 automobiles on the 50 mile stretch.

3. Let x = height from ground to bucket and $0 \le x \le 30$ ft. Divide the distance the bucket must be raised into n separate distances. At height x, to lift weight Δx the work needed is the work for the bucket plus the work for the cable or
   $300\Delta x + 3(30–x)\Delta x = (390–3x)\Delta x$ ft-pounds.

   The total work is approximately $W \approx \sum_{i=0}^{n-1}(390 - 3x)\Delta x$ and, as n gets large, the total work is

   $$W = \int_0^{30}(390 - 3x)dx = 390x - \frac{3}{2}x^2\Big|_0^{30} = 10{,}350 \text{ ft - lb.}$$

4. Company A: offers \$50,000 now and $0.05S(t)$, where $t = 0$ two years from now.
   Company B: offers \$0 now and $0.06S(t)$, where $t = 0$ two years from now.
   We can calculate the PV for the royalties on sales to two years from now, and then bring that PV back to the PV now.

	Company A	*Company B*
• amount of royalties for period t to $t+\Delta t$	$\approx (0.05)S(t)\Delta t$	$\approx (0.06)S(t)\Delta t$
• PV (at year 2) for period t to $t+\Delta t$	$\approx ((0.05)S(t)\Delta t)e^{-0.10t}$	$\approx ((0.06)S(t)\Delta t)e^{-0.10t}$

• Total PV (at year 2)	$\int_0^5 [0.05S(t)]e^{-0.10t}dt$ $= \int_0^5 [0.05(1,000,000e^{0.10t})]e^{-0.10t}dt$ $= \$250,000$	$\int_0^5 [0.06S(t)]e^{-0.10t}dt$ $= \int_0^5 [0.06(1,000,000e^{0.10t})]e^{-0.10t}dt$ $= \$300,000$
• Total PV now	$50,000 + 250,000e^{-0.10(2)}$ $\$254,683$	$300,000e^{-0.10(2)}$ $\$245,619$

Company A offers the better deal.

5. You want a value of a such that $\int_a^\infty \frac{1}{6\sqrt{2\pi}} e^{-(x-80)^2} / 2(6)^2\, dx \approx 0.10$, giving $a \approx 88$ points.

Chapter 9 Practice Test

1. By the comparison test, $\frac{1}{2+5^n} \le \frac{1}{5^n}$ for $n \ge 1$. Since $\sum_{n=1}^{\infty} \frac{1}{5^n}$ is a geometric series which converges, the series $\sum_{n=1}^{\infty} \frac{1}{2+5^n}$ converges.

2. The series of absolute values $\sum \left|\frac{\cos n}{n^2}\right|$ converges because $\left|\frac{\cos n}{n^2}\right| \le \frac{1}{n^2}$, $n \ge 1$ since $|\cos n| \le 1$ and we know $\sum \frac{1}{n^2}$ converges. Thus the series $\sum \frac{\cos n}{n^2}$ converges.

3. $\lim_{n\to\infty} \left|\frac{a_{n+1}}{a_n}\right| = \lim_{n\to\infty} \left|\frac{x^{n+1}}{x^n} \frac{\sqrt{n}}{\sqrt{n+1}}\right| = |x|$ By the Ratio Text, R = 1 and the interval of convergence is: $-1<x<1$. To examine the endpoints: If $x = 1$, the series is $1 + \frac{1}{\sqrt{2}} + \frac{1}{\sqrt{3}} + \ldots$, which diverges (p-series with $p = 1/2$). If $x = -1$, the series is $-1 + \frac{1}{\sqrt{2}} - \frac{1}{\sqrt{3}} + \ldots + \frac{(-1)^n}{\sqrt{n}} + \ldots$ which converges by the alternating series test. So the interval of convergence is: $-1 \le x < 1$.

Chapter 10 Practice Test

1. $f'(x) = -\sin x, f''(x) = -\cos x, f'''(x) = \sin x$ and $\cos(-\pi/4)=\sqrt{2}/2$, $\sin(-\pi/4)=-\sqrt{2}/2$.

$$P_3(x) = \cos\frac{-\pi}{4} + \left(-\sin\frac{-\pi}{4}\right)\left(x+\frac{\pi}{4}\right) + \frac{-\cos\frac{-\pi}{4}}{2!}\left(x+\frac{\pi}{4}\right)^2 + \frac{\sin\frac{-\pi}{4}}{3!}\left(x+\frac{\pi}{4}\right)^3$$

$$P_3(x) = \frac{\sqrt{2}}{2} + \frac{\sqrt{2}}{2}\left(x+\frac{\pi}{4}\right) + \frac{-\sqrt{2}}{2}\frac{\left(x+\frac{\pi}{4}\right)^2}{2!} + \frac{-\sqrt{2}}{2}\frac{\left(x+\frac{\pi}{4}\right)^3}{3!}$$

$$P_3(x) = \frac{\sqrt{2}}{2}\left[1 + \left(x + \frac{\pi}{4}\right) - \frac{1}{2}\left(x + \frac{\pi}{4}\right)^2 - \frac{1}{6}\left(x + \frac{\pi}{4}\right)^3\right]$$

$f(0) = cos(0) = 1$; $P_3(0) = 0.987$

2. For $x \approx 0$, $\sin x \approx x - \frac{1}{6}x^3$ and $\cos x \approx 1 - \frac{1}{2}x^2$ so

$$\frac{(\sin x - x)^3}{x(1-\cos x)^4} \approx \frac{\left(-\frac{1}{6}x^3\right)^3}{x\left(\frac{1}{2}x^2\right)^4} = \frac{\frac{-x^9}{6^3}}{\frac{x^9}{2^4}} = \frac{-2}{27}$$

3. $\cos x = 1 - \frac{x^2}{2!} + \frac{x^4}{4!} - \dots$

$$\cos\sqrt{x} = 1 - \frac{\sqrt{x}^2}{2!} + \frac{\sqrt{x}^4}{4!} - \dots = 1 - \frac{x}{2!} + \frac{x^2}{4!} - \dots \quad \text{for } x \geq 0$$

$cos\sqrt{2} \approx 1 - \frac{2}{2!} + \frac{4}{4!} = 0.1\overline{6}$

(To get a better approximation we need more terms.)

4. $\ln(1+x) = x - \frac{x^2}{2} + \frac{x^3}{3} - \frac{x^4}{4} + \dots$ with interval of convergence $-1 < x < 1$

$$\text{-ln(1-4x)} = -\ln(1+(-4x)) = -\left(-4x - 8x^2 - \frac{64}{3}x^3 - 64x^4 + \dots\right)$$

$$\text{-ln(1-4x)} = 4x + 8x^2 + \frac{64}{3}x^3 + 64x^4 - \dots$$

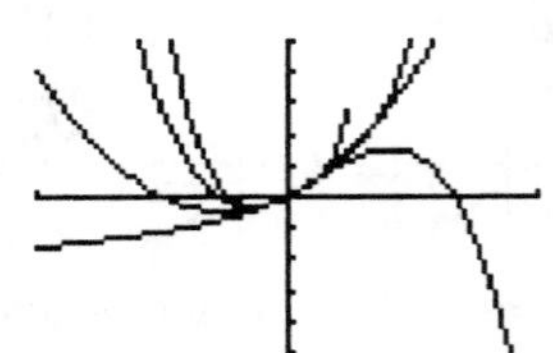

As n gets large the polynomials seem to converge on the interval $-1/4 < x < 1/4$.

5a. The Taylor expansion for $e^t = 1 + t + \frac{t^2}{2!} + \frac{t^3}{3!} + \dots$ so

$$f(t) = t\left(1 + t + \frac{t^2}{2!} + \frac{t^3}{3!} + \dots\right) = t + t^2 + \frac{t^3}{2!} + \frac{t^4}{3!} + \dots$$

5b.

$$g(x) = \int_0^x te^t dt = \int_0^x \left(t + t^2 + \frac{t^3}{2!} + \frac{t^4}{3!} + \dots\right) dt$$

$$= \left.\frac{t^2}{2} + \frac{t^3}{3} + \frac{t^4}{4\cdot 2!} + \frac{t^5}{5\cdot 3!} + \dots\right|_0^x$$

$$= \frac{x^2}{2} + \frac{x^3}{3} + \frac{x^4}{4\cdot 2!} + \frac{x^5}{5\cdot 3!} + \dots$$

5c. $g(t)=\int_0^1 te^t dt = te^t - e^t\Big|_0^1 = 1$, so $g(1)=\int_0^1 te^t dt = \frac{1}{2}+\frac{1}{3(1!)}+\frac{1}{4(2!)}+\frac{1}{5(3!)}+\ldots = 1$

Chapter 11 Practice Test

1. a. $y=e^t$

   b. $y=e^{-t}$

   c. $y=\cos x, y=\sin x$

   d. $y=e^{-t}, y=e^t$

2. a. $\frac{dy}{dt}=ky$

   b.

$$y = 2,000,000e^{kt}$$
$$1,000,000 = 2,000,000e^{k8}$$
$$k=\frac{\ln 0.5}{8}=-0.0866$$
$$y = 2,000,000e^{-0.0866t}$$

   c.

$$\frac{dy}{dt}=-0.0866y$$
$$\frac{dy}{dt}=-0.0866(1,000,000)=-86,600 gallons/month$$

   d.

$$100,000 = 2,000,000e^{-0.0866(t)}$$
$$t = 34.6 months$$

3. a.

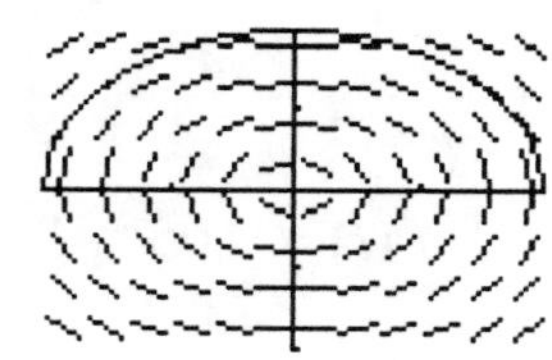

   b.

points	*slope*	*next y*
(0,2)	$\frac{dy}{dx}=\frac{-0}{2}=0$	$y_1 = y_0 + \frac{dy}{dx}\Delta x$ $y_1 = 2+0(0.25)=2$
(0.25,2)	$\frac{dy}{dx}=\frac{-0.25}{2}=-0.125$	$y_2 = 2-0.125(0.25)\approx 1.969$
(0.50,1.969)	$\frac{dy}{dx}=\frac{-0.5}{1.969}=-.0254$	$y_3 = 1.969-0.254(0.25)\approx 1.9055$

(0.75,1.9055)	$\frac{dy}{dx}=\frac{-0.75}{1.9055}=-0.394$	$y_4=1.9055-0.394(0.25)\approx 1.807$
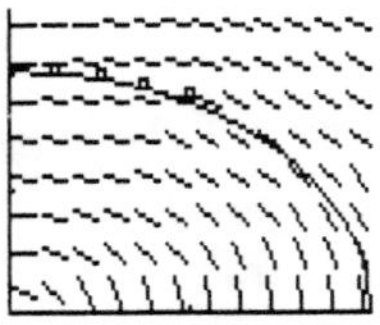		This is an overestimate since the curve is concave down.

c. $x^2+y^2=4,\ or\ \ y=\sqrt{4-x^2}$ passes through the point $x=0$, $y=2$.

$$\frac{dy}{dx}=\frac{-2x}{2\sqrt{4-x^2}}=\frac{-x}{\sqrt{4-x^2}}=\frac{-x}{y}$$

4.

$$\frac{dy}{dx}=\frac{1}{\sqrt{x}\sqrt{y}}$$

$$\sqrt{y}dy=\frac{1}{\sqrt{x}}dx$$

$$\int\sqrt{y}dy=\int\frac{1}{\sqrt{x}}dx$$

$$\frac{2}{3}y^{\frac{3}{2}}=2\sqrt{x}+C$$

$x=1$, $y=4$ gives $C=10/3$

$$y=\left(3\sqrt{x}+5\right)^{\frac{2}{3}}$$

5.

a.

i. The rate of growth is constant. $\frac{dP}{dt}=0.05$ has solution $P(t)=0.05t+C$, $t=0, P=1\ \ P(t)=0.05t+1$

ii. The rate of decay is proportional to the size of the population. $\frac{dP}{dt}=-0.05P$ has the solution

$P(t)=Ae^{-0.05t}$

$t=0, P=1\ \ P(t)=e^{-0.05t}$

iii. The rate of growth is proportional to the difference between the size of the population and the upper limit. $\frac{dP}{dt}=0.05(1000-P)$ has solution $P(t)=1000-Ae^{-0.05t}$, $t=0, P=1\ \ P(t)=1000-999e^{-0.05t}$

by separation of variables.

iv. This is a logistic model. $\frac{dP}{dt} = 0.0005P(1000 - P) = 0.5P\left(1 - \frac{P}{1000}\right)$ has solution

$$P(t) = \frac{1000}{1 + Ae^{-.5t}}, A = \frac{1000 - P_0}{P_0}$$

$$t = 0, P = 1 \quad P(t) = \frac{1000}{1 + 999e^{-.5t}}$$

b.

c.

i.. $\frac{dP}{dt} = 0.05 \neq 0$ so no equilibrium

ii. $\frac{dP}{dt} = -0.05P = 0$ if $P = 0$. This is a stable equilibrium as the population is dying out.

iii. $\frac{dP}{dt} = 0.05(1000 - P) = 0$ if $P = 1000$.
This is a stable equilibrium - the population is approaching the maximum value.

iv. $\frac{dP}{dt} = 0.0005P(1000 - P) = 0$ if $P = 1000$ (which is stable) or if $P = 0$ (which is unstable).

6. Answer:

a) Using the chain rule, we get $\frac{dy}{dx} = \frac{y(1 - x - y)}{x(1 - 0.5x - 0.5y)}$

b)

Window $0<x<4$ and $0<y<4$

c) At (3,3), $\frac{dx}{dt} < 0$ and $\frac{dy}{dt} < 0$ so both x and y decrease. In the long run, the $x \to 2$ and $y \to 0$.

d) Vertical trajectories: $\frac{dx}{dt} = x(1 - 0.5x - 0.5y) = 0$ if $x = 0$ (the y-axis) or if $1 - 0.5x - 0.5y = 0$ (the fine line on the graph below)

Horizontal trajectories: $\frac{dy}{dt} = y(1 - x - y) = 0$ if $y = 0$ (the x-axis) or if $1 - x - y = 0$ (the heavy line on the graph below).

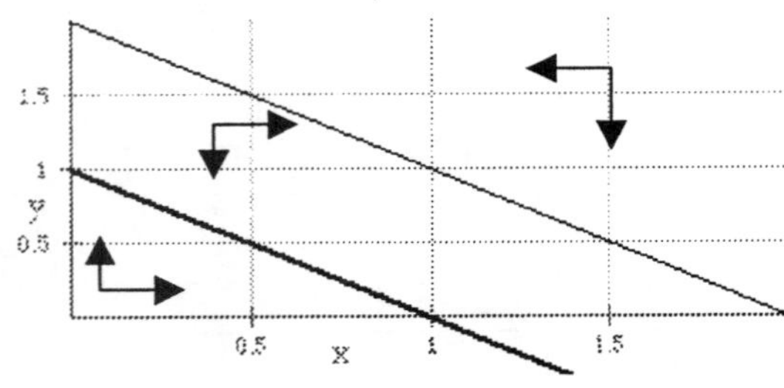

The End